FATAL TRAPS

for Helicopter Pilots

FATAL TRAPS
for Helicopter Pilots

Greg Whyte

New York   Chicago   San Francisco   Lisbon   London   Madrid
Mexico City   Milan   New Delhi   San Juan   Seoul
Singapore   Sydney   Toronto

The _McGraw·Hill_ Companies

Copyright © 2007 by The McGraw-Hill Companies, Inc. All rights reserved. Printed in the United States of America. Except as permitted under the United States Copyright Act of 1976, no part of this publication may be reproduced or distributed in any form or by any means, or stored in a data base or retrieval system, without the prior written permission of the publisher.

4 5 6 7 8 9 0 DOC/DOC 0 1 3 2 1 0 9

ISBN-13: 978-0-07-148830-3
ISBN-10: 0-07-148830-8

This book was first published in New Zealand by Reed Publishing (NZ) Ltd. © 2003 Greg Whyte. The author asserts his moral rights in the work.

The sponsoring editor for this book was Stephen S. Chapman and the production supervisor was Richard C. Ruzycka. The art director for the cover was Anthony Landi.

Printed and bound by RR Donnelley.

This book was printed on acid-free paper.

McGraw-Hill books are available at special quantity discounts to use as premiums and sales promotions, or for use in corporate training programs. For more information, please write to the Director of Special Sales, Professional Publishing, McGraw-Hill, Two Penn Plaza, New York, NY 10121-2298. Or contact your local bookstore.

Contents

Unless otherwise mentioned, the material in this book presumes that the main rotor blades on helicopters rotate counter-clockwise; generally known as 'American manufactured'. It is important to note that torque and pedal requirements will be reversed when applying any discussion to 'European' helicopters.

Measurements

1 metre = 3.05 feet

1 US gallon = 3.79 liters

1 kilogram = 2.2 pounds

Acknowledgments

I would like to acknowledge the following people:

Steve Bone: From the trivial to the critical, Steve has an in-depth knowledge of all aspects of helicopters. He did the 'raw edit' on most chapters and personally wrote many paragraphs. Vast amounts of advice, contacts and research were forthcoming, and I am hugely appreciative of the hundreds of hours Steve has given to this project.

Bernie Lewis: Apart from the fact that Bernie went through the entire manuscript in its original form at least twice (as a technical editor) and wrote the Foreword, he was hounded by telephone calls over many years. Not once was Bernie unavailable or reluctant to assist in any way that he could.

Tom McCready: Tom, currently an accident investigator with the New Zealand Civil Aviation Authority, is a talented aircraft engineer who gave up many hours to nurse me through the mechanical sections and give me the confidence to forward them to the technical editor.

Jim Cheatham: The material on wire strikes in this book is taken from an article written by Jim. He did not hesitate to give me permission to use his work. Jim is with Verticare Helicopters in Salinas, California.

Logan Sharplin: Logan is a friend who not only willingly gave encouragement and advice but also contributed a passage in the weather section about a trap that very nearly caught him – twice.

Jim Wilson: I'm sure readers will enjoy Jim's wry wit as he recounts a ditching episode he experienced. I chuckle each time I read it. His story may be humorous but there are still lessons to be learned. Jim is chief pilot at Helicopters New Zealand Ltd.

Ian McPherson: Ian has shared events surrounding the sudden loss of a tail rotor. The resulting fright comes across very clearly and gives a sobering reminder that training breeds instincts that save lives. Ian is a squadron leader with the Royal New Zealand Air Force.

Captain John Mockett: I spent countless hours in the head office of New Zealand's Transport Accident Investigation Commission and was always made most welcome by everyone from the chief executive officer to the receptionist. While I drank their coffee and used their resources, John Mockett and Joanne McMillan were never too busy to listen to any request or lend a hand.

Cliff Jenks (editor, New Zealand Civil Aviation's *Vector* magazine); Brad Vardy (editor, Transport Canada's *Vortex* magazine); Roger Rozelle (director of publications, Flight Safety Foundation). My thanks to all three for providing valuable resources. The depth, consistency and professionalism displayed in the material they publish is critical to pilot safety.

George Nadal (Australian Transportation Safety Board); Graham Liddy (Air Accident Investigation Unit, Ireland); Ken Smart, Sandra Duffin, Anna Winter (Air Accident Investigation Branch, Department of Transport, Farnborough, England); Gunnel Göransson (Board of Accident Investigation, Sweden). This book developed over six years. In that time these people answered many queries and requests. The Australian Transportation Safety Board in particular forwarded a wealth of resources after an initial short communication. The commitment to improving aviation safety by each of these bodies is awe inspiring. I am indebted to both the organizations and the individuals.

Peter Schweizer: Of all the major helicopter manufacturers, Schweizer Aircraft Corporation was the only one to have the courage to concede that there is danger if helicopters aren't flown properly. Peter didn't hesitate to assist in a project aimed at enhancing safety.

Brad Martin-Cox (Aero Products Component Services, Inc); Constable Don Bruce (photographer, New Zealand Police); Caz Caswell (UK aviation photographer and enthusiast); Callum Macpherson (editor, *Pacific Wings* magazine); Ray Deerness (New Zealand aviation photographer and enthusiast); Leigh Burney (Gusto Design & Print). My thanks to you all for helping to supply images for the book.

Brian Sutherland (Flightline Aviation, New Zealand)

Mike Feeney (independent aviation correspondent)

Reed Publishing: Thanks for sharing the vision and providing the expertise and support; a great team.

Friends and family: As I spent thousands of hours over several years researching and writing this book, those close to me often looked at me and shook their heads in wonder. But that didn't stop them from understanding and giving encouragement; thanks for that.

Foreword

I am not aware of any new way to have an accident or incident in a helicopter. They have all been tried before, but, unfortunately, people seem to keep repeating them. Greg Whyte in his excellent earlier book, *So, You Want to Be a Helicopter Pilot?*, explained the requirements and pitfalls for ab initio training. He now goes on to explain how to avoid some common and not so common pitfalls that pilots can experience throughout their flying career in helicopters.

Someone once said that a smart person learns from other people's mistakes and a fool learns from his own. This book gives you the opportunity to be smart and, after reading a variety of accident reports, they should not then feature in your future career. Hopefully it will be a wonderful career, as helicopter flying is, in my opinion, the most interesting flying of all.

I have been to a number of funerals, all of them the result of accidents similar to those featured in this book. The most traumatic of these was for some friends of mine killed in a mid-air collision on a very clear, calm day. Following this I did some research and found that most mid-air collisions seem to occur on clear, calm days when people are relaxed, sitting back and enjoying their flight. Greg features one such accident in the book.

This book covers an incredibly large range of situations and medical conditions which, combined with stress in the cockpit, may lead to a loss in concentration, perhaps a lack of situational awareness and finally to an accident. Stress in the cockpit has always been a hazard and can start simply by drinking too much coffee at breakfast prior to a long trip. After a while the bladder becomes distended and you start thinking about landing as soon as possible. However, other factors may begin to arise as you have to divert around heavy rainstorms and you drop your map in the turbulence. Suddenly, you have real stress in the cockpit and you can easily end up making bad decisions. Sometimes you can land and thus ease the situation. However, that is not always possible if you are over mountainous terrain, jungle, or water. The stress builds up too easily.

One pilot with whom I flew on a long overwater flight had the answer. He said that he 'wanted a leak' and asked me to take over the controls. I did so and a few minutes later, he said that he was now better and held up in front of us both a fragile, pear-shaped condom for me to admire! I was horrified and told him to get rid of it. He did so, tipping it out of the cockpit window, shaking it and then rolling it up and putting it back into his overall pocket! I never want to fly with him again, as I believe those things perish in the sunlight and one day he is going to have an embarrassing accident. However, I give him full marks for resourcefulness.

I started flying helicopters in 1958 and throughout the years have seen many of the accidents referred to in this book and in my early days almost suffered similar fates. For instance, one day I was tasked to go to a military unit at another airfield to pick up a S.51 (an old tandem-seated helicopter). It had several defects, including a sticky collective, but I was assured that it had been cleared 'for ferry flight only'. I was told that the center of gravity was fine, even though extra gear and covers had been stowed around my passenger in the rear seat.

The helicopter was parked about 50 yards in front and to one side of a hangar. I started up and after negotiating the various problems associated with 'ferry flight only', I was ready for lift off. Sure enough the collective was sticky and moved in short jerks. The airman in front of me was getting tired of waving his arms, indicating that I could get airborne, so at last when I was light on the wheels, I heaved on the collective to positively clear the ground. The nose promptly dropped, the cyclic came back onto the rear stop and we hurtled towards the hangar over the top of the airman who had flung himself flat on the ground! We missed the hangar by a matter of inches and managed to get back into controlled flight some two miles beyond the airfield.

My heart was pounding, my mouth was dry and I was sure I was hyper-ventilating; I was certainly going to need a runway to put this machine back on the ground. My minimum airspeed with the cyclic on the backstop was 30 knots. Fortunately, it had fore and aft fuel tanks and by the time I had flown back to base I had managed to transfer all the front fuel to the rear tank and was almost able to hover. In about one hour I had covered several pages of events mentioned in this book. I like a good adrenalin rush, as I strongly believe it is good for the system, just as Greg says, but sometimes they can be a bit severe. Caution: Do not believe what people tell you. Check it yourself.

Another incident illustrates Professor James Reason's Swiss Cheese Model mentioned in this book. I was asked to test fly a lightweight helicopter that an acquaintance of mine had imported in kit form from America. It was a beautiful looking helicopter, well engineered and assembled. The only change from the original design was that the overhead cyclic had been moved to the more conventional floor position with the approval of the designer.

On the appointed day I inspected the machine thoroughly and then checked that all flying controls were connected. Soon after starting, I checked the vibration levels and ran it up to view the tracking. It all felt very smooth and I got ready for takeoff. First of all, I checked the freedom and movement of the controls. I moved the cyclic forward and, watching the rotor disc, I saw it move down. I moved it to the rear and saw the disc move up. Great stuff! I centralized the cyclic and then moved it to the right; the disc moved to the left — to the LEFT! It was difficult to believe, so I moved the cyclic left and the disc moved to the right.

I beckoned the excited owner over and carefully explained to him that there was no way I was going to continue. It had been 40 years to the week since I started flying helicopters and I had religiously done my pre-takeoff checks prior to every flight. It was very repetitious and boring, but after 40 years I had at last found a very positive reason for having done them.

This was certainly an excellent example of Professor Reason's 'Swiss cheese model' discussed in Chapter 13. It showed:

The first slice   The original designer who approved the change.

The second slice The engineer/owner who designed the modification.

The third slice   The licensed maintainer who approved and engineered the changes.

The fourth slice The aviation authority that checked and finally certified the helicopter to fly.

The holes in all four slices were lined up for a nasty accident but, fortunately, I had done all my checks and the hole in the fifth slice did not line up with the other four, thus saving the day. Checks are stressed throughout your flying training, but I have often wondered if that is enough. In my early days of training in the Royal Air Force, they were called 'vital actions', as indeed they were. The word 'checks' seems to downgrade their importance, because they really are vital, as I found out that day.

Another modern phrase mentioned here that I have difficulty with is 'human factors'. I was always taught 'airmanship' and still believe that this is the correct word; just as 'seamanship' is still taught to sailors.

While not covering every type of accident, this book does cover a large number and, if you read it carefully, it will explain why there appear to be a greater percentage of accidents involving helicopters than fixed-wing aircraft. Pilots flying aeroplanes have other people telling them where to land, what the wind velocity is, and what the ambient pressure is. They also have the pleasure of landing on big, level open spaces with few if any obstructions. What a dull, uninteresting life they have. You, as a helicopter pilot, must rely on your own initiative and observations to accomplish a successful flight. After having possibly pumped and strained fuel into your helicopter, you take off to go to perhaps a little clearing among the scattered clouds on the top of a mountain or down in a river bed. You may have seen it only on a map but you must find it, decide whether it is safe to land, assess which way the wind is blowing, work out the power requirements for landing and takeoff and, finally, after a successful approach, land on ground that is not ideal — sloping, swampy or rocky — all the time making sure you keep the tail rotor clear of obstructions. The successful completion of such a landing is interesting and exciting. This book explains the problems involved and helps you solve them.

Greg has put an enormous amount of energy into researching this book. It is very readable and I sincerely hope it will be of great value to all helicopter pilots as well as those people who intend taking up this wonderful career.

Bernie Lewis

Introduction

The primary purpose of this book is to promote safety in rotary wing aviation by identifying and addressing the main causes of helicopter accidents. We call them 'accidents', but very few are. An accident is 'a chance event; unforeseeable'. Most helicopter crashes are extremely predictable. Someone — not always the pilot; sometimes the passenger, engineer or simply a bystander — took a chance, forgot something, ignored a principle or otherwise initiated the disaster.

Except for beginning student pilots and amateur enthusiasts, most readers will have had some training in dealing with the hazards outlined here. Each hazard is covered in such a way that working commercial pilots can understand, take note of, and incorporate the information into their everyday work habits.

There may appear to be a great many perils facing helicopter pilots. A pilot would be very unlucky to experience them all on the same day! However, it is not often that just one hazard causes an accident. The chance of encountering a hazard is greatly increased by the presence of another. For example: overloading will make the helicopter more susceptible to overpitching and vortex ring state, and make a successful autorotation a matter of much finer judgment. Problems can snowball in a heartbeat and what was intended to be 'just another sortie' quickly becomes an impending disaster. One thing not listed per se is complacency – that is the biggest killer of all.

For those readers who are not pilots but helicopter enthusiasts, there is a wealth of information in this book. The introductory paragraphs to each chapter should give you an easily understood explanation of the hazard being discussed.

It is not possible in a book of this nature to go into great depth and detail. Many topics quite easily warrant their own book. Indeed, books have been written on many individual hazards, and readers looking for further knowledge should refer to the further reading section.

Research for this book was not overly difficult as everything that is written here has appeared somewhere else at one time or another. Many parts are taken directly from excellent articles in flight safety magazines, permission for which has been kindly given by the editors. But it occurred to me that tracking down a case study, a simple explanation, a detailed explanation and some invaluable advice from experienced pilots on any one rotary wing flight hazard is very time-consuming, even for someone within the aviation arena. For other people the task would probably be impossible.

I will be satisfied if the messages contained in this book save one life. If you are a helicopter pilot you most likely have some very good friends who are also helicopter pilots. Recommend this book to them and they may be friends for a long while yet.

I would like to firmly stress that I do not sit in judgment of anyone who has had an accident for any reason. I figured out very early on that I'm a better writer than I am a helicopter pilot, which is why I now fly a laptop. Many case studies in this book leave me feeling, 'There but for the grace of God go I'. There is no doubt that on a score of occasions I have been on the threshold of an accident born of a dozen mistakes and bad calls.

Perfection makes a good target but a lousy reality. If you disagree with something written here, or you know of information, case studies, photographs or diagrams that could enhance the material in this book, please email me the details at greg@fataltraps.com. I would be eager to incorporate any improvements in the next edition.

Basic flight principles

Some books refer to 'helicopter behavior' rather than 'flight principles' and perhaps they are nearer the mark. Some helicopter maneuvers certainly seem to defy scientific explanation.

A good working knowledge of the principles of flight is required when operating helicopters, but the best weapons in your armory are preparation and judgment — a quaint old-fashioned word.

This chapter could have included a mountain of advanced calculus to truly reflect the principles of helicopter flight, but knowing how to differentiate some mind-boggling aerodynamic formulae is of little help when you are trying to resolve a critical situation. So I will keep it simple. However, make no mistake. Unless you study and understand the simple principles of flight, you are poorly armed to embark on safe helicopter flying either for recreation or as a career. Knowledge of the subject is imperative to the well-being of yourself and those who fly with you. Think of this knowledge as a form of health insurance (without the exorbitant premiums) and read on.

For readers who are non-fliers, this rudimentary understanding of the principles of helicopter flight will reveal the secrets of a number of the hazards explained in the remaining chapters of the book.

Terminology

Symmetrical and unsymmetrical airfoils

Symmetrical   Unsymmetrical

Fig 1.1 FEDERAL ADVISORY CIRCULAR 61–13B

Chord line

An imaginary line that connects the leading edge of an airfoil to the trailing edge.

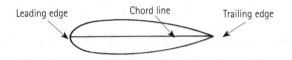

Leading edge   Chord line   Trailing edge

Fig 1.2 FEDERAL ADVISORY CIRCULAR 61–13B

Relative wind/airflow (RAF)

The direction of the airflow with respect to an airfoil.

Pitch angle

The acute angle between the blade chord line and a reference plane determined by the main rotor hub.

Angle of attack

Pitch angle

Relative wind

Fig 1.3 FEDERAL ADVISORY CIRCULAR 61–13B

Angle of attack

The angle between the chord line of the airfoil and the direction of RAF.

Lift

Lift is a force created by the pressure differential generated by the relative movement of an airfoil through the air.

$$\text{Lift} = C_L \tfrac{1}{2} \rho V^2 S$$

C_L is the co-efficient of lift and relates to the design shape of the wing, for example:

➤ High speed, as on a jet fighter .

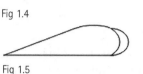

Fig 1.4

➤ High lift, as on an agricultural fixed wing .

Fig 1.5

$\frac{1}{2}$ is a mathematical constant

ρ is a Greek symbol that represents air density

V^2 is velocity (speed and direction); the squared function signifies that it is a primary factor

S is the surface area of the airfoil

Drag

Drag is the force that resists the movement of the airfoil through the air. Drag is produced any time there is movement of a body, be it an airfoil or fuselage.

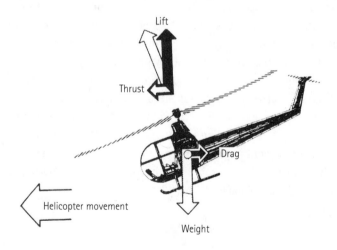

Fig 1.6 FEDERAL ADVISORY CIRCULAR 61–13B

Weight

The total weight of a helicopter is the first force that must be overcome before flight is possible. Lift, obtained from the rotation of the main rotor blades, overcomes or balances the force of weight.

Thrust

Thrust is the force that overcomes drag and moves the helicopter in the desired direction.

Rotor thrust

The same basic laws of physics govern the flight of both fixed-wing aircraft and helicopters. Both must produce an aerodynamic lifting force to overcome the weight of the aircraft. In both that lifting force is obtained from the aerodynamic reaction resulting from a flow of air over an airfoil section. On a fixed-wing aircraft, the airfoil is fixed to the fuselage as a wing, whereas the helicopter has engine-driven rotating wings. The rotor provides both lift and horizontal thrust.

While the rotor is simply an attachment that produces a lifting force, we need to consider two distinct aspects:

1. The rotor, with its behavior and characteristics, as a producer of lifting force.
2. The helicopter as a whole, with its behavior and characteristics determined largely by the rotor.

Helicopter rotors form a disc to produce rotor thrust. To understand the characteristics and control of this disc, we need to study how each rotor blade moves, and how it reacts to different control movements and airflows. This allows us to understand what happens to the helicopter when the disc performs in a manner that the laws of physics and aircraft manufacturers deem inappropriate.

Rotor blades are attached by a rotor head to a mast, which extends almost vertically from the fuselage (see Fig 1.7). Rotation of the mast allows the rotor blades to rotate. The axis of rotation is the axis through the main rotor mast. The plane of rotation is at right angles to the axis of rotation at the head of the main rotor mast. The rotor blades are connected to the rotor head at an angle to the plane of rotation called the pitch angle.

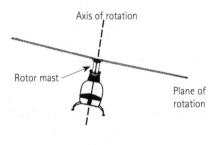

Fig 1.7

Effects of controls

The pitch angle is of primary importance. To a point, the greater the pitch angle, the greater the lift generated. If the blade is overpitched, sections of it will stall, much like the wing of an airplane will stall.

One of the main controls the pilot uses is the collective pitch lever. This increases or decreases the pitch angle of all the blades at the same time. This control, which resembles the parking brake in a car, is on the pilot's left (see Fig 1.8).

The stick in a helicopter is called a cyclic control. This also changes the pitch of the blades, but one blade at a time as it passes a certain position. If the pilot moves the cyclic control forward, the front of the disc drops and the rear rises. The helicopter then moves forward. Move the cyclic control backward and the opposite happens. Move it left to go left and right to go right.

A tail rotor is a very necessary part of the running gear for most helicopters. When the engine supplies power to the main rotors through the transmission, it applies a force. Newton tells us that for every action there is an equal and opposite

1 Cyclic control
2 Anti-torque pedal

Collective pitch lever
3 Throttle
4 Throttle friction
5 Collective friction

Fig 1.8 Controls LOGAN SHARPLIN

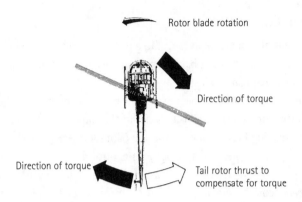

Rotor blade rotation

Direction of torque

Direction of torque

Tail rotor thrust to
compensate for torque

Fig 1.9 Purpose of tail rotor FEDERAL ADVISORY CIRCULAR 61-13B

reaction. That reaction in our case is called torque. The tail rotor's purpose is to counteract torque. Fig 1.9 shows how this works for a helicopter that has its main rotors turning anticlockwise.

For the most part, the amount of thrust required from the tail rotor depends on how much torque has to be overcome. This varies with the amount of power being used. It is also affected by the helicopter's airspeed. To turn a hovering helicopter, the pilot over- or under-compensates for the torque. A helicopter with its main rotors turning anticlockwise needs less tail rotor thrust to turn to the right. The opposite applies for a turn to the left (see Fig 1.10). For a right turn, relax the pressure on the left pedal and the torque will initiate a turn to the right.

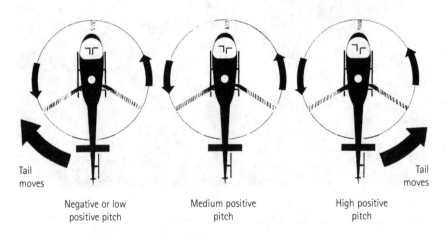

Tail
moves

Tail
moves

Negative or low
positive pitch

Medium positive
pitch

High positive
pitch

Fig 1.10 Yawing turns through tail rotor pitch variations. Note the pedal positions
FEDERAL ADVISORY CIRCULAR 61-13B

Blade forces and aerodynamics

The total effect of the lifting force produced by each rotor blade acts through the hub of the rotor and perpendicular to the plane of rotation of the rotor blades. This lifting force is known as total rotor thrust (TRT). In a calm hover, TRT provides the lifting force that overcomes the weight of the helicopter.

When looking at how lift is developed for fixed-wing aircraft, the pressure differential between the upper and lower wing surfaces is generally considered. This isn't practical for helicopters because speed and angle of attack vary greatly as each blade moves around the disc. It is usual, then, to think of rotor thrust as the reaction to acceleration imparted to air. If a mass of air is given an increase in velocity down through the disc (downwash), the reaction to the force required to do this appears in the form of rotor thrust. The greater the acceleration or the greater the mass of air affected, the greater the rotor thrust obtained.

As seen in Fig 1.11, a rotor blade moving horizontally through a column of still air deflects or displaces some of the air downward. In practical terms, helicopter rotor blades follow each other past a particular point in rapid succession. A twin-bladed rotor turning at 360 rpm has a blade passing a point twelve times every second. With the blades operating in the downwash of preceding blades, the downwash becomes cumulative and the column of air soon becomes a column of descending air. This downward motion of air is known as induced flow (IF):

IF = the downward component of airflow velocity
created in developing rotor thrust.

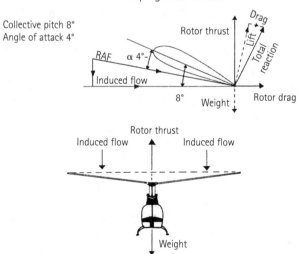

Fig 1.11 Free air hover *NEW ZEALAND FLIGHT SAFETY*

Airflow relative to the blade

Rotor blades are subjected to two airflows: one caused by the blades' horizontal travel and one caused by the downwash (IF). So, a horizontal component of airflow and a vertical component of airflow exist.

The resultant of these two flows becomes the direction of the airflow relative to the blade. The majority of airflow velocity over a helicopter rotor blade is due to the rotor's rotation (Vr). This velocity is normally large in comparison to forward speed.

RAF over a rotor blade depends on:

➤ The rotational speed of the rotor.
➤ The forward speed of the aircraft.
➤ The vertical speed of the aircraft.

Total reaction, rotor thrust and rotor drag

The aerodynamic forces acting on a fixed wing are lift, which acts perpendicular to RAF, and drag, which acts in-line to the relative airflow. A helicopter rotor blade is subject to the same aerodynamic forces, but it is not convenient to separate it into lift and drag. We are more concerned with how much of the total reaction on a blade opposes its rotation, and how much of the total reaction acts perpendicular to the plane of the disc (plane of rotation or POR). These determine how much power is required to drive the rotor at operating rpm, and how much lifting force —TRT — the rotor will produce. Fig 1.12 shows these components in comparison to lift and drag.

The component of TR perpendicular to the POR is known as rotor thrust (RT) and applies to a cross-section of the blade. This is distinct from TRT, which refers to the lifting force of the disc as a whole, or the sum of the individual sections of RT acting through the axis of rotation, usually the top of the mast.

The component of TR in the POR is rotor drag (RD) and the total of all RD components is total rotor drag (see Fig 1.12).

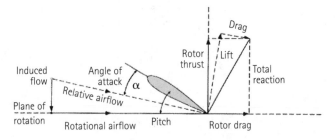

Fig 1.12 Forces acting on a rotor blade

RT and RD are in principle similar to lift and drag; both are affected by the factors expressed in the form $\frac{1}{2}\rho V^2 S$. However, the C_L and C_D (co-efficient of drag) apply only approximately to RT and RD, as the angle between RAF and POR will affect these values. The greater the angular difference, the less accurate the value of C_L and C_D when applied to helicopters. Despite that, the formulae $C_L \frac{1}{2}\rho V^2 S$ and $C_D \frac{1}{2}\rho V^2 S$ still provide a sound basis for comparing change in RT and RD with changing airflow and angle of attack.

Various forms of drag are present when the helicopter is airborne:

➤ Parasite drag is the drag caused by all the components of the helicopter not generating lift — the fuselage, skids, lights, sling loads, and so on.

➤ Profile drag is the drag caused by form drag and skin friction. Form drag is the resistance the air has to the front and rear movements of the lifting surfaces (rotors) while friction drag is the slowing down of the air as it passes over the flat surfaces of the rotors.

➤ Induced drag relates to the RT that is lost when the actual RAF is modified before it passes over the rotor. It is affected by the IF realigning the air just as it reaches the rotor disc.

Thrust distribution and coning

All the drag has to be overcome by RT. As we have seen, the effect of all RT components is the force called TRT, which acts through the hub of the rotor and at right angles to the plane of rotation. If TRT didn't act through the hub, the disc would tilt, requiring a corrective control force to restore TRT to the center of the disc. Also, as each blade bears a geometric relationship to the disc, TRT can have no residual components across the face of the disc (see Fig 1.13).

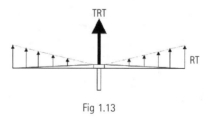

TRT

RT

Fig 1.13

If the rotor blade operated at a constant angle of attack over the full span of the blade, as in Fig 1.13, the majority of RT would be produced near the blade tips. Because the tips are traveling faster than the blade roots, they would have much more V^2 and so more lift. This is undesirable as great bending stresses would be applied to the blade in transferring the effect of RT to the hub.

To overcome this, the components of TRT are distributed more evenly, as shown in Fig 1.14. This can be achieved by:

➤ Tapering the blades (concentrating blade area, S, near the hub).
➤ Reducing angle of attack towards the tips (washout).
➤ Combining both tapering and washout.

Washout has generally been preferred as better blade characteristics can be achieved with smaller tip angles of attack.

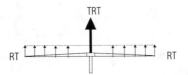

Fig 1.14 Distribution of rotor thrust across the disc

Coning

If the rotor blades are free to move in relation to the rotor hub, then RT will cause the blades to rise until they reach a position where their upward movement is balanced by the outward pull of the centrifugal force being produced by the blade rotation. The blades are now coned upward, the coning angle being the angle between the span-line of the blade and the plane in which the blade tips move (tip path plane is synonymous with POR). The greater the load (weight), the greater the RT required, therefore the greater the coning angle (see Fig 1.15).

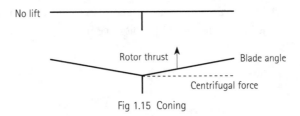

Fig 1.15 Coning

Rotor rpm limits

The transmission is limited by the amount of torque it can cope with before it exceeds design stress limitations. While the engine can actually produce more power on many occasions, the transmission simply cannot absorb that power before it is over-torqued.

A different set of boundaries defines maximum rotor rpm, but principally the upper limitation is affected by the structural fatigue factor of the components – the rotor blade itself and the sometimes intricate parts of the rotor head. A critical time to note these limitations is during an autorotation that involves low torque (no engine power) but high rotor rpm as the disc has been unloaded by a large reduction in collective pitch (refer to the chapter on engine failures).

For a variety of reasons – fatigue, aerodynamics, fuel scheduling, and so on – it is always desirable to fly within the green arc of the rotor rpm gauge. This and other important gauges are immediately visible to the pilot in the instrument panel, as shown in Fig 1.16.

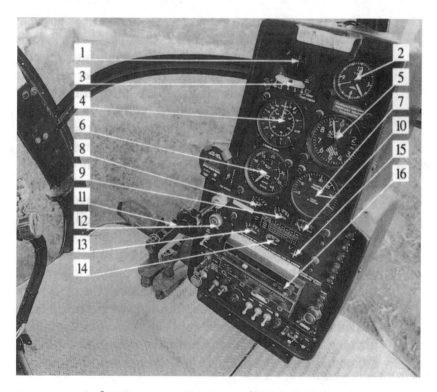

1	Compass	9	Fuel pressure
2	Clock	10	Cylinder head temperature
3	Slip indicator	11	Fuel mixture control
4	Airspeed indicator	12	Ignition
5	Altimeter	13	Oil pressure
6	Tachometer	14	Oil temperature
7	Manifold pressure	15	Ammeter
8	Fuel	16	Radio

Fig 1.16 The instrument panel LOGAN SHARPLIN

Rotor head principles

Rotor blades need to move in many ways other than rotation. They need to feather, which is altered with the collective or cyclic pitch controls, and to flap. Flapping relates to the upward and downward motions at the tip. This is to equalize the lift forces of the advancing and retreating blades and to allow for dissymmetry of lift.

Some blades also need to lead and lag, meaning they move fore and aft in relation to the rotor mast. In fully articulated rotor systems this is provided for by use of a damper, which allows the blade to pivot about a hinge point. These types of rotor-heads may also have a flapping hinge (see Fig 1.17). Two-bladed rotor systems using a teetering rotor hinge get their flapping capability from the see-saw effect.

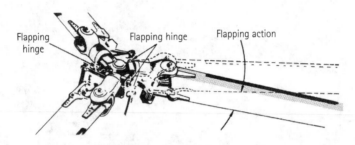

Fig 1.17 Flapping hinges FEDERAL ADVISORY CIRCULAR 61–13B

While rotor head design is developing all the time and some innovative examples are gaining certification, the two most common types seen in light helicopters are the teetering head and fully articulated.

Flapping to equality

To move the helicopter horizontally, the rotor disc must be tilted so that the total rotor thrust vector has a component in the direction required. To do this, swash plates are tilted so that the pitch angle on one side of the disc increases while, at the same time, the pitch angle on the other side of the disc is decreased. The pilot uses the cyclic stick to tilt the swash plates.

Changing the attitude of the disc changes the plane of rotation. It is important to keep the total rotor thrust magnitude steady when changing the plane of rotation, otherwise the helicopter climbs or descends. We can achieve an attitude change by adjusting the pitch angles of the blades so that they 'fly' into a new plane of rotation with no change in TRT. This is achieved by giving the blades a flapping capability.

Let's consider a blade in a hover with an angle of attack. A cyclic stick movement will decrease the blade pitch and the reduction in pitch will reduce both the blade's angle of attack and rotor thrust (see Fig 1.19).

The blade cannot maintain its horizontal flight, so its flapping capability (given by a flapping hinge, for example) allows the blade to flap down, causing an automatic increase in angle of attack (see Fig 1.20). When the angle of attack returns to α, the blade thrust returns to its original value. The blade then follows this new path until another pitch change is made with the cyclic.

While this process occurs on one side of the rotor disc, the opposite occurs on the other side; pitch angle is increased, RT increases and the blade 'flies' up to maintain angle of attack and RT (see Fig 1.21). When the pitch of a blade is changed so that the blade flaps and changes the plane of rotation, the blade is said to 'flap to equality' of rotor thrust (that is, the blade restores the original rotor thrust).

In practise, when a cyclic pitch change is made, pitch changes in tiny increments over a number of revolutions as the blade travels around the disc. So, the blades continuously 'flap to equality' as they travel through 360°.

Ground effect

A helicopter requires much less power to hover within one rotor disc diameter of the ground than at heights above that. This is referred to as hovering in ground effect (IGE) and out of ground effect (OGE). There are several reasons for this, all resulting from the interruption to the downwash velocity of air after it passes through the disc:

➤ Induced angle of attack is reduced, with the lift vector becoming more vertical.
➤ Reduction of induced drag.
➤ Induced angle of attack reduced and the lift generating angle of attack is increased.

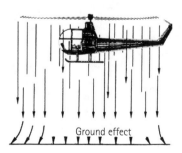

Fig 1.18 Ground effect results when the rotor downwash field is altered from its free air state by the presence of the surface FEDERAL ADVISORY CIRCULAR 61–13B

Translational lift

When a helicopter is hovering in no-wind conditions, a set amount of air is being processed by the rotor system to produce lift. If the helicopter then moves forward, the horizontal travel increases the inflow velocity and the rotor disc is supplied with a greater mass of air per unit time. The horizontal movement also advantageously affects the IF and angle of attack (see Figs 1.19, 1.20, 1.21).

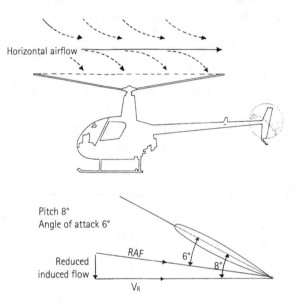

Fig 1.19 Effect of horizontal airflow on induced flow *NEW ZEALAND FLIGHT SAFETY*

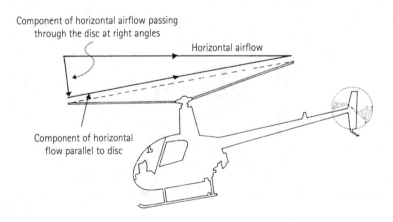

Fig 1.20 Effect of tilting disc towards horizontal airflow *NEW ZEALAND FLIGHT SAFETY*

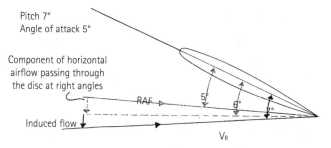

Pitch 7°
Angle of attack 5°

Component of horizontal
airflow passing through
the disc at right angles

Induced flow

RAF

V_R

Fig 1.21 Total flow through disc NEW ZEALAND FLIGHT SAFETY

This increased efficiency is called translational lift, often referred to simply as translation. Although it increases from 0 to around 55 knots, it is most noticeable at about 12 knots. The pilot has to actually reduce collective pitch and power to maintain the same rate of climb when 'going through translation'.

If the helicopter is hovering with a headwind, translational lift is already present. If the wind is over 12 knots, the helicopter can lift a greater weight for a given power setting.

Flare effect

The usual procedure for a helicopter making a normal landing is to approach into the wind, if possible, and initiate a flare to reduce airspeed. When the flare is introduced, the rotor disc tilts backward, causing TRT to also tilt backward as the horizontal component of thrust reverses. Parasite drag diminishes rapidly and the fuselage pitches nose up. TRT increases for two reasons: firstly, because the airflow relative to the disc changes with a component of the horizontal flow opposing the induced flow; secondly, the angle of attack increases. Collective pitch must be reduced to maintain the same height and rotor rpm, which naturally increases due to a force called the 'Coriolis effect'. This effect, which relates to coning, is due to the decrease in rotor drag. The cyclic must be neutralized to prevent the disc tilting back even further in sympathy with the fuselage pitching up (see Figs 1.22 and 1.23).

Forward speed reduces quickly as a result of the flare and the effects disappear. The pilot must then reverse many of the inputs by increasing collective and using forward cyclic to not only replace the lost flare effect but also compensate for the loss of translational lift.

Throughout the flare, with its changing use of power, tail rotor thrust must be controlled through the anti-torque pedals to keep the helicopter straight in the yawing plane.

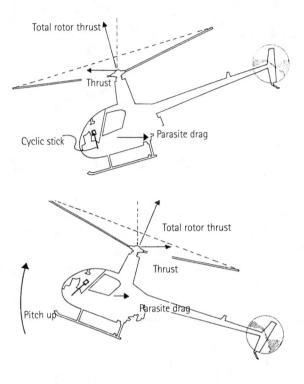

Fig 1.22 Flare effect NEW ZEALAND FLIGHT SAFETY

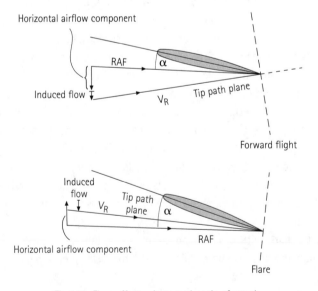

Fig 1.23 Flare effect — increased angle of attack NEW ZEALAND FLIGHT SAFETY

Performance

Helicopter performance is affected by the density of air by virtue of the ρ symbol in the lift formula (see page 17). Performance is also affected by wind velocity (translational lift) and gross weight, although many pilots are primarily concerned with how much weight they can lift given the air density and wind conditions. Performance is often considered as the amount of weight that can be held in the hover.

Air density is affected by various things:

➤ Air temperature – cooler air is much denser.
➤ Air pressure – greater at sea level than at altitude, and greater in an anticyclone (H) than in a low (L).
➤ Moisture in the air (relative humidity) – the more moisture, the less density.

A helicopter operating at sea level on a cold dry day in the middle of an anticyclone will lift a far greater weight than it will on a hot humid day in a low pressure weather system.

Flight manuals contain a myriad of data, charts and graphs that pilots can use to calculate just what their helicopter should be able to lift in a given set of conditions.

2

Vortex ring state

As shown in Fig 2.1, a helicopter under power produces vortices around the periphery of the rotor disc. This is the price of producing lift. When a descending helicopter with a tail wind slows down in forward flight, these vortices catch up with it. The rotors are then trying to channel air down using air that is already moving downward (the vortex).

If a helicopter enters vortex ring state (also known as 'settling with power') and the pilot increases power and rotor pitch angle in an effort to produce lift, the condition worsens rapidly. It is similar to stalling a fixed-wing aircraft. Pulling back on the stick doesn't help. Soon the rotors are producing no lift and the aircraft enters a very rapid uncontrolled descent.

This chapter is devoted to vortex ring state affecting the main rotor. Chapter 12, Tail rotor failures, deals with the problem of vortex ring state on the tail rotor.

Fig 2.1 Generated vortices are easy to discern in this agricultural spray RAY DEERNESS

Case study 2:1

Accident Report B/921/1036

Date: May 8, 1992

Location: Kelmscott, Western Australia, Australia

Aircraft: Aerospatiale AS 355F1

Injuries: 2 minor, 2 serious

The helicopter was owned by the Western Australia Police Department. However, because the department had no experience in operating rotary winged aircraft, it contracted a local helicopter operator to provide pilots and be responsible for the maintenance of the aircraft.

On the day of the accident, after a normal engine start, the pilot raised the collective pitch control lever to check engine power response. However, the Number 1 engine torque indication did not increase in parallel with the Number 2 engine torque indication. The pilot trimmed the Number 1 engine but the split in torque indications was still present when the collective pitch was increased. The pilot again trimmed the Number 1 engine and this time the torque readings for both engines responded normally. He then departed on the flight. The helicopter was to land on a sports oval as part of a police public relations display for school children. The oval was partly bordered by tall trees. Witnesses reported that the helicopter made a fast approach from the east and flared positively before descending rapidly and almost vertically from about 70 feet above ground level. (There was no pause between the flare and the initiation of the descent.) The initial ground impact was on the rear of the right skid and tail boom/tail rotor. The tail rotor assembly separated from the helicopter and the fuselage pitched forward before completing two rapid anti-clockwise rotations, during which the tail boom separated from the fuselage, and the main rotor blades struck the ground. The helicopter came to rest facing its original approach direction.

The pilot reported that, during the final descent to ground impact, there was a split between the engine torque gauge indications. He could not quantify the split but thought that there had been a power loss on the Number 1 engine. Neither the pilot nor any of the other occupants recalled hearing the rotor low-rpm warning horn sound at any stage during the flight. However, the horn had operated normally during the pre-flight checks.

The pilot and crewman received minor injuries and the two passengers were seriously injured as a result of the accident. During the impact sequence, the two rear-seat passengers were thrown partially outside the cabin but were restrained by lap seat belts that caught around their lower legs and ankles. The rescue of all occupants was quickly and efficiently accomplished by police officers at the site before fire took hold of the helicopter.

During the impact sequence, substantial damage occurred to the tail boom and to the main and tail rotor systems. A small fire, which began in the engine bay, could not be controlled by hand-held extinguishers and soon spread to destroy the helicopter.

Fire damage precluded a complete examination of the wreckage. However, both engines were disassembled and examined. No fault which might have prevented normal operation was found in either. Further, no fault was found in the fuel control unit and governor from the Number 1 engine (the corresponding accessories to the Number 2 engine were destroyed by the fire). The rigging of the engines individually, and as a pair, could not be checked. This was precluded by fire damage to the engine trimming control systems, including the trim linkage tree and the actuating cable between the engine trim actuator and the power turbine governor for the Number 1 engine. While the flight control system was also largely destroyed by fire, there was no indication that any failure in this system had contributed to the accident.

The pilot in command held a current Commercial Pilot License (Helicopter) and was endorsed on the AS355F1 helicopter. His total flying experience at the time of the accident was 5264 hours, of which 3463 were on rotary winged aircraft. His experience on AS355F1 helicopters was 315 hours.

The weight of the helicopter at the time of the accident was calculated as about 4784 pounds and the center of gravity was within limits. The maximum takeoff and landing weight for the helicopter was 5291 pounds.

The weather was fine at the time of the accident with a clear sky, a light southeasterly breeze, and a temperature of about 68°F (20°C).

Fig 2.2 The helicopter came to rest facing the same direction as the approach. Smoke drift indicates wind direction and strength. BUREAU OF AIR SAFETY INVESTIGATION

The pilot reported that, based on some smoke he observed in the distance, he assessed there to be no wind at the landing site. He flew the approach towards the west and described a normal approach angle and speed, terminating in a hover at about 20 feet above ground level. The pilot thought that the helicopter then experienced an engine power loss and a heavy landing had resulted. He said that apart from applying full collective pitch in an attempt to control the descent rate as the helicopter neared the ground, he did not make any other control input.

Witnesses reported that the helicopter, while on a westerly heading, performed a positive flare and immediately entered an almost vertical descent from about 70 feet above ground level. They also reported that the fuselage was randomly 'rocking' from side to side during the descent.

This information was supported by a series of photographs taken during the accident sequence by a professional photographer. Witnesses also indicated that the helicopter was descending at a higher than normal rate compared to other helicopters they had observed. Photographs of the post-crash fire indicated that a light east to southeast breeze (a tail wind with respect to the helicopter) was blowing at the time of the accident. The effect of the tail wind would have been to steepen the descent angle of the helicopter and to reduce its forward speed with respect to the local air mass.

The Accident Analysis concluded that the helicopter was in a steep descent at high-weight, low-forward speed, and under the influence of a light tail wind. Additionally, witnesses described its descent rate as high compared to other helicopters they had observed, and there was photographic and witness evidence of random rocking of the fuselage during the descent. It is probable, therefore, that the helicopter had entered a vortex ring state. By increasing the collective pitch, the pilot was exacerbating the situation.

A further possibility is that there was insufficient power available (the pilot reported seeing a split in the torque gauges) to arrest the rate of descent as the helicopter neared the ground. Had this occurred, however, the rotor low-rpm warning should have sounded. That no warning was heard indicates that rotor rpm was maintained throughout the descent and that neither engine suffered a significant power loss.

The geography of the landing area left other approach avenues open to the pilot than that which he flew. These would have permitted a more conventional and safe approach with regard to both approach angle and wind direction.

Source: Bureau of Air Safety Investigation, Australia

Case study 2:2

Accident Report B/915/1020

Date: May 12, 1991

Location: Mermaid Sound,
  Western Australia, Australia

Aircraft: Aerospatiale Puma
  SA330J

Injuries: 1 minor, 1 serious

The helicopter, with two pilots on board, was engaged in a night charter flight from Karratha to a departing liquefied natural gas (LNG) tanker to collect two marine pilots. The departure and night visual flight to the final descent point were normal.

Following an approach briefing from the pilot in command and radio advice from the ship which indicated the relative wind was from 010° at approximately 5 knots, the crew commenced the final approach from a 'gate', which is an initial approach point in level flight at 55 knots and 550 feet above mean sea level, approximately 0.75 nautical miles astern of the ship. At the time the ship was steaming in Mermaid Sound in a northerly direction at 12.5 knots, 12.5 miles northwest of Dampier, Western Australia. As the aircraft passed through the 'gate', speed was reduced to below 35 knots (minimum indicator reading) and the descent was started by reducing main rotor pitch angle and selecting the correct site picture in the windscreen. At approximately 500 feet the co-pilot, who was monitoring the instrument applications, recorded the rate of descent as 1000 feet per minute. The pilot increased the collective pitch in an attempt to reduce the rate of the descent. The corrective action had little or no effect as the rate of descent continued to increase until a slight reduction occurred just prior to impact. The accident occurred at 2133 hours Western Standard Time, when the aircraft impacted the water, rolled to the right and overturned. One of two life rafts stowed in the cabin area was dislodged and inflated. The inflated life raft provided sufficient additional buoyancy for the aircraft to remain afloat for two hours. The crew evacuated the aircraft through the co-pilot door and remained on the floating wreckage before transferring to a dinghy dropped by another helicopter. They were rescued by boat approximately 70 minutes after the accident. The aircraft sank but was subsequently recovered and transported to Karratha for inspection.

The helicopter sustained substantial damage: the main rotor blades were destroyed, the tail boom and rotor were torn off, and the main fuselage was severely dented. In addition, the co-pilot's seat collapsed and one flotation bag was torn from its stowage.

The pilot in command was aged 53 years. He held a current Senior Commercial Pilot License (Helicopter) with a valid medical certificate and was endorsed to fly Puma SA330J helicopters. At the time of the accident he had total flying experience of 17,800 hours, 2400 of which were on the Puma helicopter. His most recent night flying check had occurred on May 11, 1991. In addition, he held an appointment as a company check and training captain.

Both pilots were adequately rested prior to the flight, within their normal duty period, and had no known medical abnormalities at the time of the accident.

Both pilots had extensive experience in offshore day and night helicopter operations and, whilst they were aware of the possibility of problems caused by visual illusions, neither pilot had personal experience with them. The pilot in command had extensive experience with the vortex ring state as an instructor in light single-engine helicopters.

Weather conditions and time of the accident were consistent with the forecast and included a moonless night, no defined horizon, a strong easterly wind at altitude and a light and variable wind at sea level (dying sea breeze). The wind at 500 feet was assessed by the Bureau of Meteorology as light and variable, possibly from the southeast (beginning of the land breeze). The temperature was 83°F (28°C).

The only communication of relevance was the report from the ship to the helicopter crew indicating that they could expect a 5-knot headwind during the final approach.

The helideck was situated on the stern of the tanker behind the funnel and cabin-bridge superstructure. It met all the civil aviation requirements for a helicopter landing area for the Puma aircraft.

The helicopter contacted the water at a high rate of descent and zero forward speed. It was estimated that the rate of descent at impact was 2000 feet per minute.

Studies of accidents involving crew members with comparable experience levels (especially high levels) indicate that crew direction and supervision tend to diminish once individual members assume that they are a crew fully capable of conducting safe operations and as a result the type of detailed assistance and/or supervision that they might normally provide to less experienced crew members is not provided.

The outcome is a flattening of the trans-cockpit authority gradient as the pilot in command fails to establish full authority over the rest of the crew (approach brief, procedures, etc.) and the effectiveness of cross-checking between individual crew members is reduced.

In this accident, the co-pilot reported that, after advising the pilot of the high rate of descent and observing that action had been taken to correct it, he did not feel there was any requirement to continue to monitor the rate of descent as the pilot in command was very experienced and knew what he was doing.

With regard to vortex ring state, evidence indicates that it is widely believed within the aviation industry that this phenomenon is always accompanied by pitching and yawing and/or rolling motions that give the pilot warning of possible loss of control. This information is usually correct if the aircraft enters the fully developed vortex ring state; however, evidence was obtained from flight tests and from four experienced Puma pilots that the Puma can begin entry to vortex ring state with little or none of the expected sensory indications. A lack of indication is more likely if the airflow around the tail rotor is clear of the downwash, as would be the case if there was a tail wind or the aircraft were moving backwards.

During the course of this investigation a series of vortex ring demonstration flights in an aircraft identical to that involved in the accident were carried out. These flights showed that at the incipient stage of entry to vortex ring, a Puma, at a weight similar to that of the accident aircraft, displayed some mild random yawing (up to ± 10°) and an increase in vibration. The pitching and rolling symptoms do not become evident until about the same time as the rate of descent indication has increased markedly (to about 2000–2500 feet per minute).

It was also found that increasing the rotor pitch angle by about 1°–2° at the incipient stage of vortex ring entry resulted in an increase in the rate of descent from about 1000 feet per minute to 2000–2,500 feet per minute in about 5–10 seconds, at which point the nose tended to pitch down and random rolling occurred. Recovery was accomplished very quickly by applying forward cyclic and smoothly increasing power (rotor pitch angle) once airspeed had registered and was increasing. Height loss from initiation of recovery averaged 200–300 feet. (It should be noted that entry altitudes for these demonstrations ranged from 8000 feet down to 4000 feet.)

The demonstration flights indicated that it was possible to initiate a very high rate of descent condition, similar to that encountered during this accident, if the pilot failed to increase power sufficiently to meet the demand for increased rotor thrust required as the aircraft was flying through the entry procedures at the 'gate'. At conditions at approximately the accident aircraft's weight and density altitude, it was found that the rotor pitch angle was left at approximately 8° (a typical power setting as the entry to descent is established), with the airspeed indicating between 0 and 30 knots, the rate of descent was at or beyond 2500 feet per minute, this being the maximum on the indicator scale. In this condition heavy vibration was present, but the rate of descent reduced immediately power was increased. However, if the nose was held up such that airspeed further decreased, the helicopter entered vortex ring state accompanied by yawing, rolling and nose-down pitching with no discernible change in the rate of descent indication.

Source: Bureau of Air Safety Investigation, Australia

Case study 2:3

Accident Occurrence Number 18800739

Date: December 4, 1988

Location: near Innamincka, South Australia, Australia

Aircraft: Bell 206 JetRanger

Injuries: 1 minor

The helicopter was making an approach to a farm airstrip in remote southern Australia. The time of arrival of the helicopter was 2235 hours. The pilot in command of the aircraft had a total of 1640 hours, 184 of which were on type.

The pilot was making a night visual approach to the end of the airstrip, which was to be the helipad. The pad was illuminated by the headlights of a vehicle while lights from a house and shed provided additional peripheral light sources. The pilot reported that there was considerable turbulence adjacent to high ground near the strip. The approach was being flown at an indicated airspeed of 50 knots.

While passing over the high ground at a height of about 400 feet, the airspeed suddenly reduced to virtually zero and a large sink rate developed. The pilot attempted to go around but the helicopter continued to descend until it struck the ground in a near vertical descent.

The terrain and wind velocity were such that wind-shear over the high ground was probable. However, it was considered unlikely that this alone would have resulted in such a severe loss of airspeed. It was possible that the pilot, who had limited night-flying experience, had not noticed a decay in the airspeed prior to the sudden descent. If this was the case, the conditions were suitable for the aircraft to enter a vortex ring state. The fact that, when the pilot increased power and collective, the descent continued would also support this conclusion.

Source: Bureau of Air Safety Investigation, Australia

Principles of main rotor vortex ring state

Although vortex ring state can be entered from several flight maneuvers, the airflow conditions giving rise to its formation remain essentially the same in all cases. Vortex ring state depends on the development of two airflow systems:

1. A continuous or semi-continuous vortex around the periphery of the disc that amplifies the induced flow.
2. An increase of inflow angle in the central area of the disc which, coupled with moderate collective pitch setting, stalls the root end blade sections.

In this way, the helicopter descends with its own doughnut-like bubble of air around the disc and little effective rotor thrust.

Stylized vortex ring state

Formation of vortex ring state

Imagine a helicopter hovering OGE, as shown in Fig 2.3. The inflow angle is generally consistent along the span of the blade (even distribution of rotor thrust produces the greatest induced flow at the tips). The angle of attack is greatest at the root, reducing with washout towards the tips.

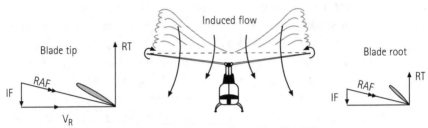

Fig 2.3 Out of ground effect hover

If collective pitch is reduced and a descent begins, there is now an upward airflow component opposed to the induced flow, as shown in Fig 2.4. This airflow quite quickly overcomes the induced flow at the root end of the blades, such that the change in inflow angle is greater than the pitch change. Put more simply, the angle of attack increases.

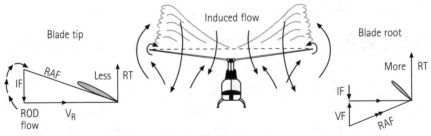

Fig 2.4 Vertical descent

Meanwhile, at the rotor tips, the rate of descent flow opposed by the induced flow tends to be deflected outside the rotor disc, then drawn inwards to become induced flow. This air reinforces the induced flow, increasing the local inflow angle. A reduction in the angle of attack results.

So, the overall flow pattern has reduced the angle of attack at the tip and increased it at the root section, but the mean angle of attack is not changed and total rotor thrust continues to equal weight.

As rate of descent increases, the rotor tips continue to become less effective, the vortices lead to increasing inflow angle and a consequent reduction in angle of attack. The burden of producing rotor thrust transfers progressively along the blade towards the root, where the angle of attack is increasing. Eventually the inner sections of the blade cannot compensate for the loss in rotor thrust at the tip sections. Total rotor thrust reduces and the rate of descent inevitably increases further. In some cases, that increase is rapid. A loss in total rotor thrust is accentuated by the stalling of the root sections when angles of attack exceed the stalling angle (see Fig 2.5).

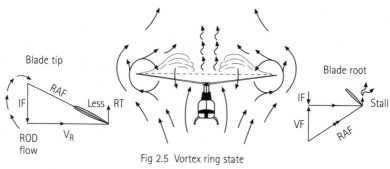

Fig 2.5 Vortex ring state

Between the tip sections, which have too small an angle of attack, and the root sections, which have too large an angle of attack, there remains an area of disc that still provides some rotor thrust at normal angles of attack. Wind tunnel experiments have shown that vortices can form and intensify in an erratic manner, subjecting each blade to large and sudden variations in angle of attack. The consequent pitch and roll is then felt. The helicopter may also be subjected to random yawing as unstable airflow from the vortex impinges on the tail rotor.

Conditions for formation
The formation of vortex ring as described here requires three conditions:

1. Minimal indicated airspeed
In theory, if the rotor disc had any component of movement in the plane of rotation, the vortex would not form as the disc would be continuously moving into fresh air, leaving the developing vortex behind. In practise, vortex ring will form at airspeeds up to 12–15 knots of airflow across the disc.

2. Rate of descent

The vortex ring state requires an airflow opposed to the induced flow. This rate of descent flow is the raw material that feeds the vortices. The helicopter's descent path need not be truly vertical. Anywhere from 70° satisfies the requirement for minimal horizontal speed.

3. Powered flight

The rotor system must be using some of the available engine power (20–100 percent) to provide the induced flow needed for vortex ring state. If there were no induced flow, the rate of descent flow would pass entirely through the disc and the offending vortex would not form. Also, collective pitch setting would be much less and root end stalling would not occur. Vortex ring state does not occur in a stabilized autorotation.

Some theorists combine the conditions of rate of descent and powered flight into one general condition – the opposition of induced flow and some other external airflow. This gives two basic conditions for vortex ring state formation:

1. Small components of airflow across the disc.
2. An opposed airflow, one component of which is induced flow, through the disc.

Do not be dismayed by theorists playing semantic games to express the same condition in different ways (wait until the serious mathematicians get in on the act!). A discussion on techniques for recovery from vortex ring makes it apparent why two neatly packaged conditions suit some theorists.

Recovery from vortex ring state

To recover from vortex ring state one or both of the conditions for formation must be removed. This involves:

1. An increase in forward speed using cyclic control.
2. Entering autorotation to remove induced flow.

Whether you chose to do one or both, a further loss of height is necessary. Gaining forward speed with cyclic control generally incurs less loss of height than autorotation. Because the inner sections of the blades are stalled, cyclic control does not provide a normal response and halt descent immediately. Despite this, gaining forward speed is the recommended immediate recovery action.

As the vortex-ring-state rate of descent may be very high (6000 feet per minute in some cases — I have seen the rate of descent needle pegged on 4000 feet per minute in a Bell UH-1), autorotation may provide an uncontrollable rotor overspeed due to the very high value of inflow angle and autorotation force.

While a state of vortex ring exists, any increase in power will only aggravate the condition by developing the vortex, stalling more of the inboard sections, resulting in an increased rate of descent. You can perhaps realize why this condition is termed 'settling with power' by white-knuckled helicopter pilots. 'Hurtling with power' would be a more apt description in some situations.

Steve Bone, consulting editor

Why avoid vortex ring state?

Four factors make vortex ring state a condition to be avoided at all costs, unless it is part of a demonstration by a competent instructor to satisfy a formal training syllabus.

1. Speed of onset

There is often very little warning of the change from a powered descent to vortex ring state. The usual warning of vibration just before the onset of vortex ring may be absent. The only indication could be a marked increase in the rate of descent from, for example, 800 feet per minute to 2000 feet per minute.

2. Control

The random pitching, rolling and yawing motions sometimes associated with fully developed vortex ring descent cannot be controlled by the pilot.

3. Recovery height

Both recovery options entail a considerable loss of height, possibly up to 800 feet.

4. Flight environment

The maneuvers most conducive to the formation of vortex ring state are normally flown at relatively low levels, allowing pilots little time to recognize the condition and perform effective recovery. Any such maneuvers must therefore be flown with great care.

Maneuvers to stay away from

The following situations can lead to the development of vortex ring state and so should be avoided.

1. Power-assisted descent with low or zero forward airspeed

The rate of descent required for vortex ring formation differs with aircraft types, but development is likely where rate of descent exceeds 500 feet per minute during a low forward airspeed descent. The situation becomes more dangerous with a heavily laden helicopter on a hot day due to the high power setting and rotor blade angles.

It is simplistic to think that power-assisted descent with low or zero forward airspeed can always be avoided. Consider, for instance, the pilot of a twin-engine helicopter during takeoff from a rooftop. Some regulations require pilots to climb their aircraft slightly rearwards so that, in the event of an engine failure, the helicopter can be flown back down to the same rooftop using the remaining engine. It is very possible that this recovery flight profile, the rate of descent and collective setting required to descend a heavily laden helicopter back to its rooftop perch could place it in vortex ring state. At best, the pilot is in for an embarrassingly heavy landing.

2. Downwind maneuvers

Downwind maneuvers – such as crop-dusting turns; downwind flares for quick stops; or quite simply, downwind approaches – are all problematic.

3. Quick stops

When a helicopter is flared in a quick stop with the disc tilting back, the horizontal airflow passing the rotor tends to come from below. If a rate of descent develops, the airflow directly opposes the induced flow.

4. Practice autorotation recovery

A recovery from a practice autorotation that increases power in the flare before leveling the helicopter creates a situation similar to the quick-stop maneuver. This does not happen when carrying out an engine-off landing – the rotor is in autorotation until the collective lever is raised to cushion the touchdown. Recovering from a slow-speed autorotation by using low forward airspeed creates a situation similar to a power-assisted descent with low or zero forward airspeed.

3

Recirculation

Recirculation refers to the reuse of air that has passed through the rotor system and returns to run the gauntlet again. Because the air is already moving when it returns, it causes a variation in the relative wind velocity (speed and direction). The acceleration applied to air that passes through the disc the first time is significant. Subsequent obstacles and courses that the air takes swirl the direction and vary the speed of what, by definition, is now 'localized wind'. As the rotor disc encounters this moving air again, whether over the entire disc or just part of it, the variations can make the helicopter difficult to control and unpredictable.

To some extent, recirculation is also part of overpitching (Chapter 7) and vortex ring state (Chapter 2). When discussing recirculation as a subject on its own, it is generally applied to a helicopter in the hover.

Finding an accident report that attributes the blame for a mishap fairly and squarely on recirculation has proven an insurmountable task. It is impossible to estimate how many accidents have involved recirculation as a contributing factor.

Principles of recirculation

Whenever a helicopter hovers near the ground, some of the induced flow is recirculated; that recirculated air increasing speed as it passes through the disc a second time. This increase of induced flow near the rotor tips gives rise to a loss of rotor thrust.

Some recirculation is always taking place when a helicopter is hovering but, over a flat even surface, the loss of rotor thrust is more than compensated for by ground effect. If a helicopter is hovering over tall grass or near a vertical surface (building, hangar, and so on), recirculation may increase and in some cases the effect will be greater than the ground cushion. When this happens, more collective pitch and power is required to hover near the ground than OGE (see Fig 3.1).

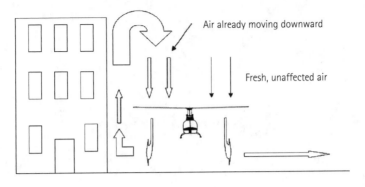

Fig 3.1 Recirculation

When recirculation occurs unevenly around the disc, the blades will flap to equality and require cyclic correction to maintain the hover. Even small pockets of swirling air can have an influence on the rotor disc. Regardless of where a helicopter is hovering, any or all of the blades (or sections) are encountering wildly oscillating portions of moving air.

Perhaps the worst scenario is when a helicopter is hovering adjacent to a sizeable obstacle like a tall building. The spent air (the downwash) has only one place to go — straight up the side of the building (Fig 3.1). At some point, it will turn around and come straight back down again. Obviously the half of the disc closest to the building will be getting far less lift than the opposite side and the pilot will have to correct with cyclic to keep the helicopter level. But this is not an unlimited option. At some point the imbalance may well cause the helicopter to be 'drawn into' the building.

I finished my commercial pilot training with Bruce Harvey, an instructor who ran a 'one helicopter, one student' operation from his farmlet in the Bay of Plenty province of New Zealand. Each morning we would wheel out the Robinson R22 Beta and then close the doors of the hangar. Some fifteen feet behind the tail rotor was the double garage, a not insignificant structure that housed the family vehicles. With less than 150 hours' experience, I used to have a tough job to maintain control of the helicopter because of the recirculation encountered from obstructions both in front and behind. It was like learning how to hover all over again!

The fact that it was not possible to find an accident where recirculation was purely to blame (although there must be one somewhere) shows that this hazard rarely traps pilots on its own. But, if another problem is being encountered, recirculation is not a welcome visitor.

4

Ground resonance

Ground resonance is a vibration of large amplitude through the helicopter, generally resounding (or echoing) between the rotor head and the ground. This can be a forced or a self-induced vibration but the helicopter must have some contact with the ground. The vibration is caused by an imbalance — the center of gravity of the rotor disc is not directly over the center of rotation.

When ground resonance is present, the fuselage rocks or oscillates. As will be seen in case study 4.1, there is often very little opportunity for a pilot to take corrective action between the time that the condition is recognized and the time it develops in full. The amplitude of the oscillation may increase to the point where it is uncontrollable and the helicopter rolls over. In extreme cases, there may be less than two seconds from onset to rolling over! It is important, therefore, that pilots are aware of the circumstances in which ground resonance can occur, and of the need to initiate appropriate recovery action immediately.

Case study 4:1

Date: August 17, 1988

Location: Alice Springs, Northern Territories, Australia

Aircraft: Hughes 269C

Injuries: 2 minor

Occurrence Report No. 8800728

The pilot was cooling the engine down at 2500 rpm after completing a circuit and normal landing. Suddenly, a hard high-frequency bounce developed and the helicopter bounced left skid low followed by a 45° yaw to the left. The pilot had tightened the control frictions after landing and he attempted to release these as he fought for control of the helicopter. He was unable to release the control frictions and the cyclic control grip was broken during his attempts to restrain the movement of this control. The engine stopped without input from the pilot and the aircraft came to rest. The period of ground resonance had been about five seconds. The aircraft was substantially damaged.

During the pre-flight inspection the pilot had noticed that the front left landing gear was slightly under-inflated, although he considered it to be within normal limits.

Source: Bureau of Air Safety Investigation, Australia

The March–April 1980 edition of the *FAA General Aviation News* made mention of a pilot in Seattle who had noticed a low oleo strut on the rental helicopter he was about to fly, but accepted the owner's word that it only appeared uneven because of the slanting ramp. As he was preparing to take off the helicopter suddenly began to vibrate so violently that he was unable to reach the key in the panel switch and turn it off. The machine flopped around on the ramp as the rotor blades were distorted by contact and chewed up the fuselage. Although the cockpit remained intact, the pilot and his passenger were both hospitalized from bruises and back injuries.

The April/May 1984 edition of *New Zealand Flight Safety* reports that at Nelson (New Zealand) a Hughes 269B helicopter with pupil and instructor on board had just alighted on the hard-standing near the fuel pumps at the airport. Normal lateral vibration commenced as the collective was lowered after touchdown but the vibration suddenly increased. The instructor immediately 'squeezed off throttle' but the engine did not respond and the vibration became violent.

The helicopter was shaken apart within about five seconds. All main rotor control rods were severed and the main rotor drive shaft became dislodged. The rotating main rotor blades struck and destroyed the canopy. The tail boom separated from the airframe.

In addition to this accident, *New Zealand Flight Safety* April 1986 tells of a Hughes 269 at Taupo, New Zealand, that touched down on a concrete pad. The pilot lowered the collective fully and rolled throttle closed. The helicopter started to rock violently. This movement rapidly increased in amplitude and the aircraft began to break up. The main rotor rpm was too low to allow the pilot to lift off and stop the ground resonance vibration.

On this occasion one main rotor blade became out of phase during engine rundown after the aircraft had alighted.

Two rotor blade dampers did not have lock-wire fitted to the adjusting nuts, allowing the nuts to back off. First stage torque measurement of these two dampers was found to be 25 percent less than that of the correctly lock-wired third damper.

Principles of ground resonance

The prime requirement for ground resonance is a source of vibrational energy – the main rotor disc. Another important requirement is contact with the ground, including a full landing. If the center of gravity of the rotor disc does not lie precisely on the axis of rotation, a wobble sets up. Because the rotor center of gravity tends to remain in the one spot, the top of the mast is forced to move around as it rotates.

Ground resonance can also be induced by the undercarriage being in light contact with the ground, particularly if the frequency of the oscillation of the oleos and/or tires is in sympathy with the rotor head vibration.

Rotor head vibration

Rotor head vibration can be caused by:

1. Blades of unequal weight or balance

Blades should be correctly weighed and balanced during manufacture but flight in icy conditions can cause imbalance if there is uneven shedding of ice on the blades. Moisture absorption and blade damage can also be causes of imbalance.

2. Faulty drag dampers

In a multi-bladed system the blades should be equally spaced, with drag dampers fitted and adjusted to ensure this. If a faulty damper allows a blade to assume a dragged position different to the other blades, the center of gravity of the entire rotor disc is displaced from the axis of rotation (see Fig 4.1). This was revealed as the main cause of two of the accidents mentioned in this chapter.

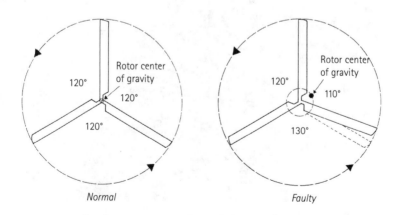

Fig 4.1 Effect of faulty drag dampers *NEW ZEALAND FLIGHT SAFETY*

3. Faulty tracking

Rotor blades are tracked to ensure that the tip path planes of all blades coincide. This is done by making adjustments to the basic pitch settings of the blades. If one blade has excessive basic pitch, its tip path plane will be higher than the others. More importantly, it will have a higher rotor drag and will therefore maintain an excessive dragged position. This causes the same out-of-balance condition as that caused by faulty drag dampers, and a roughness or vibration is apparent to the pilot.

Rotor center of gravity

Fig 4.2 Effect of faulty tracking *NEW ZEALAND FLIGHT SAFETY*

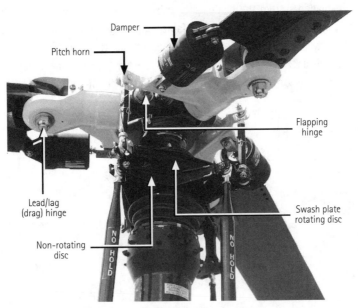

Damper

Pitch horn

Flapping hinge

Lead/lag (drag) hinge

Swash plate rotating disc

Non-rotating disc

Fig 4.3 Components of rotor dampers on a fully articulated helicopter

FAA GENERAL AVIATION NEWS

Fuselage vibration

Fuselage vibration can be caused by:

1. Mishandling of the cyclic stick during landing, which causes the aircraft to bounce from side to side.

2. A taxiing takeoff or run-on landing over rough uneven ground.

3. Incorrect or unequal tire or oleo pressure. This factor was present in at least one of the cases cited.

4. A wheel dropping into a hole or rut on landing, or deplaning personnel contacting the undercarriage when the helicopter is hovering in light contact with the ground.

Avoidance and recovery

The important thing to know is that ground resonance happens quite rarely, but often enough to make you think fast when a helicopter with a fully articulated rotor system starts to rock just a little more than you are used to.

When I was learning to fly in a Hughes 300 in 1981, I went out to start the machine and go on a solo sortie. As I engaged the rotors gradually but at a fairly low rpm the entire airframe started to rock dramatically. I disengaged the rotors again and shut down — if nothing else it was distinctly uncomfortable. I went inside and told an aging (I was 21 at the time) engineer/commercial pilot that I needed to move the helicopter because it seemed to start with ground resonance where it was. He snorted and mumbled derogatory unfathomable remarks about student pilots and stormed out of the hangar to start the machine for me. I watched with interest. He never actually said anything further when he too shut down the engine and came in search of the attachable wheels. Not only did he have a grumpy disposition most of the time, but he was also the husband of my instructor, so I didn't deem it prudent to say anything further myself.

Please note that the current Schweizer 269 series helicopters are much less susceptible to ground resonance. They are equipped with elastomeric dampers and do not need to be phased like the older Hughes manufactured 269s.

If you are an agricultural pilot operating from the deck of a flat-bed truck, or from very spongy or soft ground, you are at a high risk of experiencing ground resonance. Any surface that begins to vibrate with the helicopter can reinforce the rocking motion of the aircraft and trigger ground resonance.

Pilot technique also contributes to setting up conditions that can lead to ground resonance. Any procedure that involves prolonged single-point contact with a surface will certainly aggravate an oscillating condition, possibly to the extent of producing ground resonance. Such procedures include: landing on a slope, hovering too low over uneven terrain, making run-on landings or takeoffs over rough or uneven ground, and raising the collective pitch too slowly or hesitantly on takeoff.

The best means of avoidance is to know about the condition and whether you are flying a helicopter that is susceptible to it. If you are, ensure that your type rating has included precise details for checking drag dampers and oleo or tire pressures, and for knowing how to recognize when a rotor has started to track badly. And be firm when insisting that irregularities are corrected before you fly the helicopter! Don't accept excuses. You will be the embarrassed one sitting in a wreck wondering if all destructing helicopters make noises that loud. The ability to think after such an accident may be solely attributable to wearing a helmet: those rotors coming through the cockpit slow down for no mortal.

As for recovery, a pilot has two choices:
1. Attempt to become airborne.
2. Shut the engine down to stop the out-of-phase rotation – with any luck.

When landing, always maintain flight rpm until you are firmly on the ground. Then, if ground resonance sets in there will be enough rotor rpm to become airborne – end of problem. Set it down very gently and be vigilant. If there is not enough rotor rpm to get airborne, the best option is to shut down as quickly as possible. Any attempt to increase rotor speed also increases resonance.

5

Retreating
blade stall

The rotor blade of a helicopter gains lift by rotational speed creating airspeed over the airfoil. In a no-wind hover the airspeed is more or less the same for all blades around the disc. In forward flight, however, an advancing blade has the forward speed of the aircraft as well as rotational speed. Conversely, this airspeed is subtracted on a retreating blade.

It is possible to reach a forward speed where the retreating blade has insufficient speed (V^2) to produce enough lift, remembering that the advancing blade has a wealth of V^2. The retreating blade will stall. This problem is the primary limiter in how fast helicopters can travel. Great advances have made in blade design specifically to overcome retreating blade stall.

Case study 5:1

Accident Report

Date: June 2, 1980

Location: Vikingstad, Sweden

Aircraft: Hughes 369D

Injuries: 2 serious

This is an abridged version of the accident report. The opinions with regard to helicopter type and manufacturer's responsibilities expressed in the report are not necessarily those of the author of this book.

At 9.10am on June 2, 1980, an anti-tank helicopter leased by the Defense Forces for testing purposes, took off from Linköping/Malmen. The crew consisted of two army pilots.

The pilot in command, who occupied the right seat, was navigating after takeoff, using a map, while the pilot occupying the left seat (the gunner) was flying. The flight proceeded towards a point 2.5 miles north of Vikingstad where a military target car was parked. The car was located at the east end of a stretch of road running in a westward direction, along which the flight test was to start.

The cloud base was at an altitude of almost 330 feet above ground level and the visibility was estimated by the pilot in command to be 2.5 to 3 miles.

When the pilot in command spotted the target car the helicopter was flying on a northwesterly heading at a speed of approximately 110 knots at 160 to 250 feet altitude. He instructed the gunner to follow the previously mentioned road to the left, then continued to read the map. To follow the instruction a left turn with approximately 45° change of heading had to be made.

When the pilot in command looked up from the map he noticed that the helicopter was losing altitude with increasing left-bank angle and that there was a low rotor rpm warning indication. He also had the feeling that the load factor was high before the rotor hit the ground on the left side. The gunner also noticed that the left collective control stick, without his assistance, moved upwards during the left turn.

The target-car crew (two men) was surprised when the helicopter passed close to the vehicle in a left-hand turn at such a large bank angle (approximately 60°). From the target car they also observed that the helicopter lost altitude and collided with the ground with the rotor striking first.

When the rotor hit the ground the speed of the helicopter was approximately 100 knots, the bank angle approximately 45° to the left, the attitude nose-up and the flight direction southwesterly.

The pilot in command was 44 years old at the time of the accident and had 10 years of flight service in the army. His total flying time was 3096 hours. He had been flying the civil version of the Hughes 500 helicopter 60–70 hours and the military version approximately 8 hours.

At the time of the accident the gunner was 35 years old and had seven years of flight service in the army. His total flying time was 1813 hours. He had been flying the Hughes 500 helicopter approximately five hours.

The helicopter, a military version of the civil Hughes 500D, was equipped with dual controls, missile launchers (TOW), missile sight, autopilot and certain measurement and recording equipment.

The gunner's space in the cabin is limited by the missile sight and the left cyclic control was shorter that the right. The gunner's pedals are not adjustable because this is not possible with the missile sight installed. The left pedals are, therefore, fixed in the rear position. It is, however, possible for the gunner to make full pedal movements.

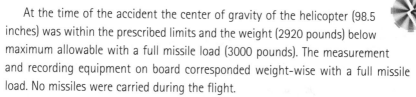

At the time of the accident the center of gravity of the helicopter (98.5 inches) was within the prescribed limits and the weight (2920 pounds) below maximum allowable with a full missile load (3000 pounds). The measurement and recording equipment on board corresponded weight-wise with a full missile load. No missiles were carried during the flight.

Analysis concluded that the left turn leading to the accident was initiated at a speed of 110 knots and at an altitude of 160 to 250 feet. According to the crew the turn was initiated gently from straight and level flight. The intended change of heading was 45° to the left and the gunner (who was flying) banked, according to his recollection, 30–35° to make a quite normal turn.

However, the turn rapidly changed character after it was initiated. The load factor and bank angle increased uncontrollably and the flight path curved sharply at the same time as the helicopter began to lose altitude so rapidly that a collision with the ground could not be avoided. The heading changed 115° instead of the intended 45°.

The way in which the left turn developed indicates that blade stall was obtained, leading to an increasing left bank angle and an increasing load factor. Since a low rotor rpm warning was obtained, the drag increase due to blade stall must have been so high that the rotor rpm was decreased from 102–103 percent (normal) to a value below approximately 98 percent. Altitude was lost since the stalled rotor could not generate sufficient lift.

The time from initial blade stall and load factor increase to impact was short (approximately six seconds). For this reason it was not possible for the crew to regain control of the helicopter. The fact that the collective moved upwards during the turn, without being pulled by the gunner, must be attributed to an imbalance in the feed-back forces from the rotor to the collective stick during the flight. The position of the collective stick in the helicopter in question is affected by a spring working on an over-center linkage in the completely mechanical control system. In this case, the collective control system was adjusted in such a way that the control stick had a tendency to move upwards if the stick motion was not prevented by manual force or by engaging the friction brake. Since the stick motion is rather slow it is not very perceptible to the pilot.

If the blade angle increases in this manner without the pilot observing it, the stall limit will be reached sooner than expected after a turn has been initiated. The collective control stick will then continue to move rapidly upwards, as experienced by a number of Swedish pilots immediately before touchdown during tests of autorotation landings. In this connection it should be observed that, if the collective control stick was raised after blade stall had been obtained, this would aggravate the situation since the increased blade angle would further the development of the stall.

Based on tests and test reports the following may be said of the handling qualities of the Hughes Helicopter: the helicopter has good handling qualities but requires more pilot attention than a Jet Ranger, for instance, which the crew involved in the accident were accustomed to flying.

The following should be observed regarding the Hughes 500:

➤ The controllability is marginal at high speeds. Although the structural limitation speed of the helicopter is 152 knots, Hughes recommends a maximum allowable speed (Vne) of 130 knots.

➤ Serious blade stall disturbances, large stick forces and flight attitude instability are obtained at 110–120 knots in turns at 30–40° bank angles at near maximum weights.

➤ The helicopter is usually flown at near maximum weight and near maximum forward center of gravity position. As a result stall is obtained at the fairly shallow bank angles.

➤ The helicopter is sensitive to uncoordinated flying. A faulty pedal depression in a normal steep turn may, for instance, result in a bank angle increase, a tightening of the turn and a loss of altitude. This is, however, most pronounced in right-hand turns.

Handling qualities of the type found in tests that limit the maneuverability of the helicopter should logically be described in the helicopter flight manual and, if required from a flight safety point of view, result in maneuver limitations and/or flight limitations. This is normally done for fixed-wing aircraft (especially military aircraft) but is, according to the board's experience, seldom done for helicopters. The reason for this is not known by the board.

From a flight safety point of view the situation is, however, unsatisfactory. It should be natural that the manufacturers and manufacturing countries' aviation authorities supply the user/operator with as much information as possible about maneuver limitation, even for helicopters.

Probable cause of the accident was reported as:

During a left turn at 160 to 250 feet altitude at approximately 110 knots and a bank angle of 30–35°, blade stall was obtained leading to tightening of the turn. Due to the stall it was not possible for the rotor to develop sufficient lift and the helicopter lost altitude. The available altitude was too low to prevent collision with the ground.

The handling quality of the helicopter in the actual case, which the crew was not made aware of and which amounted to (serious) blade stall, large control forces and instability, was probably the main contributing factor in the accident.

In addition to this the following factors may have contributed:

- The pilot flying the helicopter was occupying the left seat, the gunner's position, where limited space makes precision control difficult.
- The pilot flying the helicopter had limited experience of the helicopter type.

Source: **Swedish Bureau of Accident Investigation**

Principles of retreating blade stall

The principle of retreating blade stall is easy to grasp and requires no real in-depth explanation (see Fig 5.1).

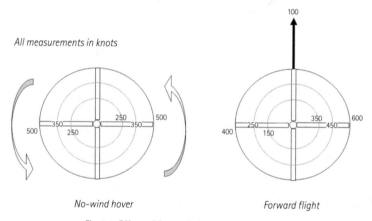

Fig 5.1 Effect of forward airspeed on blade velocity

Accidents solely attributable to retreating blade stall are rare but the danger is an appreciable one. Pilots should know this and have it in mind when the following factors are present:

1. Heavy loading on blades (weight of aircraft plus load).
2. Low rotor rpm.
3. High density altitude.

4. Steep or abrupt turns.
5. Turbulence.

All of these factors are exacerbated when operating at high forward airspeeds, as shown in Fig 5.2.

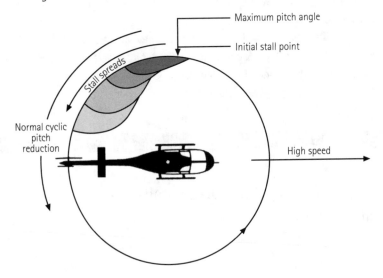

Fig 5.2 Retreating blade stall

Warning signs that a retreating blade stall may be imminent are:

1. Nose of aircraft pitches up.
2. Abnormal vibrations.
3. The helicopter starts to roll in the direction of the stalled side of the disc.

Avoidance

Retreating blade stall normally occurs when airspeed is high. To prevent it, the pilot must fly slower than normal when:

1. The density altitude is very high.
2. Flying near, at or over maximum all-up weight.
3. Flying with a high drag-generating configuration — for example, external loads, sling loads and floats.
4. Turbulence is being encountered.

Recovery

The following steps can be taken if a pilot suspects blade stall is occurring, or is about to occur:

1. Reduce power – lower the collective pitch lever.
2. Reduce airspeed – hold the nose up and maintain level flight with less power.
3. Introduce less 'G' loading when maneuvring.
4. Keep rpm to the top of the allowable range.
5. Keep the aircraft balanced – negate yaw through good pedal control.

In a fully developed blade stall, the pilot loses control. The helicopter pitches up violently and rolls to the left. In a helicopter with main rotors turning clockwise, the roll will be to the right. Still, the pilot must attempt to make the changes listed above. Each change that is achieved will shorten the duration of the stalled condition and should, with any luck at all, return control to the pilot.

6

Dynamic rollover

Dynamic rollover occurs when the skid or wheel of a helicopter is touching a fixed object, such as the ground or a structure, while a sideways force is being applied from above (the rotor mast). The helicopter pivots around the fixed point and rolls.

In one reported case of dynamic rollover, a skid had frozen to the ground in a severe frost. In another, a skid had stuck to melted bitumen. A sling rope inadvertently draped over a skid can also cause rollover.

A complete rollover of the helicopter can happen in a matter of a few seconds, even when the pilot is aware of the hazard. In fact, this hazard is the cause of so many accidents that two case studies have been included. Pilots with new ground crew are advised to have their crew read this chapter. The life they save may be their own!

Case study 6:1

Aircraft Accident Report Number 86-085

Date: October 9, 1986

Location: Motunau, North Canterbury, New Zealand

Aircraft: Aerospatiale SA315B Lama

Injuries: 2 fatal, 2 serious, 1 minor

The pilot flew ZK-HNQ in to assist in the re-floating of the fishing vessel *Howard L Shadbolt*, which had run aground two days earlier. Other helicopters had already been employed to uplift the crew following the grounding and to transfer men and equipment during the initial attempts to re-float the vessel. But a helicopter with the lifting capacity of the SA315 Lama was required to enable heavy pumps and similar equipment to be placed on board.

ZK-HNQ arrived at 1600 hours on October 6 and shortly afterwards a briefing session was conducted by the pilot for all personnel who were involved at that time with the re-floating operation. The purpose of the briefing was to explain the correct procedure for embarking and disembarking from the helicopter.

The helicopter was normally flown from the right front seat and the pilot sat in the aircraft and explained to those present (an estimated 10 persons) the procedure that was to be followed when getting into and out of the helicopter. He emphasized that no one was to jump onto the helicopter; a slow movement was required both when entering and when leaving, and he stressed that only one person was to embark or disembark at a time. Personnel were to wait until the previous passenger was positioned on the far side of the aircraft and the pilot had given a nod before climbing aboard.

During the following three days, the pilot flew numerous sorties in ZK-HNQ, ferrying personnel to and from the vessel and transporting items of equipment. The correct procedures were re-emphasized to one crew member who was inclined to climb heavily on to the skid and jump prematurely from the helicopter before landing. The operations proceeded smoothly and no difficulty was encountered by the pilot in embarking and disembarking personnel on board the *Howard L Shadbolt.*

The fishing vessel had run aground on the shingle and rock beach at the base of a line of cliffs some 295 feet high and the helicopter operation took place from a suitable paddock that bordered the nearby cliff edge.

Operations on board the *Howard L Shadbolt* could be readily observed from the vantage point of the cliffs and the adjacent paddock. Hand-held transceivers were used to maintain contact between personnel on board and those ashore for the purpose of coordinating the efforts to re-float and free the vessel.

The pilot had developed a flight pattern in which, after lift-off from the paddock, he would make a descending turn to the north over the coast and then proceed towards the bow of the *Howard L Shadbolt* to approach the heeled vessel from its lowest side. It was his practise to come in dead slow, bringing the helicopter alongside the boat's wheelhouse at a suitable height and then easing ZK-HNQ over to the right until the nearest skid was about $1\frac{1}{2}$ feet above the top of the wheelhouse itself, using a starboard ventilator mounted on the deck as a reference point.

The crewmen then boarded the helicopter one at a time from the platform formed by the top of the wheelhouse. Once a crewman had climbed onto the skid, the pilot would move the aircraft away, allowing time for the crewman to pull himself into the aircraft's rear passenger compartment and to occupy the seat on the left side, before he eased the helicopter back into position to uplift the next crew member.

On October 9 operations commenced just before daylight as it was hoped that the *Howard L Shadbolt* could be floated off at high tide, which occurred at

0900 hours. The attempt to free the vessel was unsuccessful and at about 1100 hours it was decided to shift one rope from the bow to the stern in preparation for a further attempt to pull the vessel off at the next high tide. ZK-HNQ was employed to ferry a total of five crewmen out to the vessel to perform this operation.

Following the completion of the task, the helicopter lifted off from the paddock at about 1155 hours to pick up the five crewmen. The pilot intended to make two trips, bringing back three men on the first sortie, and then returning to uplift the remaining two.

There was virtually no wind, just a very slight breeze. The vessel, which was then lying broadside to the beach, was rolling lightly at times due to the local swell, but was still fast on the bottom and was restrained from any violent motion by wire ropes which ran out to fishing vessels further offshore. The pilot estimated that vertical movement of the ventilator reference due to wave action on the hull was not more than 20 inches at the time. Two hours remained before low-water and the water depth around the hull varied between four and six feet, with the vessel lying moderately heeled to starboard.

The pilot made a normal approach toward the bow from the offshore side and could see two men waiting on top of the wheelhouse. He positioned the helicopter in the usual manner, adjacent to the radar tripod and, as he did so, the first crewman climbed onto the aircraft's skid. The pilot was just about to ease ZK-HNQ to the left to allow the first passenger to settle when, without warning, the second crewman jumped on the skid to the rear of the first crewman.

The sudden increase of mass on the skid resulted in the helicopter moving sideways to the right. At the same time, the pilot who had already applied left cyclic in an attempt to correct the imbalance, found that the control had reached its stop.

The tripod support for the vessel's radar scanner was secured to the top of the wheelhouse. The scanner itself had been removed but a steel mounting plate welded to the upper part of the tubular structure remained in place. The helicopter's sideways movement caused its right skid to contact the starboard tubular support and the aircraft moved upwards until the skid lodged beneath the extended edge of the mounting plate. ZK-HNQ then rolled rapidly onto its right side and 'spiraled' aft into the gantry structure mounted amidships. (See Figs 6.1, 6.2 and 6.3.)

During the accident sequence the helicopter's main rotor blades struck two crewmen who were standing on the deck, killing them instantly. One of the fatally injured crewmen was knocked overboard by the force of the impact. The third crew member escaped without injury.

The helicopter came to rest in a nose-down, semi-inverted attitude. The pilot,

Fig 6.1 *Howard L. Shadbolt* with Lama wreckage visible

TRANSPORT ACCIDENT INVESTIGATION COMMISSION, NEW ZEALAND

Fig 6.2 Removal of Lama wreckage

TRANSPORT ACCIDENT INVESTIGATION COMMISSION, NEW ZEALAND

who was suspended in his safety harness, managed to release himself and egress from the aircraft onto the deck. He found the second crewman who had attempted to board ZK-HNQ injured and leaning against the railing. The first crewman to climb onto the skid was lying unconscious nearby.

At the time of the accident two crew members who were to be picked up on the second trip were standing on the deck of the vessel aft of the wheelhouse towards the gantry structure; the third crew member who was to have been lifted on the first sortie was about to make his way to the wheelhouse platform.

The accident was witnessed by the two-man crew of a jet boat that was lying off the vessel after assisting with the morning operations. After maneuvering as close to the vessel as practicable, one of the crew donned a life jacket and swam to the *Howard L Shadbolt* to render assistance. At the same time, those on the clifftop who had been supervising the activity onboard had observed the accident and contacted a local station homestead by radio. The police and medical authorities were immediately advised.

The injured survivors were later lifted from the vessel by the crew of a Royal New Zealand Air Force helicopter and subsequently transferred to hospital in Christchurch.

The accident occurred in daylight at about 1200 hours.

The helicopter was destroyed. During the impact sequence the rear section of the tail boom and tail rotor installation was severed from the main structure, and the engine assembly was dislodged and fell into the sea.

The pilot in command was 36 years of age and had a total flight record of 1873 hours; 1000-odd hours had been flown in a Lama helicopter. In the 90-day period preceding the accident he had flown a total of 119 hours, 69 of which were in a Lama helicopter. The pilot estimated that over the period of four days during which he had been operating to and from the *Howard L Shadbolt* a total of 39 flights were carried out, taking personnel to the vessel or returning them from her. In addition, a further 48 flights had involved slinging equipment onto and off the vessel. All of these flights had proceeded without incident.

Using the weight and balance section of the AS 315B Lama flight manual, longitudinal and lateral center of gravity limits were calculated by estimation. The fuel quantity was estimated at 40 US gallons and the all-up weight of the aircraft at 2925 pounds immediately prior to the accident.

The aircraft was being flown with the right-hand door removed and the left door in place.

The position of the longitudinal center of gravity immediately prior to the accident was estimated as 120.86 inches aft of the datum. The position of the lateral center of gravity was estimated at 0.78 inches (positive). Both the longitudinal and the lateral centers of gravity were within the applicable limits.

The estimated all-up mass of ZK-HNQ with the two crewmen located on the right skid was 3267 pounds. The estimated position of the longitudinal center of gravity and this configuration was 117.32 inches aft of the datum. The estimated position of the lateral center of gravity was 4.05 inches (positive). In this configuration the longitudinal center of gravity was within the applicable limits but an uneven distribution of external load existed and the lateral center of gravity position exceeded the specified limits.

With regard to meteorological information there was insignificant cloud, visibility was 20 miles and there was a six-knot southerly breeze.

With regard to survival aspects of the accident; as always personnel within the vicinity of helicopter operations were at risk from the potential of the rotating rotors in the event of an unforeseen occurrence. The degree of exposure to the rotating main rotor blades of ZK-HNQ, which were under power at the time that the helicopter rolled out of control, rendered this accident unsurvivable for those who were struck by the blades. The unforeseen nature of the accident and the rapidity of the events afforded no opportunity for them to escape from the path of the blades.

Fig 6.3 Lama wreckage onshore
TRANSPORT ACCIDENT INVESTIGATION COMMISSION, NEW ZEALAND

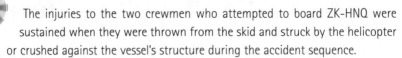

The injuries to the two crewmen who attempted to board ZK-HNQ were sustained when they were thrown from the skid and struck by the helicopter or crushed against the vessel's structure during the accident sequence.

Although all-weather protective clothing was worn by the crewmen in the course of the duties during the salvage operation — it provided protection from the elements and reduced the extent of laceration and minor injury — such clothing could not enhance materially the chances of survival in an accident of this nature.

The pilot, who was dazed but uninjured, was effectively restrained by the full safety harness that he was wearing at the time. He was also wearing a protective helmet.

Additional information contained in the report included:

The first crewman who was to be a lifted from the *Howard L Shadbolt* immediately prior to the accident was a marine engineer who acted as second-in-command of the operation to re-float the vessel and who was responsible for directing the work onboard. He had been present at the original briefing conducted by the pilot after the arrival of the helicopter and was familiar with the technique involved in boarding the helicopter from the wheelhouse roof, having been transported to and from the vessel on numerous occasions.

The second crewman was the owner of a locally based fishing boat and had assisted in running ropes to the *Howard L Shadbolt* during early attempts to free her in the days immediately following the grounding. He was normally based in Christchurch and was not in the area when Lama helicopter ZK-HNQ had arrived at Stonyhurst. Consequently, he was not present when the briefing was conducted for those who were to be transported to and from the vessel in the helicopter.

He had come back to the scene on the day following the helicopter's arrival and had observed the efforts to re-float the vessel from the clifftop during the afternoon. On that day he had accompanied the pilot of ZK-HNQ on a flight in which some equipment had been brought ashore. Later he had returned to Christchurch.

On the morning of the accident, after traveling to Stonyhurst, he had been ferried out to the *Howard L Shadbolt* in the helicopter as one of the five crewmen who were to change the position of wire rope. He did not recall receiving any safety instruction regarding the helicopter operation. Before the helicopter returned the marine engineer indicated to him to, 'Take it easy getting on, as it was easier getting off than on'. He understood that ZK-HNQ would make two trips.

The second crewman reported that he found it necessary to hold onto the radar tripod due to the slippery surface and the angle of the wheelhouse roof. As the helicopter came in, he watched the marine engineer climb onto the skid. The helicopter drifted away about a foot and the second crewman formed the impression that he had missed the opportunity of getting onto the skid. Within 'a couple of seconds' the helicopter drifted back again and he stepped onto the skid, while the marine engineer was still in the process of climbing from the skid into the helicopter. Very shortly afterwards the helicopter drifted into the radar tripod. His last recollection before regaining consciousness on the deck was seeing the aircraft's skid at the top of the tripod.

The second crewman indicated to the pilot that he had previous experience in helicopter operations. In view of this experience and the activities in which the crewman had already been engaged in connection with the salvage efforts, including a flight in ZK-HNQ, the pilot believed the crewman was aware of the procedure that had been established for boarding the helicopter from the wheelhouse roof of the *Howard L Shadbolt.*

The final analysis of the official accident report found that:

The second crewman, while not present at the briefing, was nonetheless familiar with the general sequence of operations of the helicopter but had not received specific instructions regarding the procedures established for getting into or out of the aircraft. An individual briefing covering these aspects was not provided as the second crewman had indicated to the pilot that he had previous experience in helicopter operations. The fact that he had flown in ZK-HNQ while assisting in earlier salvage attempts led the pilot to believe that he was already conversant with the procedure for boarding the helicopter from the *Howard L Shadbolt.*

It was evident that the second crewman's action in attempting to board ZK-HNQ while the first crewman was still on the skid, occurred on the spur of the moment. Not realizing the necessity for the pilot to maneuver the helicopter clear of the wheelhouse deck and allow time for the first crewman to seat himself on the opposite side of the aircraft, the second crewman attempted to board the helicopter without delay. The difficulty in maintaining an adequate foothold on the deck, which was slippery at the time and the impression that the opportunity to board would otherwise be missed, were contributing factors in his action.

The calculation of the lateral center of gravity position was based solely on the static mass of the two crewmen positioned on the right skid. The addition of the dynamic load of the crewman landing on the skid after his sudden jump would have displaced the lateral center of gravity even further. The calculation showed that the position of the lateral center of gravity moved to the right significantly beyond the permitted limits. As a result, despite the input of full opposite lateral cyclic, the pilot was unable to prevent the helicopter moving sideways to the right and contacting the tubular steel leg of the adjacent radar tripod.

The angular momentum developed about the right skid as it slid against the tripod leg and became trapped beneath the upper mounting plate, culminating in an uncontrollable rollover of the helicopter to the right while the main rotor blades were still rotating under power. These circumstances precluded action by the pilot to recover the situation.

The conclusion of the accident report stated:

> The probable cause of this accident was that the pilot in command was deprived of full lateral control of the helicopter while hovering, due to an unexpected additional load on the right skid, and was then unable to prevent the aircraft from moving sideways, contacting an adjacent obstacle and subsequently rolling over. A causal factor was that one of the persons likely to travel in the helicopter did not receive specific briefing on how to board the aircraft.

Source: Transport Accident Investigation Commission, New Zealand

Case study 6:2

Aircraft Accident Report Number 95-009

Date: June 6, 1995

Location: Mount Stevenson, South Island, New Zealand

Aircraft: Bell 206B JetRanger III

Injuries: nil

The 37-year-old pilot had three tourists on board his helicopter (ZK-HDI). They were on a scenic flight in the Mount Cook National Park and planned to include a snow landing.

Another pilot flying an AS350B Squirrel helicopter for the same company, senior in experience to ZK-HDI's pilot, had advised that the usual landing site was unsuitable for a snow landing on this day due to turbulence in a strong southwest wind. As an alternative, the Squirrel pilot landed on the northeast shoulder of Mount Stevenson, an area that he often used in these conditions. After landing and

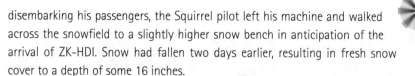

disembarking his passengers, the Squirrel pilot left his machine and walked across the snowfield to a slightly higher snow bench in anticipation of the arrival of ZK-HDI. Snow had fallen two days earlier, resulting in fresh snow cover to a depth of some 16 inches.

The JetRanger pilot noted the position of the parked Squirrel helicopter as he flew over the ridge of Mount Stevenson in a southerly direction. He assessed local wind and surface conditions and set up a left circling approach to the site. The pilot on the ground assisted in accurate determination of wind direction by throwing snow into the air as ZK-HDI approached to land.

The elevation of the landing site was approximately 6600 feet. The Bell 206 pilot had not landed at this particular site before. His proposed touchdown area lay in a shallow, slightly dished region of the snow slope to the south and well clear of the parked helicopter.

The pilot estimated the wind as 5 to 8 knots from the west. There were no significant gusts. His final approach and landing was on a northwesterly heading, resulting in a light crosswind component from the left.

As the pilot allowed the helicopter to settle onto the snow he felt the right skid sink more deeply than the left and, at the same time, was aware that he was applying considerable left cyclic control. Not being happy with the situation, he lifted off and traversed some 11 to 16 yards to the right, to an area that appeared to be flatter.

Touching down again, the helicopter felt stable. The pilot lowered the collective lever but as ZK-HDI's skids settled into the snow the right skid dropped. The pilot recalled having the cyclic control hard against his left leg. The right skid continued to go down and within seconds the helicopter rolled onto its right side. The mast and transmission assembly cracked, and the main rotor blades were destroyed as they struck the snow. The cockpit and tail rotor assembly suffered little damage but the engine sustained over-temperature indications and other damage typically associated with a power-on rollover mishap. (See Figs 6.4 and 6.5.)

Events occurred so quickly that the pilot was unable to recollect with certainty his actions at the time but believed it was likely he had raised the collective in an endeavor to lift off, as he had done previously. In fact, video tape from the camera of one passenger recorded the accident sequence and this confirmed the short time-frame of about two seconds within which the final roll developed and took place.

None of the occupants sustained any injury and all persons had been ferried to the Glentanner base within 30 minutes in the Squirrel helicopter.

At the time of the accident, a high overcast resulted in diffuse lighting rather

Fig 6.4 TRANSPORT ACCIDENT INVESTIGATION COMMISSION, NEW ZEALAND

Fig 6.5 TRANSPORT ACCIDENT INVESTIGATION COMMISSION, NEW ZEALAND

than bright sunlight. The pilot had no difficulty in establishing adequate surface references during the landing. The snow was wind-rippled and dry; despite its freshness it was well compacted and somewhat heavy in texture. The rotor downwash did not produce any blowing snow.

There was a slight slope from left to right where the pilot first touched down. The area to the right to which he then moved had appeared to be flatter but the pilot noted after the accident that it also sloped from left to right at an estimated 5° angle. The pilot did not recall any residual sideways or forward movement, as he allowed ZK-HDI to settle during the second landing, which might have induced or accentuated a tendency to roll. The skid marks observed by the recovery team confirmed that the helicopter initially had settled squarely into the snow.

The pilot's report and the circumstances surrounding the accident to ZK-HDI were consistent with the occurrence of dynamic rollover.

Source: Transport Accident Investigation Commission, New Zealand

Principles of dynamic rollover

Several factors can contribute to dynamic rollover, but the main factor is the thrust of the main rotor when it is tilted from the vertical. If the rotor disc is not tilted to the horizontal or if there is no thrust, then dynamic rollover will not occur.

Other factors that contribute are tail rotor thrust, especially when the tail rotor is located high up on the helicopter, and a down-slope crosswind. Also, for a rollover to occur, the skids must be in contact with a high friction surface. If the surface is slippery the helicopter will tend to slide and not pivot when a sideways force is applied. In addition, a helicopter with a high center of gravity and/or on narrow skid tracks is more susceptible to dynamic rollover than one with a low center of gravity and/or wide skids.

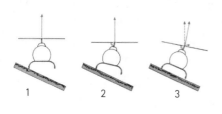

1. Helicopter in hover.
   One skid on ground.
   No roll.

2. Helicopter in controlled roll downhill until cyclic control reaches limit.

3. Any further roll is compounded by increasing side component of rotor thrust. Dynamic rollover is underway.

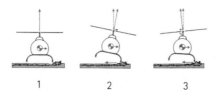

1. Helicopter in hover and drifting sideways.

2. Skid contacts obstacle and rolling moment is produced. If no cyclic correction is applied, side component of rotor thrust adds to rolling moment and dynamic rollover is underway.

3. If cyclic correction is applied in sufficient amount and before cyclic control limit is reached, dynamic rollover is arrested.

Fig 6.6 A simple illustration of dynamic rollover NEW ZEALAND FLIGHT SAFETY

Avoidance and recovery

A feature of dynamic rollover is that the helicopter's center of gravity pivots around the skid; therefore, once a roll is started it may be difficult, if not impossible, to stop. Since a pilot may not be able to stop the roll, the only way to be sure a dynamic rollover does not occur is to be aware of the problem and try to eliminate as many of the contributing factors as possible.

Avoiding dynamic rollover

When maneuvering with one skid on the ground, care must be taken to keep the helicopter trimmed — particularly in the lateral plane. If a slow takeoff is attempted and the tail rotor thrust contribution to the rolling moment is not trimmed out with cyclic, the critical recovery angle can be exceeded in less than two seconds.

Control can be maintained if the pilot does not allow helicopter control input rates to become large, and keeps the angle of bank from becoming too great. The helicopter must be flown into the air smoothly, with excursions in pitch, roll, and yaw kept to a minimum.

Level ground operations

Dynamic rollover can occur while operating on essentially level ground. If, through side drift or angle of bank, a skid is allowed to come into contact with any obstruction, then pivoting about the skid can occur. (See Fig 6.7.)

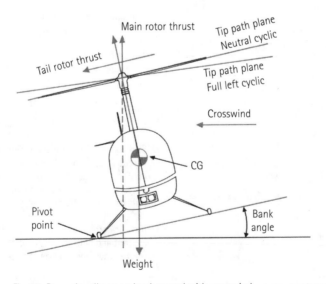

Fig 6.7 Dynamic rollover on level ground with crosswind *NEW ZEALAND FLIGHT SAFETY*

Obstructions can be as varied as tussock clumps, refueling hoses, or even simply soft ground. When taking off or landing from level ground, therefore, at the point where one skid is lightly on the ground, use the flight controls to carefully maintain the helicopter position in relation to the ground. Maneuvers should be performed smoothly and the helicopter trimmed so that no helicopter movement rates build up, especially roll rate. If the bank angle starts to increase to a large angle, say five to eight degrees, and full corrective cyclic does not reduce the angle, then collective must be reduced. This will diminish the unstable rolling movement created by the thrust vector.

Sloping ground operations

During sloping ground operations, follow the published procedures for the helicopter type. Consult the pilots' notes to determine the degree of slope that can be accommodated safely without reaching the lateral cyclic limits. The area of critical rollover will vary for different types — depending on vertical and lateral center of gravity, landing gear configuration, tail rotor thrust, and so on.

Downslope rolling tendency

Takeoff

When taking off with the helicopter facing across the slope, slowly raise the downslope skid to bring the helicopter to a level attitude, and then lift off. If the upslope skid starts to leave the ground before the downslope skid, smoothly lower the collective and check to see that the downslope skid is not caught on some object. Make the helicopter do what you want it to do before it becomes uncontrollable — accept nothing less than a vertical ascent. (See Fig 6.8.)

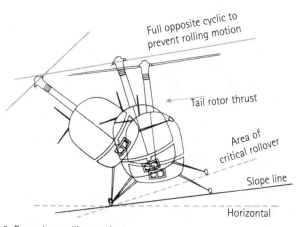

Excessive application of collective pitch lacking coordination with cyclic application into the slope. When the downslope skid is on the slope, excessive application of collective may result in the upslope skid rising sufficiently to exceed lateral cyclic limits and induce a downslope rolling motion.

Full opposite cyclic to prevent rolling motion

Tail rotor thrust

Area of critical rollover

Slope line

Horizontal

Fig 6.8 Downslope rolling motion *NEW ZEALAND FLIGHT SAFETY*

Landing

Before landing, consideration must be given to the combined effects of slope gradient, wind, load position and soil stability. When these factors have been assessed and you have decided to land, touchdown on the upslope skid first – then slowly lower the downslope skid. If the cyclic limits are reached before the downslope skid is resting firmly on the ground, return to the hover, take off, and select a more suitable landing site.

Upslope rolling tendency

During takeoff or landing, if the helicopter rolls to the upslope side, reduce collective to correct the bank, and return to the level attitude – then start the procedure again. (See Fig 6.9.)

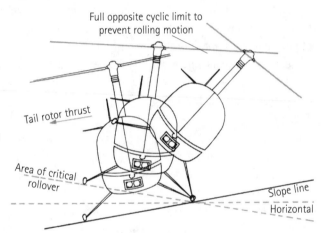

Full opposite cyclic limit to prevent rolling motion

Excessive application of cyclic into the slope, in coordination with collective pitch application. During takeoffs, this condition results in the downslope skid rising sufficiently to exceed lateral cyclic control limits and an upslope rolling motion occurs.

Tail rotor thrust

Area of critical rollover

Slope line

Horizontal

Fig 6.9 Upslope rolling motion *NEW ZEALAND FLIGHT SAFETY*

Collective is much more effective in controlling the rolling motion than lateral cyclic because it reduces main rotor thrust. A smooth, moderate collective reduction of less than about 40 percent (at a rate of less than full travel in two seconds) is sufficient to stop the rolling motion. Care must be taken not to lower collective too quickly, as this could cause the rotor blades to strike the fuselage. In addition, if the helicopter is on a slope and a roll starts to the upslope side, reducing collective too fast will create a high roll rate in the opposite direction and, when the downslope skid hits the ground, the dynamics of the motion can cause the helicopter to bounce off the upslope skid, with inertia then causing it to roll about the downslope skid and over onto its side.

Further slope operation precautions

➤ Less lateral cyclic control will be available during crosswind operations with the wind coming from the upslope side.

➤ Slope operations should be avoided in tailwind conditions.

➤ Less lateral cyclic will be available for left-skid-into-slope operations (right on helicopters with main rotor turning clockwise) because of the effect of tail rotor thrust.

➤ If passengers or additional crew members are picked up or off-loaded after landing, the lateral cyclic requirement will change and must be re-evaluated before lifting off.

Conclusion

Whether or not the helicopter rolls over depends on how quickly and accurately the pilot analyses the problem and initiates corrective action by lowering the collective. Incorrect control application can intensify the rolling motion and place the helicopter in an unrecoverable attitude.

One thing to add is that, when in a helicopter with a fully articulated rotor system and landing on a slope, as well as watching for and guarding against dynamic rollover, the pilot must also be aware of any risk of ground resonance — to have both occur in the same landing or takeoff would be quite a handful!

7

Overpitching

Overpitching is the term used to describe what happens when a pilot manages to stall all the rotor blades at the same time because there is collectively too much pitch on them for the given circumstances. It is akin to pulling the nose up on a fixed-wing aircraft until the wings stall.

Case study 7:1

Occurrence Brief 9700230

Date: January 26, 1997

Location: Geelong Heliport, Victoria, Australia

Aircraft: Hughes 269C

Injuries: 3 minor

After conducting 15 joy flights from the Geelong Heliport from about 1145 hours, the pilot flew two passengers on an advertising project. One of the passengers, a local radio announcer, was tasked to perform a live radio cross from the helicopter.

On the commencement of the last approach to land to the southeast, the pilot estimated the wind to be 15 knots from the northwest. He advised that, when on short final, 220 yards from the helipad, he noticed a sudden wind shift and the helicopter experienced loss of lift. He increased throttle and collective in an attempt to maintain height and reach the helipad. However, rotor rpm decayed and the helicopter settled into the sea 65 yards short of the helipad.

Subsequent viewing of amateur video footage showed the latter part of the landing approach to be very shallow, followed by a classic example of overpitching, with the main rotor disc coning and the rpm audio decreasing as the helicopter descended into the sea.

No fault was reported with the airframe or engine that may have contributed to the accident.

Source: Bureau of Air Safety Investigation, Australia

Case study 7:2

Accident Report LAX97TA289

Date: August 15, 1997

Location: Tahoe City, California, United States

Aircraft: Bell UH-1H

Injuries: 1 minor

The public-use helicopter was transporting fire-fighting personnel to a nearby forest fire at a small lake in the Sierra Nevada Mountains. The pilot said he calculated the allowable load for the hottest time of the day when he reported for work that morning. After dispatch for the fire mission, the pilot flew to the small lake where he determined that the area, 'appeared large enough to land and that the obstructions were 50-foot tall trees'. He made an approach over the lake and landed on the beach and was told that they were no longer needed. The pilot said that as he lifted off from the lake edge he ran out of blade pitch and available power as he tried to climb above the trees and was forced to reverse direction. At the middle of the turn the aircraft began descending rapidly toward the water, and he attempted to accelerate the aircraft and pulled power until the rotor rpm began to drop. The helicopter settled into the lake with about 10 knots of forward airspeed. The pilot said the helicopter remained upright on the water with the skids immersed and a drooped main rotor rpm. The pilot reduced collective and recovered rpm with the intention of hovering the helicopter back to the beach area. After the rpm returned to normal, he increased collective, and the helicopter rolled to the right and sank inverted. Review of the cargo manifest, load calculation sheet, and hover ceiling chart show that the helicopter was about 800 pounds over the hover out of ground effect ceiling for the temperature and pressure altitude. The density altitude was computed to be approximately 9500 feet.

The National Transportation Safety Board determined the probable cause(s) of this accident as:

The pilot's decision to attempt a takeoff in environmental conditions which exceeded the helicopter's performance capability, and his improper use of the collective, which caused a decay of the main rotor rpm, the collision with the lake, and the subsequent rollover. Factors in the accident were the high density altitude and the confined area.

Source: **National Transportation Safety Board, United States**

Case study 7:3

Occurrence Report 198803490

Date: October, 19 1988

Location: Northeast of Mount Surprise, Queensland, Australia

Aircraft: Robinson R22HP

Injuries: nil

The pilot reported that he was hovering the helicopter at about 200 feet above ground level. The wind was blowing from 90° left of the helicopter's heading at about 15 knots, with some higher gusts. As he moved the cyclic pitch control to accelerate forward, a greater than normal sink rate developed. He increased power to compensate but there appeared to be no response and he heard the low rotor rpm warning horn sound.

The pilot then fully lowered the collective pitch control and applied full throttle. He expected the engine rpm to recover quickly but the horn continued to sound. He saw the engine/rotor rpm gauge indicating 85–90 percent. At this stage he was forced to maneuver the helicopter, using all available power, around terrain and trees. A short time later he again fully lowered the collective pitch control in an effort to regain engine and rotor rpm, however, the horn continued to sound.

Because of approaching trees, the pilot was forced to land in a less than ideal area. The main rotor blades struck a small tree and the helicopter touched down firmly on the left skid causing damage to the engine frame.

No fault was found with the helicopter that might have contributed to the accident. It is possible, however, for there to have been some transient fault, the evidence of which was lost during recovery of the helicopter by vehicle over very rough terrain. The other possible cause of the loss of rpm is that the main rotor blades were overpitched when the pilot first reacted to the excessive sink.

There were a number of other aspects-that were likely to have affected the outcome. The helicopter was operating at high all-up weight, near the boundary of the flight envelope for hover out of ground effect. This meant there was little excess power available to counter the loss of lift when the helicopter was accelerated forward. By not operating the helicopter into wind, the pilot was increasing the time and power required to achieve translational lift. The maneuver the pilot was forced to make to avoid terrain and trees (using all available power) would have negated any rpm recovery achieved from the initial lowering of the collective pitch control.

In summary, while there is doubt whether there was an initial power loss, a number of factors were present which would have rendered an effective recovery difficult, if not impossible, from the ensuing situation.

Source: **Bureau of Air Safety Investigation, Australia**

Principles of overpitching

High pitch angles mean high rotor drag, which means more power is required to maintain rotor rpm. If pitch is increased without increasing power, the rotor rpm will drop (see Fig 7.1). As a result, there will be less centrifugal force and the blades will cone upwards. This coning not only decreases overall lift by decreasing the disc area, but also decreases components of rotor thrust, which further reduces the effective total value (see Fig 7.2). Now even more pitch is required for rotor thrust, and so it goes on.

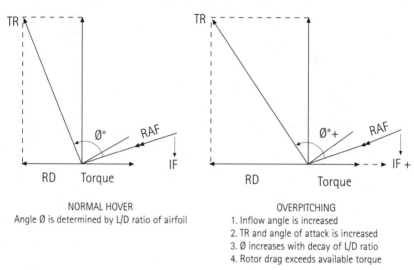

NORMAL HOVER
Angle Ø is determined by L/D ratio of airfoil

OVERPITCHING
1. Inflow angle is increased
2. TR and angle of attack is increased
3. Ø increases with decay of L/D ratio
4. Rotor drag exceeds available torque

Fig 7.1 A comparison of normal hover and overpitching

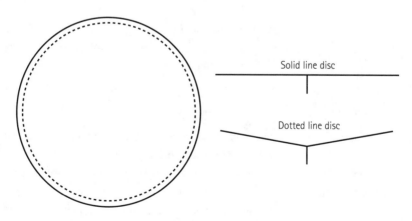

Fig 7.2 Coning reduces rotor disc size

Overpitching can occur at any altitude and at various stages of flight, but is more likely to occur when approaching hover or during a hover. If hovering is attempted with rotor rpm at too low a value, there may be insufficient power available to maintain sufficient rotor rpm. The resultant overpitching can then easily result in a crash or a very heavy landing.

There are several things that will affect a helicopter's susceptibility to overpitching. Indeed, it's fair to say that overpitching is generally the result of many of the other hazards dealt with in this book, including: lack of power due to mechanical problems, high density altitude, overloading and downwind maneuvers.

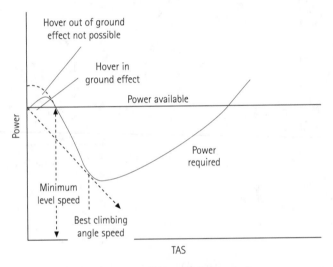

Fig 7.3 Effect of reduced power available on helicopter performance *NEW ZEALAND FLIGHT SAFETY*

Pilots must also be aware that, depending on the type of helicopter and the design limitations of its transmission/engine combination, overpitching may cause damage through over-torque, over-boosting or over-heating.

Avoidance

1. Where possible, operate into wind and never exceed the maximum all-up weight for the density altitude in which you are to operate. If starting work in the morning, use the weather forecast to predict the density altitude in the hottest part of the day and calculate your weight limitations accordingly.
2. When airborne and approaching a landing area, perform a 'power check' — an airborne assessment of the power available — before committing to a landing:

- Fly straight and level at a predetermined speed.

- Select landing rotor rpm.

- Note the manifold pressure or torque required to maintain speed.

- Maintain the rotor rpm and apply full power temporarily.

- Note the change in manifold pressure or torque.

- The difference in the two values is the power margin.

You must then judge — using knowledge, experience or calculations from the flight manual — the slow speed capabilities of the helicopter. Also to be factored in are wind effect and whether the operation is above the altitude where the rotor is most efficient.

Whilst this overall technique is standard, specific facts and figures will vary according to the situation and the helicopter type.

Recovery

The only successful recovery action is to lower the collective lever to allow rotor rpm to increase. If there is power available, an increase in throttle will be beneficial. As seen in case study 7.3, recovery is neither immediate nor guaranteed to happen in time to avoid an accident. But it is quite evident that the repeated attempts, using the correct technique, made by the pilot in that case study at least provided sufficient rotor rpm and rotor thrust to allow a final contact with the ground that could be survived by the occupants, if not by the aircraft.

Without rotor rpm, you might as well be sitting on a brick!

8

Main rotor strikes

Main rotors of helicopters are not only very strong, they are also reasonably reliable when used in conditions for which they were designed. Main rotors also move very quickly.

The speed of a rotor tip is often approaching the speed of sound! That is fast enough for the rotor to sustain damage if it comes into contact with any solid object during flight. Even objects as small as insects, heavy rain droplets and hail may damage the rotor to some extent. Larger objects – such as trees, hills, wires and so on – will, on most occasions, extensively damage rotors.

During the Vietnam War, many a Huey (Bell UH-1H) blade enlarged the size of a landing area by brute force, with some 'trimmed' branches being an inch or two thick.

In commercial operations, and in the normal course of events, contact with any item that is liable to result in damage or excessive wear to rotor blades should be avoided. Even if it doesn't cause an accident or subsequent failure, rotor blades cost a fortune in anyone's language. The owner of the helicopter will not be pleased if blades wear out before the certified number of hours.

Case study 8:1

Aircraft Accident Report Number 90–004T

Date: November 1, 1990	
Location: Fiordland, South Island, New Zealand	
Aircraft: Hughes 369D	
Injuries: 2 fatal	

ZK-HOP had been out hunting deer with considerable success for most of the morning. Because dawn and twilight are the favored times for deer to be out of the bush feeding, the crew took a break for a large part of the afternoon and left their Te Anau base at around 1700 hours for the evening shoot.

Fig 8.1 ZK-HOP ready to capture live deer
TRANSPORT ACCIDENT INVESTIGATION COMMISSION, NEW ZEALAND

Those who had been speaking and flying with the 36-year-old, 3000-hour pilot remarked that the helicopter was performing without problems and it was being flown well. Evidence confirmed that, as was their normal habit, the crew had filled the fuel tanks to capacity before departing on their last sortie for the day. This would have provided fuel endurance of around two hours, and several jerry cans were carried for extra endurance. Also, it was the pilot's practise to shut down for some time in the field prior to beginning to hunt. It is worthy of note that the 'shooter' (the crewman) had held a commercial helicopter pilot license, but now held a private license and had some 2200 hours logged. He was very experienced in venison recovery helicopter operations.

Civil evening twilight ended at 2107 hours. It was a clear night, nonetheless the failure of the helicopter to return gave rise for concern and another local operator went searching for ZK-HOP at 2300 hours without success. The police were notified at 0130 hours on November 2, and the official search located the missing machine at 0730 hours in a remote area of the Fiordland National Park with the assistance of ZK-HOP's emergency locator beacon.

The crash site was some 82 yards above the bush-line of a spur of the Cameron Mountains, at an elevation of 3250 feet above mean sea level.

With regard to specific weather around the time of the accident one must bear in mind the remote and mountainous location. The meteorological service post-cast estimated that the wind would have been 10–15 knots and dying out toward evening, with no cloud or precipitation. Little or no turbulence could be presumed.

The accident investigation report's description of the wreckage is lengthy but some highlights include:

The fuselage was lodged amongst stunted mountain beech at the edge of the bush below a steep tussock- and scrub-covered slope with numerous rocky outcrops . . . the front cockpit area completely exposed. The lower forward structure and cockpit floor had been severely compressed rearwards and upwards. The crush angle indicated that, when it first struck the slope, the fuselage was pitched markedly nose down. The extent of damage suggested that the helicopter tumbled and rolled as it descended the slope.

The aft section of ZK-HOP's tail-boom . . . was lying on a tussock ledge 60 yards up-slope from the fuselage . . . . (See Fig 8.2).

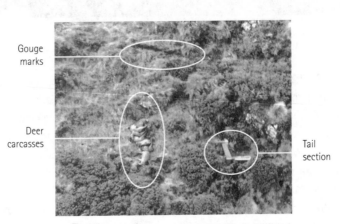

Fig 8.2 TRANSPORT ACCIDENT INVESTIGATION COMMISSION, NEW ZEALAND

The carcasses of eight deer were spread down the slope 16 feet to the north of the tail-boom . . . . The manner in which the deer were lying and the steep nature of the slope on which they had come to rest suggested that their release from the hook in this location occurred as a matter of some urgency. . . .

Approximately 23 feet upslope, directly above the spread of deer carcasses, there was a prominent gouge mark in the tussock-covered slope, clearly made by the helicopter's main rotors. The gouging of the soft soil extended over a length of some 30 feet, up to 10 inches deep and 14 to 16 inches wide. . . .

Fragmented and separated portions of the main rotor blades were recovered from various parts of the slope in an arc to the north of the fuselage descent path and up to 110 yards from the gouge marks. . . . Four of the five main rotor blades were shorn off adjacent to the blade grips. . . . (See Fig 8.3).

Fig 8.3 Chain and strop wrapped around the rotor head

TRANSPORT ACCIDENT INVESTIGATION COMMISSION, NEW ZEALAND

Portions of the tip sections of all five main rotor blades were recovered (see Fig 8.4). Each blade had sustained similar damage, curved rearwards and upwards. Each of the tips had split open at the trailing edge and contained earth and debris consistent with repeated strikes against the slope. Paint smears confirmed that the main rotor blades had flapped downwards sufficiently to strike the 'doghouse' and tail-boom at this time.

The investigation found that a single belt section of the pilot's shoulder harness had failed. The high forces of the impact had caused the metal shroud of the inertia reel to cut through the belt. Repeated wear by the belt over a long period of time had feathered the shroud to a sharp edge.

The pilot sustained severe bruising and lacerations to various parts of his body and head, and multiple rib fractures on both sides of the chest. Chest wall and lung damage resulted in respiratory failure. He did manage to exit the aircraft and manually activate the emergency locator beacon, but probably succumbed to his injuries and died one to two hours after the accident.

The shooter was not wearing any restraint; indeed it is quite possible that he was not in the aircraft at the time of the accident. It is quite common for shooters to ride on the skid of the helicopter, ready to jump onto rugged terrain to arrange the sling-loading of carcasses. The shooter died as a result of injuries sustained when falling onto rocks.

Fig 8.4 The recovered main rotors

TRANSPORT ACCIDENT INVESTIGATION COMMISSION, NEW ZEALAND

In the analysis section, the accident report noted:

... [T]he pilot had flown between 8 and 10 hours in ZK-HOP, mainly engaged in hunting activity and may have been fatigued at the time of the accident.
. . .

The extensive but localized and clearly defined gouging of the slope, and the similarity of damage to each of the five main rotors blades of ZK-HOP, indicated that all the blades had struck the slope in close succession. The location of the gouge was consistent with a main rotor strike against the slope as the pilot maneuvered the helicopter, in an approach or hover, close to and above the deer carcasses. . . .

The damage to the main rotor blades and complete loss of the tail-boom would have rendered the helicopter uncontrollable, with ensuing rotation and nose pitch down. . . .

No conclusive reason was established to explain the main rotor blade strike against the slope. . . . The damage of the main rotor blades and the melted foreign matter induced into the engine indicated that substantial power was being developed.

It was evident, however, that the steepness of the slope resulted in minimal clearance for the main rotor blades of ZK-HOP if the pilot was approaching close to, or hovering above, the area where the deer carcasses were lying, and maneuvering the helicopter to enable the shooter to hook on and then embark. While a pilot such as [pilot's name] was likely to be accustomed to maneuvering ZK-HOP with precision, the prevailing conditions suggested that an unanticipated 'sink' or wind gust may have been sufficient to erode the margins of clearance which had been allowed for and resulted in an inadvertent main rotor blade strike. The possibility that [crewman's name], the shooter, was about to board the helicopter at the time of the accident, leading to the potential for some resultant imbalance, may have been a contributory factor. Other factors such as the difficulty of judging, from the pilot's position in the left front seat, the degree of main rotor clearance available, the varying nature of the slope itself, fatigue due to the demanding nature of the task and sorties already flown, and the possibility that the accident may have occurred in fading light conditions, could also have been relevant.

Source: Transport Accident Investigation Commission, New Zealand

Avoidance and recovery

Taking off from a confined area requires extra care. If you can see space between the rotor disc in front of you and the obstacle you are trying to clear, you will out-climb it. If there is no clearance or the disc is below the obstacle and the aircraft is at full power and best-climb configuration, hastily make another plan because you are not going to clear the obstacle.

There is little more to say here — judge well, allowing as much leeway as possible.

Blade sailing — which can involve the ground in front (on some helicopter types), the tail boom or persons or objects under the rotor disc — is another cause of main rotor strikes. Because of the danger to people in the vicinity of a helicopter with rotors turning, this hazard has been dealt with in Chapter 20, Crew and pre-flight dangers.

9

Mid-air collisions

OTHER AIRCRAFT

'Maintain a good lookout for other aircraft' is pounded into the heads of student pilots from day one. Pilots have the aid of transponders, controllers and radar. The sky is, comparatively, quite a sizable place. Yet, tragically, it is not always quite big enough.

It is easy to classify mid-air collisions as rare occurrences. But there are far more than you would think and there are more to come. The main defense air crews have is the 'see and avoid' principle — and it is far from infallible.

In 1991 the Bureau of Air Safety Investigation (BASI), as it was then, published an in-depth report on the limitations of the see and avoid principle. The following excerpt explains the steps involved in this principle:

See and avoid can be considered to involve a number of steps. First, and most obviously, the pilot must look outside the aircraft.

Second, the pilot must search the available visual field and detect objects of interest, most likely in peripheral vision.

Next, the object must be looked at directly to be identified as an aircraft. If the aircraft is identified as a collision threat, the pilot must decide what evasive action to take. Finally, the pilot must make the necessary control movements and allow the aircraft to respond.

Not only does the whole process take valuable time, but human factors at various stages in the process can reduce the chance that a threat aircraft will be seen and successfully evaded. These human factors are not 'errors' nor are they signs of 'poor airmanship'. They are limitations of the human visual and information processing system which are present to various degrees in all pilots.

Case study 9:1

Aircraft Accident Report 93-020

Date: November 26, 1993

Location: Auckland,
New Zealand

Aircraft: Aerospatiale AS 355F1
and Piper PA 28-181

Injuries: 4 fatal

ZK-ENX, a Piper Archer airplane, was operated on contract to the New Zealand Police to provide a road traffic patrol service over Auckland City on weekday mornings and afternoons during periods of peak traffic. The patrol flights used the radio call-sign 'PACT 1' (Police Airborne Control of Traffic). The normal patrol altitude was 1500 feet. The pilot would transmit a brief traffic bulletin to police control once on each circuit. This information was then made available to local radio stations for public broadcast. In addition to the routine patrol, police control would, when advised of a road accident in the area, require PACT 1 to report on it. The pilot would then fly an orbit around the scene and report back on the accident and its effects on the traffic flow.

PACT 1 departed from Ardmore at 1559 hours on November 26, 1993. The pilot in command had a commercial pilot license (airplane) and had logged almost 1300 hours. He had around 700 hours experience in this role, mostly in ZK-ENX. He flew over Auckland on a normal patrol. Several road accidents or breakdowns were addressed during the next one and a half hours.

ZK-HIT, a Twinstar helicopter, was operated on contract to the New Zealand Police to provide a crime patrol vehicle over greater Auckland on weekdays. It was positioned each day from Ardmore to Mechanics Bay Heliport, one nautical mile east of the central city where the Police Air Support Unit was based. Random patrols were flown over the area throughout the day usually at an altitude of 1000 feet and the helicopter, which used two radio call-signs — 'Police One' with air traffic control and 'Eagle' with the police — was often tasked by Auckland police control using the police multiplex radio system. The police radio system also enabled the crew to monitor other police activity and to task themselves to assist.

The helicopter was crewed by a civilian pilot. On this day the pilot was one with over 13,500 logged flying hours. He had been involved in police operations in Auckland since 1989. Also on board were two police observers who were specially trained members of the Police Air Support Unit. The senior observer directed the police support function of the helicopter and operated the police radios, while the rear seat observer did the detailed navigation. The pilot was overall commander of the aircraft. Although the two police officers had no formal license rating, both initial and recurrent training qualified them for their duties. In addition to the police observer function this included helicopter groundcrew tasks, passenger briefing and safety and keeping a lookout for aircraft traffic while airborne.

ZK-HIT had landed at Mechanics Bay at 1622 hours after a routine patrol. After a break of about an hour the crew prepared to depart on another patrol. The pilot made a RTF [radio transmission] broadcast to 'Mechanics Bay Traffic' that they were lifting off and the departure time was logged in the office as 1733 hours. Shortly after, while the helicopter was climbing out, the Mechanics Bay office asked the pilot to call Auckland information to inquire about another company helicopter that they expected to arrive. sThe pilot did this and reported back by RTF. The RTF exchanges probably occupied the pilot for 45 to 60 seconds. No other RTF traffic with either 'Police One' or 'Eagle' was recorded from this flight.

Shortly after its departure to the north, over Waitemata Harbour, the helicopter was turned left onto a southwesterly heading. It was flown, climbing, across the city to near Queen Street where it was turned a further 30° to the left at about 1734:30 hours. The helicopter then continued climbing in a straight line.

At 1734:48 hours one of ZK-HIT's main rotors sheared off the left wing of ZK-ENX when the two aircraft collided. ZK-ENX then rolled and dived steeply to collide with the elevated carriageway on the interchange between the Northern and Southern Motorways. The left wing fell to lodge on the tower of a church near Queen Street.

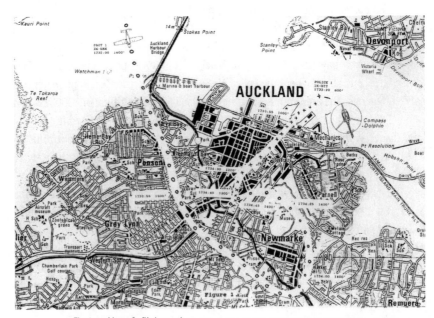

Fig 9.1 Aircraft flight paths TRANSPORT ACCIDENT INVESTIGATION COMMISSION, NEW ZEALAND

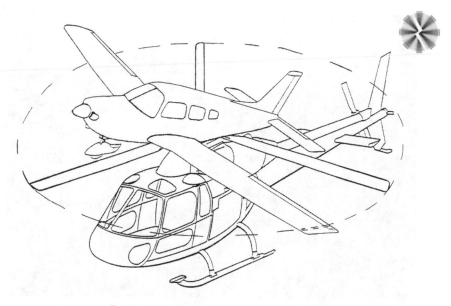

Fig 9.2 TRANSPORT ACCIDENT INVESTIGATION COMMISSION, NEW ZEALAND

On ZK-HIT the main rotor and transmission, the rear tail-boom, vertical stabilizer and tail rotor separated from the helicopter. The machine fell onto the on-ramp from Grafton Road to the Northwestern Motorway under Symonds Street Bridge. A severe fire broke out on impact. The main rotor and tail rotor section fell into a nearby cemetery.

Fig 9.3 Wreckage of Eagle One TRANSPORT ACCIDENT INVESTIGATION COMMISSION, NEW ZEALAND

Fig 9.4 Wreckage of Piper PA 28-181 TRANSPORT ACCIDENT INVESTIGATION COMMISSION, NEW ZEALAND

Such were the ground impact forces encountered by both aircraft that the accident was not survivable for any of the four occupants.

The accident occurred in daylight at a position close to the intersection of Queen Street and Karangahape Road at an altitude of about 1400 feet above mean sea level.

ZK-HIT was painted black with silver and blue trim, and had a red strobe light on the vertical stabilizer. ZK-ENX was painted white with blue, red and gray trim colors, and had strobe lights on the wing tips and a rotating beacon on the vertical stabilizer.

At the time of the accident the weather over Auckland City was fine with scattered cumulus cloud and excellent visibility. A light southwesterly breeze was blowing.

The position of the sun at the time of the accident was 247°M in azimuth and +32° in elevation.

Extensive and detailed calculations and tests were conducted in the subsequent accident investigation using radar readings and transponder records from both aircraft. Several flights were made in similar aircraft to assess views, sunlight angles, and so on. Photographs were taken to assist.

The study revealed that for the pilot of ZK-ENX the helicopter would have been located low in his central windscreen, beneath the horizon and just above the nose of his aircraft at 1734:20 (28 seconds before the collision), which was when the aircraft first closed sufficiently to be identifiable. It would have moved steadily right to become obstructed by the aircraft's nose about 10 seconds later. The study also showed that at about the same time the sun appeared in the top left of his windscreen, to move across to the right as the aircraft turned left. The dark transparent sun visor, if used, could have screened the pilot from direct light from the sun.

For the pilot of ZK-HIT, at 1734:20 the airplane would have been located in the top left corner of the left windscreen, above the horizon and within his monocular view for a few seconds only. It would have been moving aft, to go out of his view behind the windscreen and above the left window. After he turned the helicopter at 1734:30, the airplane would have moved forward but would have remained out of his view above the window and windscreen. The sun was outside his field of vision above the right side of his windscreen.

Local measurements were made in an AS350 helicopter with similar left doors and windows to ZK-HIT to determine the fields of view to the left of the observers seated in the left front and rear seats. The measurements showed that the top of each adjacent window was 14° above the horizon from a normal seated position. From the front seat the adjacent window gave a lateral view from 60° to 130° left, while the rear seat provided a view from 40° to 140° left. These fields of view would have presented no obstruction to either police observer on ZK-HIT to prevent him from seeing the approaching airplane from 1734:20 to 1734:45.

The accident report is extremely detailed, even in the summary section. For our purposes, the following phrase is overriding:

A causal factor was the inherent limitations of the 'see and avoid' concept.

Source: Transport Accident Investigation Commission, New Zealand

The see and avoid principle

The 1991 Bureau of Air Safety Investigation report mentioned earlier summarizes its findings:

Numerous limitations, including those of the human visual system, the demands of cockpit tasks, and various physical and environmental conditions combine to make see and avoid an uncertain method of traffic separation.

. . .

Cockpit workload and other factors reduce the time that pilots spend in traffic scans. However, even when pilots are looking out, there is no guarantee that other aircraft will be sighted. Most cockpit windscreen configurations severely limit the view available to the pilot. The available view is frequently interrupted by obstructions such as window-posts, which totally obscure some parts of the view and make other areas visible to only one eye. Window-posts, windscreen crazing and dirt can act as 'focal traps' and cause the pilot to involuntarily focus at a very short distance even when attempting to scan for traffic. Direct glare from the sun and veiling glare reflected from windscreens can effectively mask some areas of the view.

Visual scanning involves moving the eyes in order to bring successive areas of the visual field onto the small area of sharp vision in the center of the eye. The process is frequently unsystematic and may leave large areas of the field of view unsearched. However, a thorough, systematic search is not a solution as in most cases it would take an impractical amount of time.

The physical limitations of the human eye are such that even the most careful search does not guarantee that traffic will be sighted. A significant proportion of the view may be masked by the blind spot in the eye, the eyes may focus at an inappropriate distance due to the effect of obstructions as outlined above or due to empty field myopia, in which, in the absence of visual cues, the eyes focus at a resting distance of around 20 inches. An object that is smaller than the eye's acuity threshold is unlikely to be detected and even less likely to be identified as an approaching aircraft.

The pilot's functional visual field contracts under conditions of stress or increased workload. The resulting 'tunnel vision' reduces the chance that an approaching aircraft will be seen in peripheral vision.

The human visual system is better at detecting moving targets than stationary targets, yet in most cases, an aircraft on a collision course appears as a stationary target in the pilot's visual field. The contrast between an aircraft and its background can be significantly reduced by atmospheric effects, even in conditions of good visibility.

An approaching aircraft, in many cases, presents a very small visual angle until a short time before impact. In addition, complex backgrounds such as ground features or clouds hamper the identification of aircraft via a visual effect known as 'contour interaction'. This occurs when background contours interact with the form of the aircraft, producing a less distinct image.

Even when an approaching aircraft has been sighted, there is no guarantee that evasive action will be successful. It takes a significant amount of time to recognize and respond to a collision threat and an inappropriate evasive maneuver may serve to increase rather than decrease the chance of a collision.

In addition, keeping the windows as clear and clean as possible is a great advantage. It can save spending valuable seconds trying to ascertain if a spot in your field of vision is an Air Force jet that's going to be with you very shortly or a squashed bug.

Also, ensure that regular crew know and are frequently reminded that it is part of their duties to be alert for other air traffic, and to tell you the instant that they see anything at all. That way, the area being watched is much larger.

WIRE STRIKES

Fig 9.5 AERO PRODUCTS COMPONENT SERVICES INC.

A helicopter flying into wires is, sadly, an all too frequent occurrence. Until you have flown an aircraft it takes some believing just how hard it is to see a wire — much of the time, it is very close to impossible. So, is the main defense knowing where the wires are? Surprisingly enough, no it isn't. Many pilots have struck wires that they knew were there.

Case study 9:2

Aircraft Accident Report 99-006

Date: December 18, 1999

Location: Kawerau, North Island, New Zealand

Aircraft: Hughes 269C

Injuries: 3 fatal

On about December 15, 1999, the pilot contacted the operator to arrange the hire of ZK-HYE, a Hughes 269C helicopter, for Saturday December 18. The helicopter was based at the operator's Ngongotaha site, on the western side of Lake Rotorua. The helicopter was primarily used for local charter work and training but was available for private hire. Since the pilot was unknown to the operator's flying instructor, the pilot and instructor agreed that they would complete a short check flight in ZK-HYE before the pilot commenced the private hire.

On December 18, the pilot arrived at the operator's base at about 1030 hours, accompanied by his wife. The instructor examined the pilot's flying log book and noted that the pilot had flown about 60 to 80 hours on helicopters, mostly on the Hughes 269C. The last flight was recorded as being flown two weeks earlier, on December 4, 1999, when the pilot had completed a biennial flight review with another instructor.

The instructor and pilot discussed the intended flight. The instructor fitted the dual flight controls to ZK-HYE while the pilot 'pre-flighted' the helicopter. About 5 US gallons of fuel were added at this time. The pilot started ZK-HYE and completed the engine 'run-up' checks before lifting from the helicopter pad and departing to Rotorua Aerodrome [Airfield]. The pilot then flew a series of exercises including into wind and down wind landings, autorotations to the hover and quick stops. The instructor considered the pilot to be 'a little rusty' at first but improved during the flight to demonstrate a good standard.

The pilot flew the return leg to the operator's base and while en route he and the instructor discussed the intended private flight. The pilot advised the instructor that he was going to fly to his parents' farm about 3 nautical miles (nm) north of Kawerau and that he knew the area well despite not having flown there before. The instructor then questioned the pilot about any wires on the farm. The pilot replied that there were 'heaps' but plenty of room to land. The check flight took about 30 minutes.

The flight to the family farm took about 25 minutes with ZK-HYE initially approaching from the west. The helicopter then descended on a wide left base, flying between a cattle yard and the farmhouse. The helicopter landed in an open grassed yard near an implement shed. The arrival of the helicopter was a surprise to the other members of the family who had gathered for a lunchtime get-together. The pilot shut down the helicopter, unloaded some items and joined the other family members for lunch.

During lunch the pilot arranged to take some members of the family for a short flight before returning to Rotorua. The pilot's father and mother agreed to go first. The pilot carried out a pre-flight inspection of ZK-HYE before helping to strap in the two passengers. The pilot's father was seated on the right of the helicopter while his mother occupied the center seat. The three occupants were wearing headsets. The weather at the time was overcast with a high cloud base and a light breeze from the northwest.

The pilot's intention was to fly around the farm and return to take a second group of two for a flight. However, no arrangements were made with the second group as to where the helicopter would land or if the engine and rotors were to be kept running or stopped. The observers believed that, knowing the pilot's meticulous manner, he would most likely shut down the helicopter before off-loading the first group and boarding the second.

The pilot had some difficulty starting ZK-HYE but was successful on about the fourth attempt. The helicopter was then hover-taxied forward over a fence to a large paddock next to the yard and implement shed. The paddock was covered in flowering clover about eight inches high. A power line ran across the paddock, about 110 yards southwest of where the pilot entered. The pilot completed some hovering maneuvers before lifting the helicopter off on an easterly heading and climbing over some trees.

The farm had three power lines that ran across it. Two prominent 33,000-volt power lines ran on a northeasterly heading parallel to the main road and railway line boundary of the farm, and across an adjacent paddock to where ZK-HYE had lifted off. The two power lines were about 110 yards north of where ZK-HYE had departed. Each line consisted of three conductors or wires supported by wooden cross-arms attached to two wooden poles. The sets of poles were at about 140 yard intervals. Passing underneath the 33,000-volt power lines was a less prominent 11,000-volt power line running from the main road across the farm in a southeasterly direction, feeding the farmhouse and other buildings. The 11,000-volt power line crossed the paddock about 55 yards in from the southern fence line, or about 110 yards to the south of where ZK-HYE had departed on the flight around the farm. The 11,000-volt power line consisted of three wires supported on wooden cross-arms and single poles. No poles were located in the paddock. Instead the 144 yard span was supported by a pole near the implement shed and a second pole close to two of the 33,000-volt power line support poles in an adjacent paddock. The minimum ground clearance for the 11,000-volt power line across the 144 yard span was 21.3 feet with a pole height of about 26.2 feet.

After lifting off, ZK-HYE was observed by witnesses, including several people on neighboring properties, to climb to about 500 feet and fly around the outside of the farm in a clockwise direction. The helicopter then crossed the railway line and road to the southwest of the farm, turned left through about 270° to approach the paddock, from where it departed, from the south.

ZK-HYE then flew a 'slow steady approach, gradually descending' and turning right as it passed some tall trees near the southern corner of the paddock. As the helicopter approached abeam the implement shed it had descended to about 15 feet above ground level. The observers were aware of the power line crossing in front of the helicopter. One person believed the helicopter would pass underneath the wires while others thought it would fly over them. As the helicopter moved forward several bright flashes were seen as it contacted the wires. Several loud 'cracks' were also heard at the same time as the bright flashes. The helicopter was then observed to roll and tumble forward, impacting the ground nearly inverted. On impact a loud explosion was heard followed by an intense fire. The fire quickly enveloped the cabin area.

The wire strike and explosion were heard by several neighbors who raced quickly to the accident scene and attempted to help extract the occupants. The fire, however, prevented any rescue. Local fire and rescue services arrived on the scene

The wires are barely visible even when you are looking for them

Fig 9.6 Wreckage of the Hughes 269C TRANSPORT ACCIDENT INVESTIGATION COMMISSION, NEW ZEALAND

within about 15 minutes of the accident and extinguished the remaining fire. The duration of the flight was about six minutes.

The pilot, male, aged 37, was a licensed aircraft maintenance engineer and had commenced flying airplanes as a recreation in 1994 and helicopters in 1996.

The weather on the day of the accident was described by eyewitnesses as fine, with an overcast sky and some shower activity on the distant hilltops. The temperature was mild with a light northwesterly breeze. At the time of the accident the sun was obscured by cloud but would have been high and well to the left of the pilot as he approached to land.

The pilot was familiar with the farm, having been raised there from about the age of 11 years. A photograph of the farm taken in the early 1970s showed that the three sets of power lines were present at that time. A more recent survey photograph taken in 1987 showed that several trees had since become established to the south of the shed. The tallest of these trees was about 75 feet in height but most were 45 to 50 feet high. One of the trees was located about 35 feet from the pole supporting the 11,000 volt power line.

ZK-HYE would have been well above the 11,000-volt power line when it descended to land near the cattle yard in the morning. Nevertheless, by approaching over the power line and then turning left back towards the northwest, the pilot should have been able to see the clearly visible poles and been reminded of the existence of the wires.

Fig 9.7 TRANSPORT ACCIDENT INVESTIGATION COMMISSION, NEW ZEALAND

As ZK-HYE approached abeam the shed it would have been at about the same height as the 11,000-volt power line. The pilot probably did not see the wires and continued moving forward towards them, eventually causing the rotors to come in contact with the wires. The effect of the rotors striking the wires resulted in the helicopter rolling and tumbling forward, impacting the ground at about a 50° angle from the wires. The force of the helicopter impacting the ground inverted caused the fuel tank to rupture, resulting in an intense fuel-fed fire.

The force and angle of the helicopter impacting the ground would have disorientated but possibly not incapacitated or rendered the occupants unconscious. The location and intensity of the fire would have made any escape nearly impossible.

The flight path taken by the pilot when he approached the paddock from the south meant that the power pole on the eastern end of the 144 yard span was obscured by the trees near the implement shed. The pole at the western end of the span was some distance away and would have blended in with the poles supporting the 33,000-volt power lines in the next paddock.

Wires are very difficult to detect from the air. Pilots would normally see a supporting pole first and then try to determine the direction in which the wires ran. This would be confirmed by locating the next pole along the power line. Pilots operating at low-level will normally cross a power line at a pole to ensure a safe separation above the wires.

The eastern pole of the 144 yard span would have become visible as ZK-HYE moved abeam of the shrouding tree, about 27 yards before the wires. However, the pilot's view of the pole may have continued to be obscured by the passengers seated

Fig 9.8 The remains of the rotor head TRANSPORT ACCIDENT INVESTIGATION COMMISSION, NEW ZEALAND

Fig 9.9 Initial main rotor impact scar TRANSPORT ACCIDENT INVESTIGATION COMMISSION, NEW ZEALAND

to his right. Because the pilot was not able to clearly see either pole he was deprived of a reminder of the presence of the wires across his view.

As ZK-HYE approached the wires, the height of the helicopter would have meant that the wires would have been cast against the background of the trees at the far end of the paddock. The weathering of the wires had resulted in a dulling of their color. This may have made the wires less conspicuous against the trees and therefore harder to detect.

As the pilot prepared to land he would have been aware of the group of observers waiting by the paddock fence. The pilot was now probably looking forward and down, with his attention focused on flying the helicopter towards the landing spot. Maintaining a hover-taxi height of about 15 feet, the wires would have slowly moved above the pilot's field of view, reducing the possibility of detection.

As ZK-HYE slowly approached the wires there was ample opportunity for the pilot to take avoiding action before striking the wires, had he seen them. There was no evidence that the pilot had taken any avoiding action before the helicopter struck the wires. Further, there was no evidence to indicate that the passengers either saw the wires or warned the pilot of their proximity.

The flight path flown by ZK-HYE would suggest that the pilot did not remember the existence of the power line, despite having flown over them on the approach that morning. However, the track flown on arrival that morning was well to the south of the accident site, and the pilot may not have been concerned about the power line at that time.

Source: Transport Accident Investigation Commission, New Zealand

One could be forgiven for focusing on the pilot's lack of helicopter experience but he was also a fixed-wing pilot and that is all flight experience. It is not at all uncommon for pilots to fly into wires that they absolutely knew were there. And experience is no real factor. Some years ago a pilot in Wellington (New Zealand) — a true hero of many rescues and well over 10,000 hours logged — was helping police track an escaped prisoner and inadvertently hovered into major transmission wires that he himself had help to erect across a gully. It would seem that often the comfort of knowing that wires exist is so reassuring that the perils move to the subconscious.

For years people have clamored for suspended wires to bear colorful markers. The problem with that is the probability of resulting complacency — every unmarked wire could catch pilots out. To think that all the wires that people might fly into could be located and marked is unrealistic. Also, would you lay money that a marked wire would never get hit by an aircraft? I think it would be a dangerous wager.

While fitting wire cutters to helicopters has saved many lives, they are definitely not infallible. Pilots have to hope that the wire they don't see is caught in the coverage area and is of a gauge that the cutter will actually cut. Any wire that just catches a skid, main rotor or tail rotor, most likely with the helicopter in descent, is still potentially lethal. But anything to reduce the percentages is well worthwhile.

The answer could rest with global positioning systems, which are becoming more accurate, affordable and common. Surely it is possible to gather wire location data,

Fig 9.10 Wire strike kit on a Bell 206 AERO PRODUCTS COMPONENT SERVICES INC.

area by area, and hook that up to an alarm system in the aircraft. The pilot could have the ability to program in further wires. There are many cases where farmers have told pilots of wires — physically shown them in fact — only for the pilot to collide with them hours later. Of course, you would still be in danger from temporary wires that people string up. Tom McCready of New Zealand's Civil Aviation Authority is working on a warning system based on this concept.

Avoidance

The following material is from an article entitled 'The easiest accident to avoid' by Jim Cheatham from Salinas, California.

All wire strikes are pilot error!

Sound pretty radical? I'll admit it's a bit judgmental . . . and sarcastic and flippant. But it is essentially the truth.

I've hit several wires in my career. Lucky for me, they were all of the small copper variety one sometimes encounters at embarrassing moments in ag [agricultural] flying and none of my encounters resulted in a crash. But they were all pilot error. I've logged over 18,500 hours so far and very little of it was at an altitude above 500 feet. I'm an expert on this, so I ask you to please listen to this obvious stuff.

Okay . . . I've said that wire strikes are the easiest accident to avoid. . . . But if it's so damned easy, why do helicopter pilots and their passengers die from hitting wires every year?

Airline pilots will, of course, agree that all wire strikes are pilot error. These guys fly all aloof and stately, from airport to repetitious airport, leaving contrails in the stratosphere while speaking to air traffic control in those insider tones that most helicopter pilots never quite master. They needn't fly low to do their job. A few helicopters also inhabit this sanitary world of concrete ramps and pilot's lounges with private lines to weather and computer flight plans. They also fly repeatedly from 'A' to familiar 'B' using 'L' charts and approach plates while under the watchful radar eye of big brother.

But most of you guys find my statement a little harder to take, right? You fly agriculture, fire fighting, external loads, line patrol, aerial camera work, in mountainous terrain and doing all the other thousand or so jobs that helicopters do. You get 'down and dirty' in the course of your work. You're at low level where the wires are. You get real busy and have a lot on your mind, without benefit of a three person crew. 'Cockpit resource management' means just you by yourself trying to do a good job. They can't blame a guy if a wire jumps out and gets him when he's busy doing a job under tough conditions, right? Well, yeah, but let's think about this. Does it really matter whether or not they blame you if you are dead? The real question is, 'could this accident have been avoided?' And the answer to that question is always 'Yes'.

I'm trying to help you younger pilots learn to do your job while avoiding wires and hopefully survive to at least social security time.

Of course, those of us whose jobs take us close to the ground under conditions of high mental workload are the ones who are most at risk and therefore have the need to make special mental preparation. We must train our minds to see the world a little differently than nature prepared us for. The normal human eye to brain connection tells us that if we see nothing in a place, then nothing is there. The human animal is designed to respond instantly when threats occur in nature. The trouble is, wires don't occur in nature. Helicopters are not a natural phenomenon either, so we must train ourselves to see the world as technology has modified it. We must re-think the assumptions that control our judgment in the same way we re-train ourselves to trust our instruments and not our inner ear when flying in Instrument Meteorological Conditions.

First you must convince yourself of the fact that wires are usually unseen. Pilots almost never hit the wire they see. We are very good at avoiding what we see. We hit the ones we don't see, and even worse, the ones we didn't even think about. If you assume that your eye will detect a wire and give you enough warning to react, I ask you please, don't take any passengers . . . They don't deserve to die from your ignorance. What you must do is avoid flying through spaces where wires likely could be. A wire cannot 'jump out and get you'. They are passive, inanimate objects. They won't hit you; you must invade their space in order to hit one. Wires are found only in predictable places.

Wires are subject to gravity, and so are found only below and between objects to which they can be attached. It is impossible for them to be above the imaginary line between the tops of the highest objects in the area.

If you first assume that a wire is any place where it could be, you will get in the habit of visualizing wires along the imaginary line between the tops of any two objects or structures. I call this the inferred wires. You must look for them and prove to yourself that they do not exist before invading

Fig 9.11 The remains of a Bell 205 with stainless steel spray tanks after it hit high tension wires. The pilot knew the wires were there but was distracted looking for another wire that was harder to see.

their space. Of course, you relieve yourself of this burden if you fly above the inferred wires.

Do not assume, just because you can see wires at a height below the inferred wires, that there are not more, less visible, somewhere above the ones you see, but below the inferred wires (static lines or another line connected to a higher structure).

What about at night?

How do you avoid hitting something when you can't even see the towers, poles and buildings? Of course, most of us are smart enough to figure this one out. Fly high. And remember, if you are depending on the idea that high objects have red lights or strobes on them . . . keep in mind that, except near airports, there is no legal requirement to light objects less than 500 feet above ground level and, in fact, many dangerous things are lurking if you fly below 500 feet above ground level at night.

Does all this sound very basic and boringly obvious? Of course it does, but how often have you failed to heed the obvious? Some obvious things you can ignore at no particular peril. This is not one of them.

Let's review some of this obvious stuff:

When can we hit a wire?

Only when flying low. If you don't have a need to be low, you can solve the whole problem by simply flying high.

Wires are never present in the absence of something to which they could be attached. They are found directly between objects that can be seen in normal daylight, so you can easily visualize the space between structures as being dangerous. Also, the area near a structure (guy wires to the ground) is dangerous. The key word here is 'between'.

Rivers and canyons are especially dangerous. If you like to fly down rivers and canyons without knowing for sure what is there, you will eventually hit a wire.

If you really need to fly between some towers, trees, etc., how do you determine for sure whether a wire is present? Hey, I said this stuff was obvious, didn't I? You are flying a helicopter, aren't you? They can hover, can't they? Go Slow! Take a close look and check it out before you commit yourself. Examine the structures for tell-tale connectors.

These are most of the real obvious ways you can avoid hitting a wire. If you follow my advice, I can almost guarantee that you will not become a wire-strike statistic. But if you choose not to heed it and go blasting carelessly into the unknown? Try praying; that might work.

After you hit a wire

Sounds gruesome? Sorry, but maybe I can say something here that will save somebody's life. I know it has saved mine. I had the benefit of breaking into ag flying by being around some old timers who had made a lot of mistakes and seen a lot. I was able to watch them and listen and thereby learn. A lot of pilots today don't get the benefit of this kind of upbringing and it's a shame.

Let's say you are flying low, doing your job. You get momentarily distracted and you hit a wire you had forgotten about. What happens next?

Is your life now over? Maybe, MAYBE NOT.

Do you close your eyes and wait for the end? HELL NO!

Maybe serious damage was done to the aircraft and MAYBE NOT.

How do you know if you can still control the aircraft? KEEP FLYING IT!

What's cool?

Helicopter pilots have hit wires and come to a stop, then backed away from the unbroken cables and flown away with the cockpit half destroyed. They have flown through wires, breaking all of them without a scratch on the aircraft. They have been spun around and flipped over, yet managed to regain control and land. Helicopters have landed with a half mile of wire trailing from the skids. There are too many of these kinds of stories to recount, but there is a common theme to each and every one. The pilots kept their cool and did not quit flying the aircraft!

It is a recurring theme that sometimes a pilot will strike a wire and then crash, but the aircraft had been intact and apparently controllable after the wire strike. Why then, did it crash? Usually because, in the trauma of the emergency, the pilot quit flying the aircraft. Maybe his instructor never reminded him that Yogi Berra's words applied to aviation as well as baseball: 'It ain't over 'til it's over'.

Make yourself this promise . . . 'If I am going down, it won't be because I quit trying. I'll try everything in the book to control my aircraft, all the way, no matter what.'

Keep your cool.

BIRD STRIKES

Compared to fixed-wing aircraft, particularly jets, helicopters don't go very fast and are less susceptive to bird strikes — although they still happen, of course, as case study 9.3 shows. Case study 9.4 shows that the threat of a bird strike can be more dangerous than an actual incident.

Case study 9:3

Accident Report FTW97LA354

Date: September 20, 1997

Location: Houma, Louisiana, United States

Aircraft: Bell 206L-3

Injuries: 1 minor

At 1615 Central daylight time, N210PH collided with terrain following a loss of control during cruise flight. The commercial pilot received minor injuries and the helicopter sustained substantial damage. Visual meteorological conditions prevailed for the repositioning flight that departed Fort Jackson, Louisiana, at 1547 hours, under a company VFR flight plan.

An interview, conducted by the investigator-in-charge, revealed that the helicopter was in cruise flight at 800 feet above mean sea level about 14 miles from Houma when the pilot heard a loud bang and the helicopter started 'shaking violently'. Cyclic inputs did not correct the attitude of the aircraft. The pilot, with seat belt and shoulder harness secured, was thrown about the cockpit as the helicopter pitched down and to the left. The pilot lowered the collective full down and rolled the throttle to the idle position. The helicopter entered a 'near level attitude and a slight left bank'. At about 15 feet above the ground, the pilot applied full 'aft cyclic and pulled pitch to try to cushion the landing'. The helicopter touched the ground in a slight left sideways movement and bounced once before coming to rest.

An examination of the helicopter revealed that a bird had struck and sheared the cyclic (yellow) push pull tube. The aft portion of the tail-boom was bent and the right horizontal stabilizer winglet was separated from the helicopter.

Source: National Transportation Safety Board, United States

Case study 9:4

Accident Report FTW94FA158

Date: May 16, 1994

Location: Tulsa, Oklahoma, United States

Aircraft: Bell 47G-2

Injuries: 1 fatal

At 1502 hours (Central daylight time) the Bell 47G-2 was destroyed when it impacted terrain during an uncontrolled descent. Visual meteorological conditions prevailed.

Eye witnesses were interviewed and thirteen witnesses' statements were submitted. The consensus of these statements revealed:

The accident helicopter was following a lead helicopter (both helicopters were en route to a helicopter repair facility for maintenance to be performed on the accident helicopter).

Five witnesses reported hearing a loud 'pop', 'exploding' or 'backfire' noise.

Two witnesses indicated that there was a power loss.

Eight witnesses saw something separate from the helicopter.

One witness saw the helicopter tilt and go sideways. 'I could see the top of rotors from my view point which told me the copter was completely sideways', he wrote. Another witness verbally reported seeing the helicopter roll 90° to the left, and said he also saw the top of the rotor disc.

Five witnesses said the helicopter went inverted, four witnesses said the helicopter entered a spin or spiral and another witness said it 'tumbled'.

The helicopter impacted the back yard of a residence causing minor damage to the chimney structure. Two of the three longitudinal tail boom trusses were found severed. Approximately 240 feet away the left synchronized elevator and end cap were located. These were the only parts that were found to have separated from the helicopter. One of the main rotor blades had an indentation on its leading edge near the blade tip. Measurements made on this indentation match those made to the elevator end cap.

A metallurgical examination of the fractured surfaces concluded that they were the result of overload forces. Measurements taken from a three-dimensional scale drawing of the Bell 47G-2 indicate that in order for the main rotor blade to contact the synchronized elevator, the rotor system would have to tilt aft approximately 18°.

According to a report by Bell Helicopter's Rotor System Design Group, static stops limit the angular movement of the rotor mast. In a static state, the rotor blades will clear the tail-boom by about 14 inches. The report noted:

> In trimmed, high speed forward flight, the rotor would have about 8° of collective pitch and 8° of cyclic pitch, and the rotor flapping would be near 0°. However, an unusual and violent maneuver such as maximum rate drop of the collective pitch with full after cyclic pitch applied simultaneously could command after flapping exceeding the stop-to-stop clearance. Under this circumstance, heavy flapping stop contact could occur, and with large mast and blade bending deflections a blade strike on the tail boom is possible.

Additional Information: The lead helicopter pilot said that shortly after departing the airport, he was forced to bank sharply to avoid a flock of birds. He radioed this information to the second helicopter pilot but received no reply. When he turned around, the second helicopter was not in sight.

Source: National Transportation Safety Board, United States

There are definitely areas and times of the year that one has to be more mindful of birds. Some birds are a little more likely to cross your flight path than other species. There are species that, during the nesting season, will attack even the biggest of slow-moving trespassers and can put a big dent (or worse) in your tail rotor.

I once heard of a Hughes helicopter brought out of the sky by a duck, according to the pilot's account. Nobody ever did find one at the scene and rumor has it that they never found any fuel either. 'Duck' would be the appropriate word, apparently.

Case study 9.4 could be summarized as:

Birds were in the flight path and reflex avoidance actions (control inputs) by the pilot exceeded the limitations of the teetering mast rotor system of the helicopter he was flying.

This was a tragedy because there is a very, very high likelihood that receiving one or even multiple actual bird strikes on a slow-moving aircraft like a helicopter would never amount to the total loss of control that mast bumping or main rotor/ tail boom strike surely will.

Be mindful of the limitations of your aircraft, even when there is an apparent threat – what seems to possibly be a cure could actually be far worse than the disease.

Issue 4/95 of Transport Canada's *Vortex* magazine told how the pilot of a Bell 206 flying at night over a heavily populated part of Kelowna, British Columbia, was at 1200 feet and 80 knots when he noticed a dark object pass on the left side. A split second later there was an explosion-like noise followed by a severe blow to the pilot's head and face. Though cut about the face and almost unconscious, the pilot managed to successfully land at the nearby airport. The average western grebe, which is what had hit the right-side door and window frame about 12 inches down from the top of the door, is about 25 inches long with a wingspan of around 34 inches. It most likely weighed two to four pounds.

The pilot stated that:

Due to force of habit from military flying days, I had my clear visor down. Had I not been wearing my helmet with the visor down, I certainly would have been incapacitated, with most likely a fatal crash. At the very least I would have lost my right eye.

The editor of the magazine rightly intimated that a baseball hat and sunglasses wouldn't offer much protection at all and stressed the sagacity of wearing a helmet.

10

Mast bumping

A rotor mast is the upright part that connects the transmission to the rotors. It supports the fuselage of the aircraft and is a key piece of carefully manufactured equipment. Helicopter rotor blades must be able to flap; it is an integral part of rotary wing aerodynamics. But there must be a limit to the amount the blades can flap. On a teetering rotor system there are ways to exceed these limits — briefly.

To exceed the flapping limit is to initiate mast bumping — repeated contact between the static stops and the mast. Mast bumping during flight generally results in the failure of the mast, the departure of the rotor blades and a near vertical plummet to the ground, mostly with fatal consequences.

Case study 10:1

Aircraft Accident Report Number 91–001

Date: January 1991

Location: North Auckland, North Island, New Zealand

Aircraft: Robinson R22

Injuries: 2 fatal

Just after noon one day in early January 1991, the 31-year-old private license holder was flying over the North Auckland province on a cross-country to Ardmore Airfield, Auckland City.

The aircraft had been refueled prior to departure. With one female passenger, no concerns existed with regard to center of gravity limits. The mid-summer weather at the time included an unusually strong inversion layer between 3000 and 8000 feet; cloud was variable, scattered to broken thin stratocumulus with a base 2500 to 3000 feet; and visibility was hazy but in excess of 16 nautical miles. A steady southwesterly surface wind of some 20 or 25 knots prevailed in the area. All in all, not a bad day to be flying.

HDC's altitude for cruising that day is unknown, however, the pilot decided to descend to between 50 and 100 feet above ground level to over-fly a rest area on the highway. He either did this to see and greet two friends who were making the journey by motorcar, or, with the 20 to 25 knot crosswind, he may have been trying to minimize the headwind component to reduce flying time.

The rest area was on the crest of a ridge that ran across the aircraft's track and rose some 100 feet above the general ground level to either side. The pilot may have climbed the helicopter as he approached the ridge and then endeavored to descend to his former altitude as soon as he had passed over the crest and/or he may have encountered an up-draught created by the upward slope of the ground beneath the machine in relation to the prevailing wind. For whatever reason, it seems that the pilot eased the cyclic control forwards to regain his selected height above the ground.

Fig 10.1 Mast bumping accidents tend to be catastrophic

TRANSPORT ACCIDENT INVESTIGATION
COMMISSION, NEW ZEALAND

This forward movement of the cyclic unloaded the main rotor disc, resulting in negative G. With the disc thus unloaded, the effects of the tail rotor would create a roll to the right. It would appear that the pilot attempted to level the aircraft with left cyclic. The main rotor blades flapped downwards in excess of 50° from the horizontal and struck the aircraft canopy. The blade hub bumped the mast and the main rotor head assembly, along with a section of shaft separated in-flight over a point immediately past the rest area on the crest of the ridge. The severity of the impact and subsequent fire made this accident unsurvivable for either pilot or passenger.

The pilot had one more written examination to pass before a commercial pilot flight test. He had a little experience in fixed-wing aircraft but was concentrating on helicopters. He began his training in a Robinson R22 on July 4, 1989 and made his first solo (helicopter) on August 6 after 19 hours of dual instruction. On August 28, 1989 a 1.1 hour dual flight was recorded as 'Low G recognition and recovery'. In November 1989 he had 42.3 hours on type but then completed his private license

Flapping hinge   The remainder of the rotor mast

This is what 'bumped the mast'   The rotor blade root

Fig 10.2 TRANSPORT ACCIDENT INVESTIGATION COMMISSION, NEW ZEALAND

training in a Hughes 269. He later gained a Bell 206 rating but had accumulated only two hours prior to the accident. His total logged flight experience was 213 hours.

The operator stated that he reminded the pilot of the hazards of low G or negative G flight prior to departure.

Subparagraph 3.3.6 of the aircraft accident report states:

The probable cause of this accident was that the pilot failed to recognize that he had inadvertently entered a low G flight regime that caused the aircraft to roll. Consequently he endeavored to right the aircraft by applying left cyclic without first restoring positive loading to the main rotor blades. Other factors were the pilot's inexperience on type and a lack of awareness of the helicopter's vulnerability to low G flight.

Source: Transport Accident Investigation Commission, New Zealand

Principles of mast bumping

On a multi-bladed rotor head with flapping hinges there is a 'droop stop' that is hit when limits are exceeded. Instead of being termed mast bumping, this is called droop stop pounding and does not often have the cataclysmic results that mast bumping does.

Mast bumping is most often caused by attempting to maneuver in situations involving low rotor thrust. Pushing the cyclic forward, as the pilot did in the case study, meant that upward centrifugal force opposed the weight. Naturally, rotor thrust diminished markedly, perhaps to zero or even a negative value. When the tail rotor thrust caused the helicopter to bank, the pilot corrected with cyclic but, as there was no rotor thrust, there would be no reaction to the input. More and more input would result in the tip path of the rotor teetering accordingly. In such situations, aerodynamic forces are strong enough to supersede mechanical limitations and mast bumping occurs.

Avoidance

When flying a two-bladed helicopter with a teetering rotor system, avoid situations that cause low amounts of rotor thrust, particularly negative G. Even when practicing autorotations, don't drop the collective too quickly as this can create a low-thrust configuration.

Landing on slopes can also involve low thrust components. When firmly placing the upslope skid, the ground suddenly supports an increasing amount of the aircraft weight and required rotor thrust decreases – but you still want the helicopter to 'fly' until the downslope skid is also on the ground.

If the cyclic is not centered during start-up or shut-down, mast bumping is possible. Accordingly, cyclic should be centered when at ground idle (see case study 20.9).

The center of gravity can play a part in how much cyclic input is required to achieve a desired lateral movement. If center of gravity is near the limits (or outside them), the additional cyclic input required might be just enough to initiate mast bumping where it otherwise wouldn't be a factor.

Mast bumping is not something that occurs only in flight. If a two-bladed teetering head rotor helicopter is parked in wind (it doesn't have to be strong) and the blades are see-sawing up and down, they can be actually bumping the mast at the extremity of their travel. These helicopters need to have their main rotors secured in all but nil wind conditions (see Figs 10.3 and 10.4).

Fig 10.3 The front main rotor tie-down on a Bell 212 STEVE BONE

Fig 10.4 The other main rotor on a Bell 212 tied to the tail boom STEVE BONE

Recovery

If the mast has already bumped, chances of recovery are negligible (and that's the optimistic viewpoint). With mast bumping, successful avoidance is the only guaranteed recovery. However, rather than just sitting there and watching it all happen, moving the cyclic rearward immediately may load the rotor disc. Then center the cyclic laterally and land immediately.

11

Engine failures

The easiest way to grasp the concept of how a helicopter might glide without engine power is to think of how that marvel of nature, the sycamore seed, floats. There is little more to it, although a helicopter weighs much more and requires good control.

Unless recovering from an engine failure in the hover, which is probably better described as a rough cushion if you are quick enough to react, true autorotation is best set up with forward airspeed. Recoveries do work flying backwards, but it is a most uncanny sensation. The rotors are driven by the very descent we are trying to control.

Following an engine failure, a freewheel unit in the transmission automatically disengages the drive. This prevents the slowing engine from dragging the rotor to a stop. However, there is no longer an engine driving the rotor. To maintain rotor rpm fully lower the collective immediately. This reduces the angle of attack and so reduces drag. The forces of the autorotation that drive the rotors begin and are so powerful that, unless the pilot reintroduces some pitch, the rpm may well cause an over-speed and component failure. The simplest definition of autorotation is:

A condition without engine power where rotor rpm is maintained due to airflow up through the rotor disc. This airflow is called 'rate of descent flow'.

The helicopter will not glide as far as a fixed-wing aircraft without power. But the fixed-wing must land with forward airspeed above stalling speed and therefore requires a large area of flat obstruction-free ground to avoid damage from the ensuing roll-out. The helicopter is fully controllable, can be autorotated into a very small area and can touchdown at zero speed. A flare (briefly pulling up the nose of the helicopter) at the bottom of the descent gives enough of a final burst of energy to the rotor system to enable a comfortable landing. If the landing area is flat and level, all onboard can walk away from an undamaged helicopter.

Case study 11:1

Aircraft Accident Report
Number 93-005

Date: March 17, 1993

Location: Fox Glacier, South Island, New Zealand

Aircraft: Aerospatiale AS 350D

Injuries: 2 serious, others minor

ZK-HGV was based near Fox and Franz Josef Glaciers on the West Coast of New Zealand's South Island. The first flight of the day had been to both glaciers and had included a snow landing above Fox Glacier, at an elevation of about 6500 feet above mean sea level. During the pre-flight and the first flight that morning the pilot had noticed nothing unusual in engine start parameters or performance. He refueled the helicopter at one location, flew to another and left the engine running at ground idle.

Due to the size of the next group, the subsequent flights were to be performed by ZK-HGV, another AS 350 and a Bell 206. The pilot rostered to ZK-HGV for this flight was 25 years old and had almost 1900 hours of flight time logged, 234 in the Squirrel helicopter. He carried out routine preliminary checks on the machine, confirmed the fuel contents as 40 percent of total capacity, assisted the passengers to board the helicopter and briefed them with regard to the proposed flight. A slight delay ensued to coordinate departure with the other helicopters.

The intention was to largely repeat the first flight of the day but weather conditions were such that a lowering cloud base may well have precluded some of the higher elevation operations and therefore plans were to some extent left open-ended.

Lift-off and climb were normal and the pilot of ZK-HGV, which was the leading helicopter, followed the promulgated route skirting the southern side of the Fox River valley, then climbed in an easterly direction above the lower part of the glacier towards a prominent feature on the northern side known as 'Victoria Falls'. At a climb airspeed of 90 knots the pilot took particular note that all indications were normal. Turbine outlet temperature was 720°C at 90 percent torque. The all-up weight of the aircraft was approximately 220 pounds below the maximum authorized weight of 4300 pounds.

The helicopter was flying 1100 to 1200 feet above ground level when it suddenly yawed to the right and the main rotor rpm warning horn sounded. The pilot corrected the yaw and lowered collective without delay to commence an autorotative descent. He cancelled the warning horn and was adjusting main rotor rpm when engine power briefly returned but ceased completely within a few seconds.

The pilot of the following helicopter had observed ZK-HGV yaw to the right and realized that an engine problem had occurred. He maintained radio contact with the pilot of ZK-HGV who had advised him of the sudden power loss. The second pilot kept the helicopter in view to monitor its progress during the emergency and to assist as required.

Initially, the pilot of the stricken aircraft considered descending to the river flat at the foot of the glacier but it was evident that this area could not be reached safely. At this time the helicopter was above Victoria Flat, a flatter region of the glacier which, although severely undulating and crevassed, presented the best available landing site. The pilot flew ZK-HGV in a right descending turn of about 270°, selecting the smoothest contours for a landing across the glacier toward the north.

The pilot was well aware of the broken uneven nature of the icy surface and endeavored to ensure that the helicopter touched down with skids level. Firm contact was made in a level attitude with some residual forward speed. ZK-HGV remained upright but, as it slid over the frozen surface, it struck a ridge of ice, fracturing the skids and disrupting the lower structure of the cabin. About 36 feet further on it entered a shallow crevasse and came to rest, rolled onto its left side.

The right cabin door was dislodged during the latter stages of the ground slide and the passenger seated in the rear, nearest the door, fell from the helicopter. The passenger received bruises and abrasions but was otherwise unharmed. No conclusive

Fig 11.1 Failed components
TRANSPORT ACCIDENT INVESTIGATION COMMISSION, NEW ZEALAND

Fig 11.2 TRANSPORT ACCIDENT INVESTIGATION COMMISSION, NEW ZEALAND

reason was obtained to explain why this passenger, who recalled tightening her lap-belt before the landing, was thrown clear at this time. The seatbelt anchorages were intact. The tongue and buckle assembly, which was open when inspected, operated normally. The pilot, who was restrained by a lap-belt, sustained a compressive spinal injury and other minor injuries but was able to vacate the cabin without undue difficulty.

The operator had had the foresight to stow survival equipment in both the left- and right-side lockers to ensure availability in the event of the fuselage lying on one side. It was this precaution that enabled the pilot and uninjured passengers to readily access the equipment and make the two most seriously injured passengers as comfortable as possible. An added bonus was that all the persons on board ZK-HGV were dressed in warm clothing with suitable footwear.

Evacuation of all parties from the scene was efficiently carried out by helicopter after due medical precautions had been taken.

In the subsequent accident investigation it was found that heavy spline wear on the fuel pump driveshaft had caused it to fail. There had been no warning symptoms, the pump had not reached the total hours permitted between overhauls, and there was no maintenance inspection or special procedure which would have enabled the operator to detect or anticipate abnormal spline wear on the fuel pump driveshaft assembly between prescribed overhaul periods or while the pump was in service.

Source: Transport Accident Investigation Commission, New Zealand

Case study 11:2

Accident Report MIA01FA004

Date: October 14, 2000

Location: Pembroke Pines, Florida, United States

Aircraft: Robinson R-22

Injuries: 2 fatal

When the pilot experienced a loss of engine power, he entered autorotation but failed to maintain rotor rpm and the proper rate of descent. The pilot had accumulated 67 hours in helicopters, of which 18 were as pilot in command. The pilot's last known autorotation was four months before the accident.

The National Transportation Safety Board determined the probable cause/s of this accident as:

The pilot's failure to maintain rotor rpm and the proper rate of descent during a forced landing/autorotation resulting in an in-flight collision with terrain. Contributing to the accident was a loss of engine power due to the failure of the No. 3 exhaust valve for undetermined reasons, and the pilot's lack of total experience in the R22 and recent experience in autorotations.

Source: **National Transportation Safety Board, United States**

Principles of autorotation

Autorotation in still air

In powered flight, rotor drag is overcome with engine power. When an engine fails or is deliberately disengaged from the rotor system, some other force must be used to maintain the rotor rpm. To achieve this, the helicopter is allowed to descend and the collective lever is lowered fully, so that the resultant airflow strikes the blades in such a manner that the airflow itself provides the driving force. When the helicopter is descending in this manner, the rate of descent becomes the power equivalent and the helicopter is in a state of autorotation. In practise, most auto-rotations are carried out with forward speed, but to best explain why the blades continue to turn when in autorotation, we will consider that the helicopter is autorotating vertically downwards in still air. Under these conditions, if the various forces involved are calculated for one blade, the calculations will be valid for all the other blades. The various airflows and angles which will be referred to are shown in Fig 11.3.

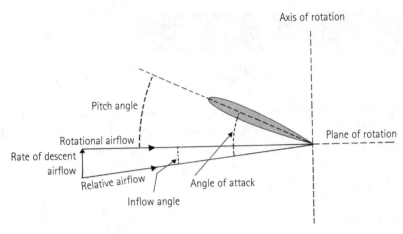

Fig 11.3 Autorotation – terms used *NEW ZEALAND FLIGHT SAFETY*

Note that the inflow has been determined from the blade rotational velocity and the airflow resulting from the rate of descent. This is not strictly true, as the action of the blades slows down the rate of descent airflow, which in effect produces an induced flow, making the inflow angle smaller than has been shown. For the purpose of our discussion, however, it will be assumed that the inflow angle is as shown in Figure 11.3, and the fact that it is smaller and how this affects the blade will be considered later.

Autorotative force/rotor drag

Consider a rotor blade made up of three sections – A, B, and C (Fig 11.4). The direction of the relative airflow for each section can be determined from the rotational velocity and the rate of descent of the helicopter. The rate of descent will have a common value for each section but the rotational velocity will decrease from the tip towards the root. Comparing sections A, B and C, the inflow angle must therefore be progressively increasing (Fig 11.5). The pitch angle is also increasing because of the wash-out incorporated in the blade. The blade's angle of attack is the pitch angle plus the inflow angle; therefore the blade's maximum angle of attack will be at the root.

If the angle of attack for each section of the blade is known, the lift/drag ratio for these angles of attack can be ascertained by referring to the airfoil data tables and, by adding lift and drag vectors in the correct ratio, the position of the total reaction can be determined (Fig 11.5).

Relating total reaction position to the axis of rotation (Fig 11.5a) at section A, the total reaction lies behind the axis; at section B, it is on the axis; and at section

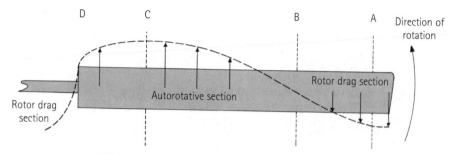

Fig 11.4 Autorotation — Blade section NEW ZEALAND FLIGHT SAFETY

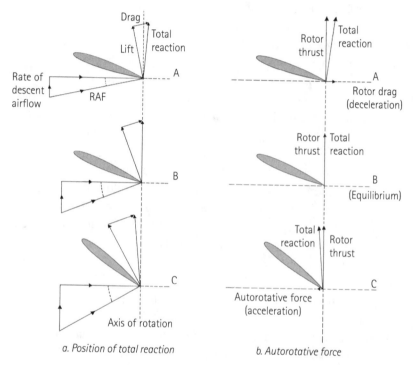

a. Position of total reaction   b. Autorotative force

Fig 11.5 Autorotation — position of total reaction and autorotative forces
NEW ZEALAND FLIGHT SAFETY

C it is in front of the axis. Having determined the position of the total reaction, it can now be considered in terms of rotor thrust and rotor drag (Fig 11.5b). At section A, the condition is the same as in powered flight and the component of total reaction in the plane of rotation opposes rotation and is continually trying to decelerate the blade. At section B no part of the total reaction is acting in the plane of rotation and it is all rotor thrust; at section C the component of total reaction in the plane of rotation assists rotation and is continually trying to accelerate the blade. Under these conditions it is no longer referred to as rotor drag, but as the autorotative force.

The section of blade producing an autorotative force will be accelerating the blade, while the section producing rotor drag will endeavor to slow it down. To maintain a constant rotor rpm, the autorotative section must be sufficient to balance the rotor drag section of the blade, plus the drag set up by the ancillary equipment, tail rotor shaft and tail rotor, all of which continue to function in autorotation.

In normal conditions, with the collective lever lowered, the blade geometry is such that the autorotative rpm is in the correct operating range, provided an adequate rate of descent exists. If the lever is raised during autorotation, the pitch angles increase on all sections (Figs 11.4 and 11.5). Section B will tend towards A and section C will tend towards B; thus the autorotative section moves outwards. However, section D at the root becomes stalled and the extra drag generated causes a decrease in the size of the autorotative section and therefore rpm decreases, stabilizing at a lower figure. This continues with further raising of the lever until such time as the blade is no longer able to autorotate.

Autorotative descents from high altitudes or at a high all-up weight lead to high rates of descent. Inflow angles will be higher and autorotative sections will be further outboard on the blades; rpm will be higher in autorotation under these conditions. It should be noted, however, that descent into more dense air decreases rate of descent and rpm for a constant lever position.

Rate of descent

If the engine fails during a hover in still air and the collective pitch is reduced, the helicopter will accelerate downwards until such time as the angle of attack is producing a total reaction to give an autorotative force to maintain the required rotor rpm and a rotor thrust equal to the weight. When this condition is established, the acceleration will stop and the helicopter will continue downwards at a steady rate of descent. If some outside influence causes the angle of attack to increase, there will be an automatic reduction in the rate of descent, the reverse taking place if the angle is decreased.

Autorotation with forward speed

Unlike a vertical autorotation in still air, the rate of descent during an autorotation will initially decrease with increases in forward speed. Beyond a certain speed, however, the rate of descent will start to rise again. This variation in the rate of descent with forward speed is due to the changing direction of the relative airflow which occurs throughout the speed range in autorotation.

Relative airflow — vertical autorotation

Consider a helicopter of given weight which requires a mean angle of attack of 8° to provide the required rotor thrust and autorotative forces to maintain it in vertical autorotation. Then assume that this angle of attack is obtained when the rate of descent is 2000 feet per minute. If the inflow angle is determined from the rate of descent and a mean rotational velocity, it will be found to have a value of, say, 10° (Fig 11.6a), but because the action of the blades slows down the airflow coming from below the disc, the actual inflow angle will be less, say only 6° (Fig 11.6b). If the mean pitch value to the blade is 2°, then the angle of attack will be 8° — which is the angle required. So a 2000-feet-per-minute rate of descent is required by this particular helicopter to produce an inflow angle of 6°.

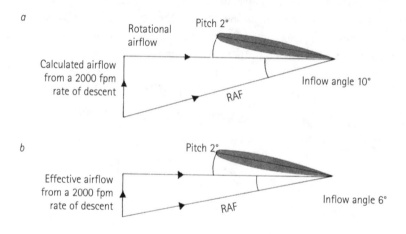

11.6 Inflow angle and rate of descent relationship NEW ZEALAND FLIGHT SAFETY

Relative airflow — forward autorotation

In determining the direction of the relative airflow when the helicopter is in an autorotation with forward speed, three factors must be taken into account. We will consider these factors individually, then collectively.

Factor A

To achieve forward autorotation the disc must be tilted forward. If the effective airflow from rate of descent (Fig 11.6) remains unchanged, then the inflow angle must decrease (Fig 11.7). The angle of attack — and therefore the rotor thrust — must also decrease, thus causing an increased rate of descent.

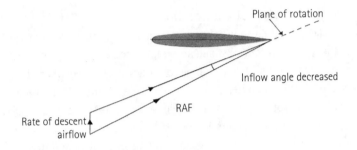

Fig 11.7 Inflow angle — Disc tilted forward *NEW ZEALAND FLIGHT SAFETY*

Factor B

When the helicopter is moving forward, the disc will be subjected to a horizontal airflow in addition to the descent airflow. Because the disc is tilted to this horizontal airflow, it will further reduce the inflow angle (Fig 11.8). The angle of attack is further decreased resulting in a further increase in rate of descent.

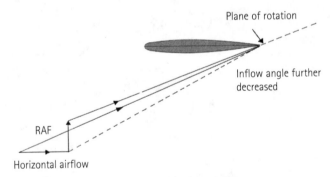

Fig 11.8 Inflow angle — Effects of horizontal airflow *NEW ZEALAND FLIGHT SAFETY*

Factor C

When the helicopter moves forward, the disc is moving into air which has not been slowed down by the action of the blades to the same extent as it is when the helicopter is descending vertically. The effective rate of descent airflow will therefore increase, resulting in an increased inflow angle (Fig 11.9). The angle of attack and rotor thrust increases, giving a decreased rate of descent.

Combined effect

At low forward speed only a small tilt of the rotor disc is required, and the effect of

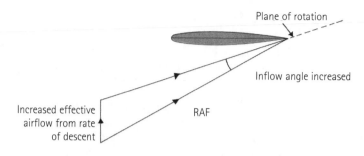

Fig 11.9 Inflow angle — Effects of forward speed *NEW ZEALAND FLIGHT SAFETY*

factor C will be greater than the combined effects of factors A and B, so the inflow angle will increase. The angle of attack, and therefore the rotor thrust, will increase, and the rate of descent will decrease. As the rate of descent is reduced, the inflow angle will decrease, and the rate of descent will stabilize again when the angle of attack is such that the value of rotor thrust equals the weight. As forward speed is progressively increased, the effect of factor C will continue to increase the inflow angle, but, as with the induced flow in powered level flight, its effect is large initially but diminishes with increasing forward speed. As the disc has to be tilted more and more to overcome the rising parasite drag from the fuselage, the combined effects of factors A and B rapidly increase with forward speed. A forward speed is, therefore, eventually reached where the combined effects of factors A and B equal C, and balance out. When this happens, the helicopter will be flying at the speed which gives minimum rate of descent. Beyond this speed, the effects of factors A and B will be greater than factor C, the inflow angle will therefore reduce, and the required rotor thrust can only be obtained from a higher rate of descent.

Rate of descent requirements in autorotation

During autorotation, a rate of descent is required to:

➤ Produce a rotor thrust equal to the weight.
➤ Provide an autorotative force for the selected rotor rpm.
➤ Produce a thrust component equal to the parasite drag.

If these three components are plotted against forward speed, the graph would be similar to the one showing the power requirements for level flight (Fig 11.10).

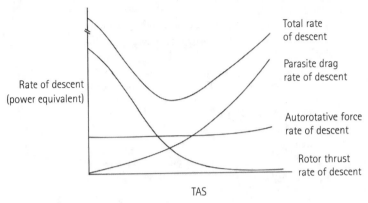

Fig 11.10 Effect of forward speed on rate of descent *NEW ZEALAND FLIGHT SAFETY*

Best speed for autorotation

Each type of helicopter has a minimum rate of descent speed in autorotation. Know this speed — it is critical. In some types, a maximum indicated airspeed (IAS) is also specified for autorotation. This maximum speed restriction is necessary because of excessive rate of descent and low rotor rpm. One particular type of helicopter has a minimum rate of descent speed to 48 knots IAS and at this speed the rate of descent is less than 1500 feet per minute (fpm). The same helicopter has a maximum autorotation speed of 98 knots IAS where the rate of descent is 6000 fpm.

Normal autorotation speed may need to be varied depending on conditions, such as all-up weight and wind velocity. With a head wind of, say 25 knots, it may be necessary to increase airspeed so that a reasonable forward speed is achieved when flaring to reduce rate of descent. A higher all-up weight will increase rate of descent and rotor rpm, therefore a slight increase in airspeed may be desirable. The correct airspeed for autorotation descent should be minimum rate of descent speed for ideal conditions, increasing by no more than 10 knots for the worst conditions. A turn during autorotation has the effect of increasing all-up weight, thus increasing the rate of descent and rotor rpm. A slight increase in collective pitch may therefore be required to keep these values within limits. These factors are applicable for a standard autorotation in flat pitch with the collective right down. But let us now look at positioning the helicopter for an actual autorotational descent to a particular spot.

Autorotation for range

The horizontal distance traveled can be lengthened considerably by increasing airspeed 10–20 knots and reducing rpm to approximately a quarter up from bottom green rotor rpm. While this effectively increases rate of descent, the higher airspeed has the effect of reducing angle of descent.

The actual airspeed gain will be governed by the type of helicopter and rotor system, its minimum rate of descent speed, and the prevailing wind conditions. In a typical light helicopter with a two-blade high-inertia rotor system and a minimum rate of descent of, say 48 knots, the still air range speed would be about 60–65 knots, or 70 knots near bottom green rotor rpm with a head wind of approximately 20 knots. With low-inertia rotor systems the minimum rate of descent speed is usually higher, therefore, range speed is increased accordingly by an amount similar to that outlined for high inertia rotor systems.

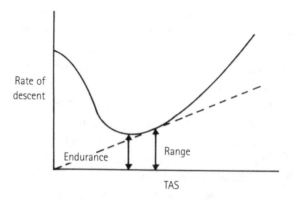

Fig 11.11 Range and endurance NEW ZEALAND FLIGHT SAFETY

During initial application of autorotation for range, you will likely experience the opposite effect of an apparent loss of range, as airspeed is increased to the desired range speed. This will soon diminish as rotor rpm is reduced and the effect of extended range becomes evident.

At some point before the flare, however, it is essential that airspeed is reduced to minimum rate of descent speed and rpm increased to near top green rotor rpm. This is achieved by initiating a slight flare to reduce speed (which also has the effect of increasing rotor rpm) and at the same time lowering collective fully down. In a helicopter with a low-inertia rotor system this action can be left to just before a normal flare, since the rotor rpm will rise rapidly. With a high-inertia rotor system, it takes longer to increase rotor rpm, therefore corrective action must be initiated earlier.

Do not fall into the trap of having to execute an excessive flare in order to reduce a high rate of descent. Although this will increase rotor rpm considerably, any benefit gained from the maneuver is lost as a result of the abnormal amount of pitch forward necessary to achieve a level touch-down attitude.

Remember, a flare will increase rotor rpm; a pitch forward will reduce rotor rpm.

Autorotation for minimum range

An opposite situation applies when a suitable landing area is closer than normal autorotation descent angle. In this case, a steeper angle of descent can be achieved by reducing airspeed. This is quite safe providing the speed is kept out of the range which could produce power settling (vortex ring state).

While this condition is not likely to result in a full down collective situation, with near zero airspeed the raising of collective can produce a vortex ring state very quickly. So, reduce airspeed if necessary but still maintain a minimum of 10–15 knots IAS. Recovery from this slow airspeed must be commenced early as it may take 300–400 feet in height to reach minimum rate of descent speed.

Possibly a better way of reaching a landing site almost below the helicopter is to execute a 360° turn. This enables the landing point to be kept in view and also allows for adjustment of the turn to reach the required spot. However, a height of about 800 feet is normally required for such a maneuver. The actual height loss can be reduced substantially by turning in the same direction as the rotor rotation. For example, turn right in a Lama or AS350 Squirrel; turn left in 'US types'.

The profile of an autorotation descent can be likened to a funnel angled about 30° from the vertical with the neck on the landing area. The lower side of the funnel represents the angle of autorotation for range and its upper side the near zero airspeed angle of descent. As long as the descent profile remains within the two sides of the funnel, the helicopter will go through the neck and minimum rate of descent speed will take it to the desired touch-down spot.

The flare

Having established a stable autorotation, the next phase of the maneuver is the flare. The main purpose of the flare is to reduce both forward speed and the rate of descent. It can also be used in a limited way to make good an alighting point and, if necessary, to restore rotor rpm.

The flare effect following autorotation will be exactly the same as for the flare in powered flight. Rotor rpm will rise because the increased inflow angle will cause the autorotative section to move towards the tip, and the increased thrust will reduce the rate of descent as long as the flare effect lasts.

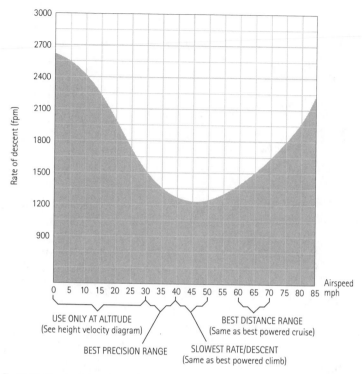

Fig 11.12 Steady state autorotation rate-of-descent graph for a typical light helicopter

NEW ZEALAND FLIGHT SAFETY

The autorotative landing

The flare from minimum descent speed should be a steady, progressive maneuver to reduce the rate of descent and forward speed. As mentioned earlier, this will increase rotor rpm and a very slight tug on the collective at this point may be possible. The flare should be initiated early enough so as to arrive at the landing spot in a level attitude. An excessive nose-up attitude not only requires more height for adequate tail rotor clearance, but also produces a reduction in rotor rpm in the resulting large roll forward.

Actual touch-down must be made with skids level and straight and, if possible, before translational lift is lost. This can be achieved by a small run-on landing if ground conditions permit, or a zero groundspeed landing if the wind strength is 15 knots or higher. This principle dictates an in-to-wind landing regardless of wind strength. Even with a 5-knot wind, translational lift is lost at a ground speed of 10 knots landing in-to-wind, or at 20 knots groundspeed landing downwind.

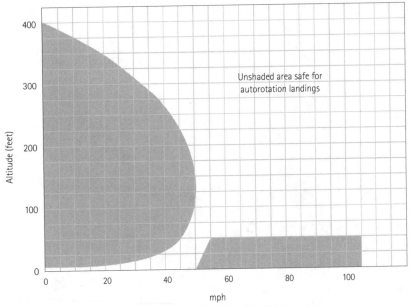

Fig 11.13 A typical airspeed versus height limitation chart *NEW ZEALAND FLIGHT SAFETY*

Avoid areas

There are certain areas of the flight envelope of a helicopter from which it is impossible or dangerous to carry out a successful autorotation.

Establishing a fully developed autorotation after an engine failure will inevitably involve a loss of height. The extent of the height loss will depend on the airspeed, aircraft weight and the density altitude at the time that the engine failure occurs.

From the lowest forward speeds, the height loss necessary to establish full autorotation will be greatest. This is because, not only will the lever have to be lowered fully and rapidly to maintain rotor rpm, the helicopter will also have to be put into a descent in order to accelerate to the optimum autorotation speed (see Figs 11.11 and 11.4). At higher airspeeds, the kinetic energy of the helicopter can be used to assist in maintaining rotor rpm, and achieving optimum airspeed will require less acceleration — or even deceleration. In these circumstances, therefore, the height loss will be much less.

However, at low level and high forward speeds, it will be necessary to flare the aircraft in order to avoid hitting the ground at high speed. During the speed reduction it may be necessary to manipulate the lever in order to prevent the rotor over-speeding. This flare may require a steep tail-down attitude and, bearing in mind that an immediate height loss is inevitable, at very low level there may be insufficient time for the pilot to carry out a flare, or insufficient room for it to be completed. These factors are taken into account in producing the low-level high-speed 'avoid' area on the aircraft flight envelope graph (Fig 11.13).

The other 'avoid' area on the graph, the low-speed area, can be explained as follows.

At very low forward speeds and close to the ground, it is possible to cushion the helicopter onto the ground using primarily the kinetic energy of the rotor system. This is achieved without entering stable autorotation. As height above the ground increases, however, a point will be reached at which the rotor energy is insufficient to cushion the descent onto the ground, and thus it will be necessary to achieve a developed autorotational state. This height is depicted as the lowest point on the low-speed avoid area graph. To this must be added the height loss that will be incurred in establishing the autorotation, and this then gives the upper limit of the low-speed avoid area.

In Fig 11.13 we show a specific example of an avoid area graph (commonly known as 'dead man's curve').

Operational considerations

Some pilots consider that a helicopter has less performance in an actual auto-rotation than in practise with the throttle closed. As long as the needles are split in practice autorotation, there is no technical reason for any difference in performance. But the circumstances at commencement are very different.

Consider, first, the practise situation. Having decided to carry out an autorotation descent, you become mentally prepared and position the aircraft at a selected height. At about the right angle you lower the collective fully and then close the throttle. Rotor rpm, airspeed, rate of descent are all good and from then on it is plain sailing.

Now take the actual emergency situation. You are climbing out or cruising along with the collective half up under your arm and with your thoughts some-where else. You are probably not as high as you may have been if there was any thought that the engine may stop. Without warning everything goes quiet, or the horn starts blowing. In the split second it takes to decide what is happening, rotor rpm is now going back through 70 percent because of the high collective position and associated drag. You lower the collective, maybe still not right down, and have a quick look round for a suitable landing area. By now you have lost 200–300 feet, with rotor rpm still very low and not coming up very fast. Your selected landing spot proves to be a bit too far away and you arrive short, too fast, and with rotor rpm too low. The result is inevitable.

Many tasks performed by helicopters have to be conducted fully at low level, and sometimes within the height/speed avoid area ('dead man's curve'). This should be accepted only if the operational requirement dictates. But careful thought should be given to planning so that the time spent in these critical areas of operation is kept to an absolute minimum. (See Fig 11.13).

To a helicopter pilot used to operating predominantly at low level, an en route climb to a safe autorotation height of 500 or 1000 feet may seem quite unnecessary and time-consuming. In fact, with a reasonably high performance helicopter there will be little or no extra time involved since the reduction in ground speed during track climb is compensated for in the descent.

Let us consider a suitable, safe transit height for a well-loaded turbine-powered helicopter on a short distance flight from A to B. A climb to, say, 1000 feet at 80 knots IAS will take about two minutes. A descent over four minutes will make up for this earlier reduced ground speed. By cruising at 1000 feet above ground level, the autorotation range from any point on track is like an inverted funnel and this allows you to reach any point within the funnel. If there is a tail wind, an engine failure will require 200 feet to stabilize and a further 400–500 feet to turn into wind. This leaves only about 300 feet to vary the approach path and make a satisfactory landing.

Conversely, if you are cruising at 500 feet above ground level in a tail-wind condition, an unexpected engine failure will probably leave you with insufficient height to turn into wind and maneuver the helicopter to a suitable landing area. You will probably not be able to make a successful landing downwind in other than a very slight breeze.

It is therefore apparent that a cruise height of 2000 feet will provide a funnel area of possible landing sites many times the size of the options at 500 or 1000 feet. In the event that a suitable landing area is either not available or just out of autorotation range, you will have to adapt your mind quickly and firmly to an alternative course of action. It is recommended that you control the rate of descent at the correct speed and flare to zero speed on the best available landing area, even if this happens to be the softest looking tree! If it is on sloping terrain and the landing approach is transverse to the slope, kick the nose uphill just before touchdown. A controlled crash landing such as this is preferable to arriving too fast, too low, and with rotor rpm too low.

To fly helicopters without regularly practicing autorotations is a little like playing Russian roulette. If you haven't practiced the skill and the engine stops (it usually does so unexpectedly), you have loaded a few more chambers in the revolver. If you are a high-time pilot your natural instincts may see you walk away, perhaps from a wreck. But if you have less than 1000 hours, chances are you won't be this lucky unless you are habitually practicing and improving engine-off talents. Much of this is the skill of good judgment; something that doesn't always come naturally.

The time you need to react to an engine failure is small, particularly in a helicopter with small blades that don't carry much inertia. Rotor rpm will begin dropping immediately and dramatically. Your left hand should never leave the collective and don't be shy to push it toward the floorboards whenever the nose gives an unexpected twitch; yaw is the first indication that the engine isn't delivering full power. If the cause of the twitch turns out to be a bit of unforeseen turbulence, that's fine; better safe than sorry.

Twin-engine helicopters are not immune from problems either. In fact you could argue that, with two engines strapped on, you are twice as likely to encounter engine problems.

Although not as vulnerable as a single-engine machine in various flight regimes, the twin-engine model does have similar susceptibilities. Even now, there are not too many twin-engine machines that can hover OGE at maximum all-up weight on one engine if the other has failed. So, at those crucial parts of the flight envelope – just after takeoff and on approach to land – the twin has similar problems. Even with one engine still operating, a twin-engine aircraft cannot sustain lift at low airspeeds.

With two engines you are certainly better off in cruise flight should one of your engines fail, as long as you handle the failure appropriately.

Case study 11:3

Twin engine failure

Date: Mid 1990s

Location: Unknown

Aircraft: UH-60A Black Hawk

Injuries: 4 fatal

In the mid-1990s, two UH-60A helicopters (Black Hawks) took to the air at 2120 hours on the first leg of their flight. The crews were using night-vision goggles and there was ample starlight for these to be effective.

Flying over water at 700 feet and 140 knots, the crew of one helicopter (codename 'Chalk 2') saw sparks coming from the exhaust on the lead aircraft's Number Two engine. Chalk 2 made a radio transmission advising Chalk 1 that their Number Two engine was on fire. There was no response but a few seconds later Chalk 1 turned toward the shore in a slow descent. Shortly afterward the controlled descent increased rapidly. Just before impact the Chalk 1 crew abruptly raised the nose of the aircraft but it was too late. With a descent rate of 1200 feet per minute and a ground speed of 214 knots, the 16,382 pound Black Hawk hit the water tail-low, rolled onto its left side, appeared to explode and then sank in 140 feet of water.

It was found that the engine failed because a stage-one turbine blade on the gas generator had fatigued; not an unheard-of event.

But the crew did not take the proper measures when faced with the failure of one engine. Sometime after the crew was told that the Number Two engine was on fire, one of the pilots pulled off the throttle (power control lever) which shut down the Number One engine. Now it was a total power-off situation, but the pilots did not adjust the flight configuration to establish autorotation and failed to maintain rotor rpm.

Source: *Vortex*, 4/98

Case study 11.3 serves to underline that engine failures require accurate judgment and procedure, even for helicopters with two engines. Mistakes can be easy to make but four lives is an expensive price to pay.

The United States Army conducted an extensive review of procedures and did a study that involved 4000 multi-engine aviators. The results were quite surprising: 70 percent believed the potential was there to shut down the wrong engine, and 40 percent confirmed that during simulated emergencies they had become confused. Half of those confused pilots admitted they had, in fact, shut down the wrong engine or moved the wrong power control lever! This revealed that there was a one in six chance that the crew of a two-engine aircraft would react incorrectly when one engine failed.

Subsequently, it was decided to do more crew coordination training, expand the training, and improve procedures and checklists with regard to malfunction analysis and emergency procedures.

One emphasis the review identified concerned procedures. All procedures are subordinate to the primary function of a pilot — fly and control the aircraft.

12

Tail rotor failures

The tail rotor is an anti-torque mechanism that controls a hovering helicopter in the yawing plane. When more power is applied, more tail rotor pitch (and consequently thrust) is required to keep the nose straight. Bear in mind, that for the tail rotor to deliver more thrust, it requires more power. As pilots increase the power setting and generate more torque, more tail rotor thrust is required. This process can snowball to the extent that there is insufficient power available for the desired maneuver. Pilots commonly call this 'running out of tail rotor'.

Even though this seems very straightforward, other factors make it more complicated. The rpm of helicopter tail rotors varies with make and model, but a figure of 1600 to 3500 rpm is realistic. A piece of machinery moving at that speed, and having variable forces and settings, is an advanced and finely tuned device.

Being right at the back of the helicopter, the tail rotor is the furthermost thing from the pilot and, in the heat of a tricky operation, it can be forgotten. A tail rotor blade striking an object, no matter what size, can do incredible damage. Grass, wires, fences, animals, people, bushes, trees, buildings — you name it; somewhere in the world someone has hit it with a tail rotor. Whatever is hit, maintaining control of the helicopter can be very difficult if not completely impossible.

Of course tail rotors can suffer mechanical failures: the engineer leaves something out during re-assembly, a component just ceases to work, or there is too little maintenance. A mechanical failure is the subject of case study 12.1.

When a tail rotor fails for whatever reason, things get out of balance very quickly and compounding factors can happen almost instantaneously. The tail rotor gearbox, a solid and reasonably weighty object, often tears itself loose. Located at the end of the tail boom, quite some distance from the center of gravity, its loss rapidly moves the C of G some distance forward. As a result, there could well be insufficient aft cyclic available to keep the nose level. Such a sudden shift, together with a resulting control input from the pilot, may well cause a main rotor blade to flap downward and strike the tail boom or cockpit. In a teetering rotor system it may also cause mast bumping and subsequent loss of the main rotors. Whatever happens, losing the tail rotor is nobody's idea of fun!

Case study 12:1

Aircraft Accident Report Number 91-008

Date: March 17, 1991

Location: Castle Point, North Island, New Zealand

Aircraft: Hughes 369HS ZK-HZG

Injuries: nil

The 38-year-old commercial helicopter pilot was completing a task which had taken him around two hours. He was lifting bales of seaweed from an isolated beach and sling-loading them to a collection point suitable for ground transportation.

In the hover, while detaching the last of the 25 loads, the pilot sensed a transient 'buzz' from the tail rotor area. He attributed this to the effect of a crosswind gust. The 'buzz' disappeared with slight application of yaw pedal in either direction.

On the transit back to the pickup point, the pilot positioned the helicopter parallel to and about 200 yards seaward of the shoreline to facilitate a right-hand circuit onto the beach.

At about 200 feet altitude with an indicated airspeed of 60 knots on the downwind leg, the pilot felt a brief high-frequency vibration throughout the airframe, followed by a loud noise. Both yaw pedals 'kicked back' momentarily and then 'went slack'. The helicopter's nose pitched down sharply, giving the pilot the impression that the helicopter was going inverted.

At this time, the attention of the coordinator on the beach, also a commercial helicopter pilot, was attracted to the helicopter by a loud noise. The witness observed that the tail rotor, apparently intact, had separated from the aircraft and was 'half spinning, half fluttering' into the sea.

The pilot lowered the collective pitch lever, at which point the aircraft began to rotate to the right. In an attempt to raise the nose he applied full aft cyclic, which was partially effective only. He raised the collective, accepting that the aircraft was going to continue to rotate, to explore the possibility that it might assist in raising the nose. The rotation was so violent that he hit his head on the side of the cabin and the top of the instrument panel. He was restrained by a full harness and was wearing a safety helmet.

The helicopter descended, still rotating uncontrollably, toward the beach and the high ground beyond. The pilot stated that his intention throughout these revolutions was to arrive on dry land, avoiding ditching at all costs.

As it crossed the beach the aircraft's rate of rotation appeared to decrease as it slowed and the direction of travel changed away from the high ground and back to seaward. At that stage the pilot closed the throttle and pulled full collective. The helicopter landed on some rocks with sufficient sideward motion to roll it onto its side. The rotor blades struck the rocks and 'thrashed themselves to pieces'.

The pilot stopped the engine by closing the fuel valve and vacated the wreckage, shaken but uninjured. The Hughes 369 was destroyed.

The approximate position of the tail rotor was marked by a crayfish pot and float but, despite a number of attempts, it was not recovered.

Examination of the wreckage revealed that the tail rotor, its driveshaft and a substantial portion of the tail rotor gearbox housing had parted company from the aircraft. The remaining portion of the gearbox was that which housed the input driveshaft, with its crown wheel and associated bearings.

Laboratory examination of the available evidence was largely inconclusive; however, it is interesting to note that the tail rotor had suffered a strike some 192 hours' flying time prior to this accident. The gearbox had been dismantled, examined and rebuilt before being restored to service. The loss of a gear tooth subsequent to a tail rotor strike is a known phenomenon in Hughes 369 helicopters, the first 300 hours after the strike being the 'danger period'.

Fig 12.1 Remaining portion of the tail rotor gearbox

Findings in subparagraphs 2.1 to 2.3 of the accident report stated:

The probable cause of the gearbox failure was a gear tooth separating from the output drive pinion and lodging between the two gear wheels, but other causes could not be eliminated as the remainder of the tail rotor assembly was not recovered.

The resultant loss of anti-torque effect, coupled with the sudden change in center of gravity due to the departure of a significant mass from the tail, rendered the helicopter uncontrollable.

The pilot's lack of injury was attributable to the helicopter's low touchdown speed and wearing a full harness and safety helmet.

Source: Transport Accident Investigation Commission, New Zealand

Flight Lieutenant Ian MacPherson shares his tail rotor incident:

I recall this story more than eight years after the event but it feels just like yesterday.

Joining the Royal New Zealand Air Force in 1988 was the realization of a dream. Having gained my private pilot's license a few years earlier I found the discipline, high standards and relentless emergency training somewhat of a shock. In hindsight I believe this training may have saved my life and the lives of all those on board my aircraft. This is my account of the day the tail rotor of the UH-1 Iroquois helicopter on which I was a very junior captain failed.

With only 301.4 hours on the Huey I was a junior bograt and was seldom authorized as captain; normally the tasks the squadron flew would require the better-qualified pilots to occupy the right seat. However, in order to gain experience we were occasionally sent out on day navigation exercises, as was the case on 29 April 1991. As Number Two in the formation we were positioning downwind left hand for runway 03 at Wigram. The rejoin was briefed to be a run-in low level on 03 followed by a one second 270° right break, to terminate outside Number Four hangar. All was going as planned except for Number One's VHF radio, which had failed minutes before; my crew were therefore doing all the joining calls for the formation.

In mid-downwind and only one and a half miles from Wigram we felt and heard a high frequency vibration through the airframe. It's funny how often you hear, feel or smell something in the cockpit but after consultation with the crew the problem appears to have been a figment of your imagination. In the hope that this was perhaps one of those times, I sheepishly asked the question, 'Can you guys hear or feel that?' To this they both replied over the intercom, 'Sure can'. There was little doubt in our minds that, with Wigram so close, we should turn left, leave the formation, and land as soon as possible. Once clear of the other aircraft and with the whole airfield in our sights we had a minute to address the issue; strangely, however, the vibration had gone away. All of the flight controls were responding normally, all of the instruments were normal, and since the noise too had gone I felt no cause for alarm. My old course-mate in the left seat had his hands full operating two radios. I don't think he was particularly worried either. Why should he be? The noise was gone.

As we crossed the perimeter fence Number Four hangar came into sight. I certainly felt a slight sense of relief as home was now only 300 yards away. However, we were far from out of the woods. As the Huey approached translational, I introduced collective to arrest the rate of descent and the aircraft made a sudden and violent yaw to the right. I had never seen anything like this in all of the training I had received. One thing did seem obvious; pulling up the collective had caused this immediate problem so the sensible act was to put it back down. I did this and fortunately the rotation stopped at about 110° and then the nose came left again to settle at 60°–70° out to the right. It came as no great surprise that the vibration was back and with far more vengeance this time. Some height and speed were lost but we were still crossing the ground at approximately 20 knots and descending through about 40 feet.

Acting more on instinct I tried to introduce power and increase airspeed. The aircraft responded but was flying at an alarming attitude, the left skid was very low and the nose wavering between 70° and 90° out to the right. Despite reducing the rate of descent, thereby delaying impact with the ground, efforts to climb proved fruitless as the aircraft threatened to rotate through 90° every time power was increased beyond a critical point.

The co-pilot and crewman automatically performed the critical actions of a mayday call and securing the passengers in their crash positions.

The point of impact was quite obvious, about 100 yards away. As the aircraft approached 10 feet and was still crossing the ground at about 20 knots I had no choice but to treat this as a 'low power tail rotor emergency'. For a helicopter like the Huey with an anticlockwise rotating main rotor, a low power tail rotor emergency means the tail rotor is not producing the thrust required for a given power setting and, as power is introduced and/or airspeed is reduced, the nose rotates to the right. The only corrective measure is to close the throttle, which eliminates the torque; the nose then rotates left for a matter of seconds, during which time the aircraft should be run on while the skids are aligned with the direction of travel.

During training we would never practice such an extreme low power case; I had not seen this maneuver done with the nose beyond 20°–30° to the right. I briefed the crew of my intention to close the throttle to flight idle and proceeded to do so. The aircraft yawed to the left but didn't quite reach the direction of travel when suddenly the rotation reversed and again the nose was rotating to the right. Now we really were committed; the rotor rpm was reducing rapidly and the rotation was accelerating through 90° to the right.

Again instincts took over and I attempted to reduce ground speed to zero, as the aircraft would surely turn over if we hit with any sideways movement. The left toe then kissed the ground but didn't grab, allowing a further 15 feet of flight before contacting again, this time with more vengeance. As the skid tore into the soft topsoil the remaining sideways movement caused the aircraft to rear up on its left skid, all the time rotating to the right. The last of the energy was dispelled as the left heel also penetrated the topsoil and the aircraft, now completely out of control, threatened to roll over.

Fortunately the dynamic rollover effect ceased before the center of gravity exceeded the limit of the left skid and the aircraft ungracefully fell back down, for the first time contacting the ground with both skids. The aircraft was now facing 180° opposite to the approach heading.

The crewman quickly vacated the aircraft to check for the cause of this hair-raising ride. The co-pilot and I looked anxiously at each other while we secured the engine and turned the electrics off. The main rotor was still winding down when the crewman returned to inform us that the tail rotor was not turning at all and that he could see the failed tail rotor drive hanger bearing. This was the first time during this 50-second ordeal that any of us could think clearly enough to acknowledge the fault and its seriousness.

I flew the Huey for a further five and a half years, qualified as a fixed-wing and helicopter instructor and now have a total of 2500 flying hours, 1800 of which are on helicopters. I have never forgotten the time my tail rotor failed, and I doubt I ever will. Surprisingly, I don't think I would do anything different if it happened again. Perhaps I wouldn't have flown a normal approach but again the noise was gone and there were no signs of a serious problem — how often do you hear a noise that turns out to be nothing?! I attribute our instinctive handling of this problem to the excellent instruction I received during my training. Quite simply, when you fly helicopters there are some things that need to be instinctive. One of them is the initial actions required when you lose yaw control or tail rotor effectiveness.

Loss of tail rotor effectiveness

Sideways movement, shuddering in the tail rotor pedals and loss of directional control to the right are all symptoms of the condition commonly known as loss of tail rotor effectiveness (LTE). In this state the airflow through the tail rotor is reduced — the combined effect of the actual wind, the direction of movement of the helicopter and/or the turbulent air emanating from the rotor disc. As a result, the tail rotor can no longer supply sufficient thrust to counter the turning effect caused by the rotation of the main rotor system, and the helicopter will yaw to the right. If the effect is severe enough, application of left pedal may aggravate the situation.

Case study 12:2

Accident brief

Date: July 1989

Location: Parafield Airport, South Australia, Australia

Aircraft: Hughes 300

Injuries: unknown

The pilot of a Hughes 300 aircraft was approaching Parafield Airport in South Australia. The pilot had 387 hours of flight experience, 25 of which were in the Hughes 300 aircraft.

The pilot was practicing hovering at various altitudes. With the helicopter heading 320°, the pilot commenced a slow descent from about 15 feet. The wind was 310° at 10 knots gusting to 20 knots. While descending to about three feet, the helicopter began a slow yaw to the right through 30°. The pilot elected to let the yaw continue and carry out a 360° pivot turn.

After turning through about 100°, the pilot stopped the descent by increasing power and pulling on collective. Coincidentally, the pilot reported that the helicopter then rapidly increased the rate of turn despite the application of full left anti-torque pedal.

After about one turn, the pilot said he lost all control and, following about another three to four turns, the helicopter crashed onto its side at about 45° angle of bank. During the sequence several main rotor blade ground strikes occurred. The pilot said that he was also aware of over-controlling on the cyclic and pulling on collective at one stage in his efforts to control the helicopter.

The investigation could not find any mechanical cause for the loss of control. Both the pilot and passenger reported that at no time did the passenger touch the controls or have his feet near the anti-torque pedals.

In addition to the fact that the relative wind was favorable to the formation of vortex ring in the tail rotor, the tail rotor was also moving to the left in the right pedal turn.

At the point where power was increased, the tail rotor was in a critical position with the relative wind from 250° and strength from the left of between 10 and 20 knots. If the tail rotor was close to a vortex ring state, increased power would, through the torque effect, produce a marked right yaw. Due to the vortex ring, left pedal input would have no effect.

Source: Bureau of Air Safety Investigation, Australia

Case study 12:3

Accident brief

Date: May 1996

Location: Western Australia, Australia

Aircraft: Robinson R22

Injuries: unknown

The pilot, who was engaged in aerial mustering, was flying the helicopter sideways 15 feet above the ground as he attempted to herd a bull into a paddock. He felt a shudder through the tail rotor pedals and the helicopter immediately began to rotate to the right. Full left pedal did not stop the rotation. After $2\frac{1}{2}$ turns the pilot was able to land the helicopter, cushioning the touchdown with the collective control.

During the landing the tail rotor struck a fence. A ground witness reported that the tail rotor system appeared intact until the collision. The engine had already stopped when the pilot attempted to shut down after landing. The fuel tank contained sufficient fuel and no water was evident. Wind strength at the time of the accident was reported to be 5 knots.

Source: Bureau of Air Safety Investigation, Australia

Principles

Loss of tail rotor effectiveness is very well outlined in this passage from the Australian Transportation Safety Board's Accident Report on Occurrence Number 200003293:

> The phenomenon of LTE, also known as unanticipated right yaw, has been identified as a contributing factor in several helicopter accidents. According to United States Army testing, OH-58 series helicopters (the Bell 206 series is the civilian variant) have proven in the past to be susceptible to LTE under certain low-speed maneuvers. LTE is not related to a maintenance malfunction and is associated with single main rotor, tail rotor configured helicopters. LTE is a result of the tail rotor losing aerodynamic efficiency due to a combination of several factors. Those factors include main rotor vortex interference and tail rotor vortex ring state (related to airflow disruption over the tail rotor), helicopter weathercock stability, and the loss of translational lift. The regimes in which LTE may be encountered include low airspeed (less than 30 knots) when translational lift is lost or reduced, high power, and in the case of the United States-designed helicopters (note: this indicates helicopters with main rotors that rotate counter-clockwise), operating in a left crosswind or tailwind or with a high yaw rate to the right.

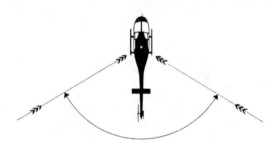

Relative winds of more than 5 knots from within this sector
will cause the aircraft to weathercock into wind, either
left or right, depending on the relative wind direction

Fig 12.2 Weathercock stability *NEW ZEALAND FLIGHT SAFETY*

There is greater susceptibility for LTE on US-designed helicopters in right turns and more so in right turns over water. This is especially true during flight at low airspeeds when the pilot is looking out the right window (not viewing the instrument panel) and is unaware of the airspeed dropping to a low value. The turn is commonly done with reference to the ground where the pilot attempts to keep a constant groundspeed by referencing ground cues. Flying over water, the pilot does not have the visual cues available as when flying overland.

In turbine-powered helicopters, the frame of reference for the engine power governor is the main rotor rpm (Nr) with reference to the airframe. Once the helicopter begins spinning rapidly to the right, as during the onset of LTE, the governor will sense a false increase in Nr and reduce fuel flow to the engine in order to maintain what it believes to be a constant Nr with reference to the airframe. Any reduction in Nr will result in a corresponding reduction in tail rotor rpm, with an associated reduction in the effectiveness of the tail rotor.

AUSTRALIAN TRANSPORTATION SAFETY BOARD

Tail rotor vortex ring state

Being the same phenomenon as vortex ring state in the main rotor, tail rotor vortex ring state has the same result: uncommanded directional travel; in the main rotor the direction is downward, in the tail rotor the direction is to the left (assuming counter-clockwise rotating main rotor).

1. Wind from between 210° to 330°

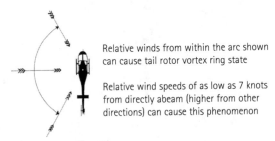

Relative winds from within the arc shown can cause tail rotor vortex ring state

Relative wind speeds of as low as 7 knots from directly abeam (higher from other directions) can cause this phenomenon

Fig 12.3 Tail rotor vortex ring state *NEW ZEALAND FLIGHT SAFETY*

Winds between 210° to 330° promote the development of tail rotor vortex ring state. As the flow passes through the tail rotor, it creates thrust to the right (causing airflow to the left). A left crosswind will oppose the airflow. That opposition causes erratic unsteady airflow into the tail rotor, and this causes the vortex ring state to form. The resulting oscillations in tail rotor thrust cause variations in yaw.

Rapid and continuous pedal movements are necessary when hovering helicopters in a left crosswind. The pilot must compensate for the rapid changes in tail rotor thrust demands. A high pedal workload with tail rotor vortex ring state is normal. The condition presents no real problem unless corrective action is delayed.

When the generated thrust is less than the thrust required, the helicopter will yaw to the right. Pilots hovering in left crosswinds must concentrate on smooth pedal coordination and not allow excessive right yaw to develop. Without such action, the machine can rotate into a tailwind position where weathercocking accelerates the rate of turn to the right.

2. Wind from 285° to 315° between 10–30 knots

Winds speeds of between 10 and 30 knots from 285° to 315° cause the main rotor vortices to be blown into the tail rotor. The tail rotor will be operating in very turbulent air.

Relative winds of 10–12 knots from within the sector shown can cause main rotor disc vortex to interfere with tail rotor efficiency

Fig 12.4 Main rotor disc vortex interference NEW ZEALAND FLIGHT SAFETY

During a hovering turn, tail rotor thrust reduces as the tail rotor reaches the area where the main rotor disc vortices are channeled by the wind. The tail rotor disc encounters turbulent vortices, which result in an increase in the angle of attack of the tail rotor blades. This in turn gives an increase in thrust, requiring the pilot to add right pedal (reducing thrust) to maintain the turn rate.

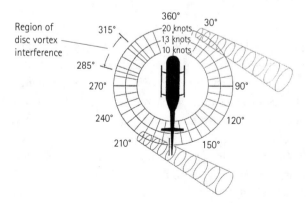

Fig 12.5 Main rotor disc vortex interference NEW ZEALAND FLIGHT SAFETY

As the tail rotor clears the affected area, its angle of attack suddenly reduces. This causes a reduction in thrust and right yaw acceleration begins. As the pilot was adding right pedal to maintain the right turn rate, the sudden yaw can be unexpected. If uncorrected, the yaw will develop into an uncontrollable rapid rotation about the rotor mast.

Avoidance and recovery

A sound knowledge of the forces involved in loss of tail rotor effectiveness and how they combine is imperative. Couple this with a high level of concentration when controlling the aircraft, particularly in correcting uncontrolled yaw in any direction (principally the 'with torque' movements), and you are well on the way to being on top of the problem.

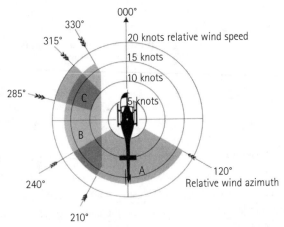

Fig 12.6 Combined relative wind velocity diagram for weathercock stability

NEW ZEALAND FLIGHT SAFETY

Extensive studies in the mid-1980s revealed that LTE was not a 'tail rotor stall' caused or aggravated by the application of left pedal. Testing determined that full left pedal (to counteract the yaw) and forward cyclic (to increase speed) invariably stops the unanticipated right yaw (anticlockwise main rotors). Reducing collective will also assist in arresting the yaw rate, but will most likely cause an excessive rate of descent. Near the ground, a large increase in collective to prevent or cushion contact with the ground may again aggravate the undesired yaw. The pilot must decide on reduction of collective with the height available for recovery in mind.

If the undesired yaw cannot be stopped and ground contact is unavoidable, an autorotation may be the wisest option.

13

Mechanical failures

A helicopter is an intricate piece of machinery. Turbine engine parts rotate at tens of thousands of rpm with clearances of just a few thousandths of an inch. Bearings, control linkages, simple rivets and seemingly a million other vital parts must be regularly inspected, maintained and, from time to time, replaced by qualified personnel. Such personnel are given different titles (mechanic, engineer, and so on) in different countries so here they will be referred to as maintainers. The original titles are retained in the case studies.

The part schedule for the average helicopter can be a seemingly endless, mind-boggling list of technical jargon that only an aircraft maintainer can understand. Many parts have to be adjusted to a precise torque setting, a lock-wire must go here, a seal must go there — so critical are the hundreds of maintenance tasks that each must be 'signed off' twice: firstly by the person performing the task and then by another qualified person after checking that the task has been done correctly.

In an ideal world, aircraft maintainers work in a tidy clean maintenance hangar with tools neatly arranged on a shadow-board. The only interruptions or distractions are for refreshments delivered during regular breaks. The maintainers work for a realistic number of hours per day, with plenty of rostered rest days. The spotlessly clean helicopters on which they work are delivered on time, with advance notification of all the faults the operator has noted since the last scheduled maintenance inspection. The maintainers are able to give the operator a very good realistic estimate of the amount of time the inspection and servicing is going to take. No pressure exists to have the helicopter serviceable earlier as the operator has scheduled his workload to accommodate the required maintenance. Of course, the word budget is never used; all replacement parts are bought new and availability is never an issue.

Case study 13:1

Accident Report Number A00Q0046

Date: April 27, 2000

Location: Beloeil, Quebec, Canada

Aircraft: Bell 206B JetRanger III

Injuries: 2 fatal

Around 1735 Eastern Standard Time, the aircraft, with the pilot and one aircraft maintenance engineer (AME) on board, was making a Visual Flight Rules flight to check the transponder. About five minutes later, after the pilot advised the area control center he was returning to the airport, the main rotor separated from the mast and the blades penetrated the cockpit. The helicopter crashed on its back in a plowed field 1.2 nautical miles northeast of the point of departure. A fire broke out after impact and destroyed the aircraft. The two occupants sustained fatal injuries.

In January 2000, the aircraft owner, who held a private helicopter pilot license, had hired a commercial pilot for additional safety while flying and to oversee the day-to-day operation of the aircraft. The commercial pilot was expected to ensure all work listed in the purchase service contract was completed within a reasonable time. The deadline for completing the work was set at May 4. The registration certificate was issued on March 7. Transport Canada then issued a certificate of airworthiness on April 20 after a compliance inspection.

On April 25, the AME started to complete the work listed in the purchase service contract and to correct the deficiencies noted by the owner on pleasure flights on April 20 and 24. These deficiencies included the defective transponder, a leak in the ceiling, and corroded washers on the droop restrainers. A nut screwed onto the top of the mast secures the droop restrainers and the rotor head in place. A mechanical lock bolted to the droop restrainer plate is secured with a lock-wire and prevents the mast nut from unscrewing in flight. After the mast nut is installed, an independent inspection is required. The work must also be entered in the aircraft logbook and signed off by either two AMEs or one AME and a qualified pilot.

At the request of the AME, an apprentice AME removed the droop restrainers and the mast nut from the aircraft, then stripped and primed them. The next day, April 26, the AME's partner, the owner of the maintenance facility, noticed that the apprentice AME had not used an epoxy primer, and he asked him to strip the parts again so he could paint them on the evening of April 27. At that time, no flights were scheduled for April 27. After the droop restrainers were stripped, they were placed on a tool box beside the aircraft; the mast nut and its securing mechanism were placed on another work table. The investigation could not determine whether the pilot was advised that the droop restrainers had been removed.

On April 27, the day of the accident, no other work was scheduled to be done on the helicopter. The AME was doing administrative work, and the apprentice AME was working on an aircraft beside the subject Bell 206. However, the pilot, who arrived at the hangar around 0930, asked the AME to work on the helicopter in preparation for some flights scheduled by the owner for April 28. Consequently, the AME had to interrupt the job he was doing and devote the remainder of the day to servicing the accident helicopter.

After replacing the transponder, the pilot and the AME pushed the helicopter out of the hangar around 1500 to find the water leak. Around 1730, the pilot started the aircraft and hovered it. The aircraft landed a few minutes later so the owner of the maintenance facility could approach the helicopter and talk to the AME. The helicopter then took off toward the northeast around 1735 to check the transponder. At 1737, the pilot called the Montreal control center to transmit his intentions. The flight determined that neither the transponder nor the altimeter was functioning. At 1740, the pilot advised that he was returning to Beloeil Airport; it was the last message received from the helicopter. Radar recordings indicate that the aircraft was orbiting left when it vanished from the screen. Since the transponder was not working, the aircraft altitude was not displayed on the screen.

The wreckage trail was on a track of 350° magnetic. The first debris, small fragments of Plexiglas from the bubble and the cockpit interior finishing, was found 1200 feet south of the main wreckage. Several other parts were strewn about in the field, between the wreckage and the south end of the debris area. The two blades were found attached to the main-rotor hub about 400 feet southeast of the aircraft.

Examination of the hub revealed that the mast nut, the droop restrainers, and the spacer that replaces the droop restrainers when they are not installed were all missing. The internal threads in the holes where the droop restrainers attach to the hub were intact, and the examination revealed no attachment bolt debris. Examination of the main rotor mast and the head trunnion indicated that the damage was caused by a vertical movement of the hub. The two pitch control rods failed in overload. Shortly after the occurrence, the droop restrainers and the mast nut for the helicopter were found in the maintenance hangar at the same location where they had been left the day before by the apprentice AME.

No entries concerning the removal of the mast nut and the droop restrainers were made in the aircraft technical log, open job lists, inspection sheets, or worksheets. As a rule, the pilot checks only the logbook before a flight to ascertain the condition of the aircraft. The logbook for the helicopter was found in the wreckage area. The investigation was unable to determine whether the pilot knew the mast nut had been removed. It was determined that the apprentice AME and the owner of the maintenance facility forgot, before the aircraft took off, that the mast nut had been removed.

Some companies place a warning flag in the cockpit and/or on the fuselage to indicate that the aircraft is not airworthy and that maintenance work is in progress. This practise is not required by law in Canada. In this occurrence, neither the AME nor the apprentice AME followed this practise: there was no visual indicator that the mast nut was not in place.

The AME was licensed and qualified to service the accident helicopter. He received his AME license (helicopter) on April 4, 1985 and formed [company name] in 1995. He was the only AME in the company, and he supervised the apprentice AME, who had four years' experience. Although the investigation did not precisely determine the AME's workload, it was established that he had been especially busy in the months preceding the occurrence. In addition to working weekdays, he worked on weekends and had practically no days off during this period. The AME worked an average of 12 hours a day.

While Canadian aviation regulations do not specifically require a pre-flight check, the flight manual, which includes the pre-flight checklist, is approved by Transport Canada and is required for the efficient and safe operation of the helicopter. Chapter 2 provides a detailed description of the pre-flight check and states that the pilot is responsible for determining whether the helicopter can complete the flight safely. The pilot should climb onto the cockpit roof to check the main rotor head and hub. From the roof, it would be obvious if the mast nut was missing: the mast threads and opening would be visible. In fact, the droop restrainers and the mast nut can also be seen from the roof.

The analysis of the accident report included:

A description of the work to be done should have been recorded on one of the documents, as required by regulation to advise maintenance personnel. Maintenance personnel could have referred to the documents and could have prevented the aircraft from taking off. However, the three persons who could have performed the work (the apprentice AME, the AME, and the maintenance facility owner) were aware that the mast nut had been removed and was to be painted on the evening of the occurrence. It is unlikely the AME thought that the mast nut had been installed by the apprentice AME or by his partner because the AME had received no notification or indication that the work had been completed. Consequently, it is reasonable to believe that the three persons who took part in the removal of the mast nut, and who were present when the aircraft took off, did not remember that the mast nut was in the hangar. It is unlikely that the helicopter would have taken off without the mast nut if a document had indicated the work that remained to be done and if the AME had consulted that document before the flight. There was no indication that the pilot or the AME consulted the aircraft documentation before the flight. It would have been unusual for the pilot to consult the maintenance documentation.

The occurrence flight expedited the work to be done on the helicopter because it was not anticipated by the maintenance personnel. It had been decided that the droop restrainers and the mast nut would be painted that same day. It seems that, after a schedule change, the work methods of the maintenance personnel did not enable the AME to be aware of the airworthiness status of the aircraft at all times.

It is likely that the pilot was not aware that the mast nut had been removed. Given that no visible warning device was placed in the cockpit or on the aircraft, there was nothing to tell the pilot that the aircraft was out of service. A visual aid such as a warning flag or sign, while not required by regulation, would have alerted the flight crew to the danger. Also, the missing mast nut undoubtedly would have been noticed by the pilot if a pre-flight check had been done as specified in the aircraft flight manual. Consequently, it is reasonable to conclude that the pilot did not climb atop the aircraft and did not examine the rotor head. Even if the restrainers were visible from the ground, noticing that something is missing is probably more difficult than noticing that something is present. The AME responsible for the maintenance was on board the aircraft when it took off and had worked with the pilot during the hours preceding the flight; this certainly gave the pilot a false sense of security.

Findings as to causes and contributing factors

1. The main rotor head separated in flight because the mast nut was not in place.
2. The helicopter took off without a mast nut.
3. The pilot did not check the rotor head before the flight.
4. Maintenance documentation did not indicate the mast nut had been removed.
5. No visible device was placed in the cockpit or on the aircraft to indicate that the helicopter was out of service.
6. The three persons who participated in the removal of the mast nut were present when the aircraft took off. None remembered the mast nut was not in place.

Source: Transportation Safety Board of Canada

Case study 13:2

Accident Report MIA00GA057

Date: September 20, 1999

Location: Fort Myers, Florida, United States

Aircraft: Hughes OH-6A

Injuries: 1 serious

According to witnesses, the aircraft flew over the landing area and was downwind when it suddenly turned to the right and descended at a high rate of speed, in a nose-low attitude, and impacted a building. Post-crash examination revealed that the lateral control rod end fitting had separated from the control rod, and the fitting's bearing did not move freely, when compared to a normal rod end bearing. In addition, the rivet that normally passes through the lateral control rod and rod end fitting was missing. Dark and white deposits were found in the threaded region of the rod and rod end fitting, and laboratory analysis of those deposits revealed that the substances were aluminum oxides.

Maintenance records show that the control rod had been inspected and modified when the helicopter was first acquired, and had been delivered to the Lee County Sheriff's Office at about 3546 total flight hours. Following receipt of the helicopter the operator's maintenance log did not indicate any specific maintenance entries related to the control rods. At the time of the accident the helicopter had accumulated 4419.4 flight hours.

The National Transportation Safety Board determined the probable cause of this accident to be improper maintenance inspection of the helicopter by unknown maintenance personnel. A worn control rod bearing continued in service, and subsequently separated, resulting in an in-flight loss of control and a crash.

Source: **National Transportation Safety Board, United States**

MAINTENANCE RELIABILITY

With a malfunction of a helicopter shortly after a scheduled inspection (they will inevitably happen sooner or later), many pilots have said to their maintainer in despair, 'It was going great until you guys touched it'. Helicopters, like any other machine, will be unreliable at the most inconvenient time despite your best efforts. Pilots and aircraft maintainers often develop a love-hate relationship with their helicopters and each other. One frustrated pilot was heard to utter, 'If you engineers would leave these helicopters alone, they'd go much better'. He's right to a point. Maintainers refer to this reliability as one side of the 'bath-tub curve'.

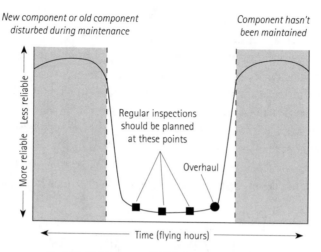

New component or old component disturbed during maintenance

Component hasn't been maintained

Regular inspections should be planned at these points

Overhaul

More reliable ← → Less reliable

Time (flying hours)

Fig 13.1 Maintenance reliability (bath-tub curve) TOM MCCREADY

The left-hand side of the bath tub is the higher than normal unreliability rate of a component because you have disturbed it during the process of inspecting, removing or overhauling. This can involve something as simple as adjusting the set of bellows on a fuel control unit. The next obvious question that all pilots will ask is, Why disturb it then?

The answer is simple. If you don't look at the component, you will eventually reach the right-hand side of the bath-tub curve and have failures anyway. The failures and problems encountered on the left side of the bath-tub curve are often only irritating events such as o-rings leaking from not settling in or fine adjustments to something that is not quite right. On the far right of the curve the consequences are generally far more serious, with actual failures and potentially disastrous results.

In between the two steep sides of the bath tub is the more desirable area of just occasional unreliability — they haven't invented the perfect machine yet! At the right of this area is the inspection or overhaul point, where the component is disturbed and so starts the left side of unreliability again.

Technology continues to advance in terms of materials and assembly techniques, and also with regard to inspection processes (non-invasive surgery in medical jargon). This advance is seeing vibration analysis, oil sampling and various forms of monitoring and recording black boxes that allow the maintainer to be confident of the actual condition of a component without having to take it apart to look at it. As an example, in some late-model helicopters an onboard computer will record, store and compare the running temperatures of many components during various operational situations. A rise in temperature of only a few degrees, a minuscule percentage of the historic temperature, may be a very early indication of a problem that can be solved in a timely fashion or of an impending actual component failure. While an early indication of a problem may mean a repair can be carried out at minimal cost, the real value is in preventing an expensive and possibly dangerous failure later on. Without these modern tools, the only clue may have been a temperature gauge briefly flickering out of the green operating range before the actual failure occurred.

With these advantages inspection and overhaul times are being stretched out. We still need both sides of the bath tub, but the bottom of the tub is getting wider and cheaper.

PILOT INSPECTIONS

When a commercial helicopter is put through its certification process, one of the key elements is the establishment of a maintenance program to specify how the helicopter will be maintained and therefore provide years of trouble-free operation. The program takes into account fatigue life and normal operational use. Most pilots are familiar with a maintenance program because, when the helicopter is returned to the maintainers, this program is what is 'signed-off' step by step to ensure that a 100-hour inspection is completed properly.

But that is all maintenance stuff, right? Wrong! The maintenance program or instructions for continued airworthiness includes pilot inspections, usually in the form of pre-flight or after-flight checks. These are intended to keep an eye on the helicopter parts during its operation between maintenance inspections. The bond between a pilot and a maintainer can, at times, be strained. Pilots essentially want to go flying, and are frequently eternal optimists, while maintainers are often faced with the unenviable task of telling pilots that they can't.

When it works correctly the relationship between a pilot and maintainer is best described as a partnership. The maintainer looks after the best long-term interests of the helicopter and crew in terms of reliability and safety, whereas the pilot (naturally enough) focuses more on the operational side and has a natural desire to always have the helicopter available for work to please and impress customers. Why then would a pilot want to go looking for trouble on an inspection? It could mean passing on bad news to customers. This is one reason many pilots fall into the trap of conducting quick gloss-over inspections that involve checking the oils and not much else — a 'kick the tires and light the fires' attitude.

This can work well for many years but consider the well-known model produced by Professor James Reason. This is a way of looking at a number of factors represented as flat plates with holes in them. Plates might belong, for example, to each of the following: the chief executive officer (CEO) of the company, the operations manager, the accountant, the chief maintainer, the maintainer, the helicopter, the individual parts of the helicopter, the pilot. The holes are created by a myriad of things: from domestic disputes to indigestion; money concerns to a hangover; corrosion to metal fatigue. The plates are lined up side by side and can rotate to any position. If the holes in all the plates happen to line up, an accident usually occurs. You can see clear through from the CEO to the accident on the other side.

Let's follow an example through the various plates. If the CEO concentrates on the smooth running of the business and leaves the chief maintainer to worry about maintenance, there's a hole you might label 'indifference to maintenance'. The accountant knows the maintenance section is understaffed but persuades himself it will get through, as the alternative is another salary eating into company profits: another hole. The chief maintainer knows that staff are overworked and shortcutting procedures to some extent but is afraid to say anything for fear of losing his job. This hole might be 'avoids issues; turns a blind eye'. Consequently, the maintainer does a poor inspection (this hole is 'overworked'). Luckily, however, the pilot does a very thorough inspection. All the holes don't line up and the leaking hydraulic fitting (or whatever) is found before it becomes a major problem.

Likewise, events can take place during operations that mean a thorough pilot's inspection can find an unscheduled maintenance problem before it causes an accident. For example, a load is heavier than planned or a landing thump on the ground is heavier than intended, but nothing is reported. A prudent pilot, however, finds a crack in the airframe before it jeopardizes the safety of the helicopter.

Pilots need to realize that their inspection, be it after-flight or pre-flight, is the last line of defense in detecting a fault before it causes an accident. This is even more important when a helicopter is used by numerous pilots.

Become familiar with your helicopter, follow the required inspections in your flight manual (see Fig 13.2 for an example) and spend time discussing the idiosyncrasies of your helicopter with your maintainer. Pilots can learn a lot from their maintainers, especially about particular things to watch out for.

A pilot's inspection must be done in a progressive and logical manner. Plan beforehand, do it the same way each time, and don't rush the job. Don't let anything interrupt you. Many people believe mobile phones cause a great many accidents by interrupting checklist tasks. Get an answer service and leave the phone in the helicopter when you need to pay attention.

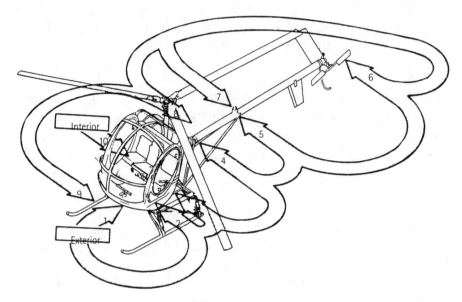

Fig 13.2 Pilot pre-flight checks SCHWEIZER AIRCRAFT CORP

With the pre-flight inspection, this is generally what pilots do. But have you ever considered the logic and obvious advantages of doing an after-flight inspection? After you have shut down the helicopter for the day, put everything away, cleaned the screen, finished the paperwork, and the helicopter has cooled down, why not go out and do the bulk of your pre-flight inspection? Perhaps the maintainer hasn't quite gone home yet or, if you're at a remote base, you can telephone your aircraft maintainer and discuss a problem you have found. Wouldn't it be far better to find a problem now than just before you want to fly the aircraft, perhaps early the next morning, when the customer is already waiting?

Speaking of early the next morning, you could sleep an extra 20 minutes by doing the bare basics before you leave the night before. Take care that your helicopter is secure and can't be tampered with overnight. Ask your maintainer to give you a list of things you can check in an after-flight inspection and leave out of the next pre-flight.

If you find a problem during the operational period between maintenance inspections, don't just nurse the aircraft to the next maintenance – tell the maintainer in advance. This allows time to research the fault, to make provision for any replacement parts and to allocate extended time to repair the problem. There is nothing worse for a maintainer than a helicopter landing outside the hangar for a standard 100-hour inspection (in by 8 a.m., out by 5 p.m.) only to be presented with a long list of other 'little problems' that clearly will require another two or three days to fix. This impacts on the resources available for the next helicopter/s scheduled on the following days, and is the beginning of a cycle of high stress that ends up with shortcuts being taken to save time.

This is an avoidable problem if maintenance is seen as a partnership between pilot and maintainer. Such a partnership results in a better operation for all concerned. Let's face it – the pilot is the last line of defense. If that fails, the pilot is the first to get hurt. So look after your own interests by giving your aircraft a good check and by improving your relationship with the maintainer.

In the interests of a good relationship with the maintainer, make sure your machine is clean when you deliver it for a scheduled maintenance inspection. Cracks are hard to see through a layer of agricultural spray or exhaust carbon. Time will be wasted if dust, dirt, cigarette butts and empty soft-drink cans have to be moved to get at critical components. Your maintainer will get the impression that you are a shoddy operator – sometimes those attitudes are contagious!

Finally, and also looking after your own interests, always make your most thorough inspection immediately after maintenance, no matter what your approach to pilot inspections. This is the time that offers the best opportunity for all the holes in the plates to align.

Making decisions about mechanical problems in-flight

Have you ever thought about what are you going to do when a serious mechanical problem develops during a flight? Do you have a good grasp of what the problem is and what is likely to happen next? Is there time to get some advice via the radio? Whatever course of action you resort to, there is one key obligation that overrides all other considerations: the preservation of life – your own and that of any passengers! That being the aim, what are other factors that can come in to the decision-making process? Some of them may include:

➤ Is there a co-pilot or someone in the seat next to you who can help in any way?

➤ What are the emergency procedures for this situation?

➤ How experienced are you?

➤ How much confidence do you have in your ability to successfully autorotate?

➤ How hostile is the environment where you will have to land? Is it so hostile that attempting to nurse the helicopter to safety is the only option?

➤ What is your height above ground/sea?

➤ Are you wearing a life jacket?

➤ Do you have a survival pack?

➤ Have you made a mayday or pan call?

➤ Is the emergency locator beacon armed?

➤ Transponder to emergency mode? Should it be activated now?

➤ Have you briefed your passengers?

Given half an hour and a committee to thrash out all the options, you would probably come up with the right answer. Unfortunately, it's noisy, lights are flashing, horns are blaring and people are looking genuinely apprehensive. You have seconds to make a decision. While doing so you must control the aircraft. It's no time to reflect on your initial decision to take up flying as a career, you can't change that now. Sitting comfortably, reading this book, can you visualize this?

There are glaringly obvious things that you can embark on right now — education, practise, training, prior thought.

THERMAL RUNAWAY

Nickel-cadmium batteries (commonly referred to as Ni-Cads) have become very popular in helicopters because of their ability to accept high charging rates after a heavy discharge period associated with a prolonged start sequence.

This is a desirable characteristic as it allows the battery to restore itself quickly; is suitable for helicopters flying short sectors and multiple starts; and meets the needs of turbine engine starters, which require a lot of electrical energy. Ni-Cads can discharge at the same high rate as lead-acid batteries but without the voltage drop associated with the latter. A starter motor that sounds as if it is dropping rpm during a prolonged start is a sign of a voltage drop.

Along with the wonderful improvements that Ni-Cad batteries give us, there is a downside. During a difficult and prolonged start, a high current is drawn off. The battery, especially the middle cells, get warm, giving them a lower voltage and lower internal resistance. Once the engine is started and the generator is online, a high current is put back into the battery. Normally the current reduces as the battery charge is regained but, if some cells have a temperature imbalance, the current will continue to rise. This causes a rise in temperature, which in turn causes a further rise in current. Soon the battery is accepting all of the charging current from the generator. This condition is known as thermal runaway, and can cause so much heat that the battery explodes.

Avoidance

How does a pilot reduce this risk?

1. Ni-Cads require more maintenance than lead-acid batteries. Get them serviced on a regular basis (six monthly is common) and carry a spare so that you are not tempted to try one start too many on a tired battery.
2. Use a ground-start unit whenever possible to preserve the installed battery.
3. Install a battery temperature monitoring unit and use it.
4. If your battery has discharged overnight, do not jump start the helicopter with a ground-start unit or vehicle. The generator will rapidly charge your flat installed battery, setting up a perfect opportunity for thermal runaway. Change the battery.

Recovery

When you have done everything to avoid thermal runaway, but still your battery temperature gauge suddenly pegs at panic stations, land as soon as practicable, shut down the machine, call your maintainer and go to the nearest café. There is nothing you can actually do. Don't be a hero and try to remove the battery as this can lead to a very bad burn and intense pain. Unstable Ni-Cad batteries can and do explode. Buy an extra coffee for the maintainer because it will be quite some time before the aircraft is safe enough to approach.

HYDRAULIC JACK STALL

This subject has been included here largely for want of a more suitable place. The phenomenon predominantly occurs in AS350 helicopters and is not a failure as such, more a hydraulic control design that may overload in certain flight regimes, loadings and resulting aerodynamic forces. The accident report in case study 13.3, which follows, explains it like this:

> The AS350 type helicopter's flight controls are hydraulically boosted by a single hydraulic servo system. This servo system supplies hydraulic power to reduce the operational loads of the cyclic, collective and directional control systems. Under normal flight conditions the servo system provides adequate power to overcome the aerodynamic forces encountered and the controls 'feel light'. When maneuvering this helicopter type it is possible to load the rotor disc to a point where the servo system is not able to overcome the

aerodynamic forces encountered and 'feedback' may be felt in the cyclic control system. This will be accompanied by an increasing heaviness of the controls which, if not corrected by reducing the severity of the maneuver, will result in a hydraulic jack stall, referred to as 'servo transparency' by the manufacturer. At this point the controls will become very heavy and difficult to move.

In this helicopter type, jack stall (servo transparency) is an aerodynamic phenomenon that can occur when the helicopter is flown outside its normal flight envelope and subjected to positive maneuvering (g-loading). It results in un-commanded aft and right-cyclic and down-collective motion accompanied by pitch-up and right roll of the helicopter. The maneuver, often abrupt and a surprise to the pilot, tends to be self-correcting since the rapid loss of airspeed due to the pitch-up and down collective causes an equally quick reduction in feedback forces. The maneuver, though uncomfortable, is therefore always short-lived. However, height loss during the recovery phase may be critical if jack stall occurs when the aircraft is at a relatively low height above the ground or water.

Case study 13:3

Aircraft Accident Report Number 94-022

Date: October 11, 1994

Location: Needle Rock, Coromandel Peninsula, New Zealand

Aircraft: Aerospatiale AS 350B

Injuries: 2 fatal, 1 serious, 2 minor

After a busy morning of productive flying, ZK-HZP was at Matarangi and in need of fuel for the remainder of the day's schedule, including a scenic flight for the employees of the company who had chartered the aircraft. The nearest fuel source was at Pauanui, a coastal resort some 10 minutes flying time to the south. As the aircraft was to be empty, the pilot elected to take the five employees along on the round trip rather than to make a separate flight later in the day. The pilot was 35 years of age with 607 hours total experience, 501 in helicopters and 47 of those in the AS350 type.

The round trip flight was intended to take a total of 30 minutes. Choosing a coastal route via Opito Bay, the pilot asked his passengers if they would like a closer look at Needle Rock, a rocky outcrop 252 feet high protruding from the sea one nautical mile from the mainland, a prominent scenic attraction. The passengers agreed to this suggestion.

Fig 13.3 Needle Rock with a navy recovery vessel on station

The pilot descended ZK-HZP toward Needle Rock from the northwest at an airspeed of some 110 knots and entered a right-hand turn. He intended to fly around the seaward side of the rock, from north to south, and then back to the mainland at 500 feet.

Passenger statements indicated that, during this maneuver, the helicopter descended to a height below or level with the top of the rock. One passenger recalled that he could see the mainland through the hole in the rock and another estimated that the helicopter descended to within 130 to 165 feet of the sea. The pilot could not be specific regarding an altimeter reading at this stage of the flight as his attention was primarily directed toward Needle Rock.

The pilot recollected that, during the turning maneuver around Needle Rock, he believed the helicopter had effectively ceased to descend. As the turn to the right continued around the southeastern face of the rock (the lee side) the pilot felt the controls suddenly 'lock up'. The helicopter rolled to the right and impacted the water on its right side in a nose-down attitude.

The occupants survived the impact but two of the passengers drowned. One had sustained serious back injuries and the other minor injuries. The pilot was seriously injured and the three surviving passengers received minor injuries. They found themselves floating close to the wreck along with the body of one of the passengers but the fifth passenger was still in the inverted helicopter. Realizing that one of the

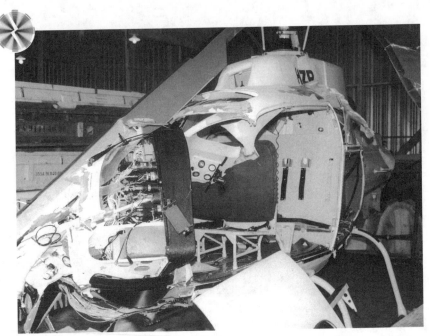

Fig 13.4 ZK-HZP TRANSPORT ACCIDENT INVESTIGATION COMMISSION, NEW ZEALAND

passengers was unaccounted for the pilot dived back into the wreckage to search for him. He briefly located the passenger but was unable to free him before the helicopter sank. Despite their injuries the survivors managed to swim to a nearby rocky ledge that jutted out from Needle Rock.

With the aid of a passing fisherman in a boat, the coast guard and a rescue helicopter, the survivors were rescued from the ledge some two and a half hours after the accident.

The Royal New Zealand Navy recovered the wreckage from 100 feet of water about 55 yards from the southeastern side of Needle Rock on October 13, 1994.

Examination of the wreckage showed that the helicopter had been destroyed during the impact with the water. The damage to the rotor blades, rotor head and transmission was consistent with a sudden rotor stoppage and the tail boom had separated on impact. The extensive damage to the nose area and right side of the helicopter showed that it had struck the water nose low in about a 90° banked attitude with a high rate of descent and at a high forward speed.

Extensive examination of the helicopter's control system, including the hydraulic system, showed no evidence of any pre-impact failure. Each of the hydraulic components was individually tested and the three hydraulic accumulators were found to be charged to normal operating pressure. The main hydraulic system filter

Fig 13.5 The main hydraulic system filter TRANSPORT ACCIDENT INVESTIGATION COMMISSION, NEW ZEALAND

had been in service for about 300 hours out of a 400-hour servicing cycle and was thought, by the pilot, to contain a higher than normal level of contamination thereby having the potential to cause the fluid flow to be restricted abnormally. Examination and differential pressure testing of the filter by the Royal New Zealand Air Force at Woodbourne showed no evidence of any blockage or restriction and the contamination was found to be usual debris from the hydraulic system itself.

An electrical hydraulic pressure dump switch situated on the end of the collective control lever was found to be in the normal open position and functioned correctly when tested. All switches and controls were in the normal position for flight and further examination of the helicopter showed no evidence of any malfunction that might have contributed to the accident.

During normal operation a failure in the hydraulic system causing a hydraulic pressure loss will be indicated to the pilot by illumination of a warning light and activation of an audible alarm (horn). The available evidence indicated that these warning systems had been functioning normally up to the time of the accident and the pilot recalled that he was not alerted by either the horn or the light at any stage throughout the flight.

Throughout the morning, and at the time of the accident (1351 hours), a strong westerly flow covered the region, with the wind strength being reported as 25 to 35 knots. Significant turbulence and down-draughting was reported to exist in the lee of the hills and the eastern side of the Coromandel Ranges. A local pilot reported severe turbulence in the Opito Bay area at about 1430 hours. The pilot of the rescue helicopter experienced severe buffeting of his helicopter in the lee of Needle Rock during the rescue.

ZK-HZP's weight was calculated to have been around 4131 pounds, 168 pounds below its maximum allowable. It was also calculated that, if the helicopter had descended from 800 feet off the coast of the mainland to below the height of Needle Rock, a rate of descent in excess of 1000 feet per minute would have been necessary in the time available.

During the turning maneuver the pilot recalled having focused his attention principally on Needle Rock and the surrounding expanse of water, not the horizon. Consequently he may not have immediately perceived a higher rate of descent developing and/or an increase in the angle of bank. Such factors, combined with the helicopter's heavy weight and high forward airspeed would have contributed to a high rotor disc loading in the helicopter. This high disc loading would have been accentuated if the helicopter flew into turbulence on the lee side of the rock.

Notable sections of the findings portion of the accident report read:

2.12 The available evidence indicated that during a turn to the right around Needle Rock, the pilot of ZK-HZP encountered hydraulic jack stall, also known as servo transparency, or its incipient stages.

2.13 The height of the helicopter above the water at the time of the occurrence provided little opportunity for the pilot to recognize the problem and recover control of the helicopter before impact with the sea.

Source: Transport Accident Investigation Commission, New Zealand

Avoidance and recovery

The problem is virtually self-correcting. Setting up loadings conducive to hydraulic jack stall should be avoided at all times, but the most important time to observe prudence is when flying at low levels as there is generally a loss of height during the recovery phase.

Pilots rarely fly at speeds beyond Vne (Velocity Never Exceed), but do occasionally induce jack stall as a result of excessive maneuvering.

The ability of the helicopter to sustain positive Gs is highly variable. At light weight, on a cold day, at sea level, and at a moderate airspeed of 80 to 100 knots, a helicopter might have the capability to execute a 3-G or greater maneuver without reaching jack stall. Under these conditions, it is very difficult and certainly uncomfortable to even demonstrate jack stall. The aircraft that can pull 3 Gs at moderate speed might, at Vne, maximum weight, and high density altitude, be able to sustain only 1.3 Gs before the onset of jack stall. If jack stall is unexpectedly encountered, the remedy is the pilot's natural reaction; if it occurs during a maneuver, decrease the severity of the maneuver! It is rarely the result of turbulence in level flight; however, if it is induced by turbulence, decrease airspeed by immediately lowering the collective. A lower collective, even at high airspeed, will provide increased margin from jack stall even while the airspeed is decreasing.

Jack stall is a normal aerodynamic phenomenon, which, if understood, can be avoided. Should it be unexpectedly encountered, knowledge of its cause can be helpful in avoiding undue concern.

SOURCE: JAKE HART (DIRECTOR OF OPERATIONS FOR EUROCOPTER IN GRAND PRAIRIE, TEXAS), *VORTEX*, 1/94

BOGUS OPERATIONAL PARTS

Killing for money is not new, nor is it uncommon. Some criminals use weapons such as guns and knives. Others use aircraft parts. Because so many of the certified parts of helicopters are critical, and their failure catastrophic, there is a huge chance that dealers in bogus parts will be a link in the chain that kills sooner rather than later.

There are many ways a part becomes bogus. The part involved in case study 13.4 was taken from a wreck and remanufactured, quite illegally. Some people manufacture bogus parts from scratch, very convincingly simulating all the inspection markings, serial numbers and the like. Others run a part for many more hours than they log for it, then sell it as having done hundreds or even thousands of hours less service. All scams have two things in common — they make or save money, and they kill people. Those responsible don't care that you are the person who pays the ultimate price or that you are giving your loved ones a treat the day the part fails.

Don't think it's just the criminals that are at fault here. When it comes to writing checks for parts, operators can sometimes not ask as many questions as they should. Sure, it's obvious that this is false economizing, for the aftermath of a serious accident doesn't come cheap, especially if the history of a part involved is questionable. But this doesn't stop some operators buying a 'bargain' part. Many of these operators and the dealers from whom they buy have reputations for dubious deals, so keep your ear to the ground. Avoid flying, as a pilot or passenger, with anyone who has a tarnished reputation.

Case study 13:4

Aircraft Accident Report
Number 95-017

Date: October 25, 1995

Location: Opotiki, North Island, New Zealand

Aircraft: Robinson R22 Beta

Injuries: 2 fatal

The 27-year-old 700-hour commercial helicopter pilot and his crewman were returning to their Opotiki base from a morning of venison recovery in ZK-HKM. They had talked to a third person before departing the hunting area. This was the driver of a truck supporting the operations. The driver stated that prior to departure the pilot had refueled the helicopter and completed a thorough pre-flight inspection. Apart from an insignificant 'drip' or 'blob' of oil around the engine compartment, no abnormalities or problems were mentioned to the driver by the pilot, and the driver stated that he did not notice anything unusual with the helicopter; that it sounded normal when it departed.

At about 0915 hours witnesses observed the helicopter in cruise flight at a safe height near a highway. A number of the witnesses saw the helicopter pass almost directly over them and they did not notice anything unusual with the machine.

A short time later these same witnesses heard a loud 'bang' and observed pieces flying off the helicopter. At the same time it was seen to pitch steeply nose down and fall onto the highway. Some witnesses thought they heard two loud 'bangs', and several of them believed the main rotor blades had stopped turning before the helicopter struck the ground. Several said that following the 'bang' or 'bangs' the helicopter went quiet, as though its engine had stopped.

An intense fire erupted shortly after the machine struck the road. No immediate assistance could be given to the occupants due to the fire and the ammunition that began exploding. In any event, the accident was not survivable and both occupants would have died on impact.

The weather was not a factor as it was calm, clear and warm with no rain.

A scene examination by accident investigators revealed a trail of debris some 175 yards long. Significantly, the first item on the trail was a piece of tail rotor blade that had separated from its root fitting. This piece comprised the inner third of the blade and it had been cut through at an angle of approximately 45°, apparently by a main rotor blade. The outer two-thirds of the blade was found about 142 yards from the main wreckage (see Fig 13.6).

At varying locations lay:

➤ The complete empennage, broken at its attachment fitting to the tail boom.

Fig 13.6 Tail rotor blade (inner third) TRANSPORT ACCIDENT INVESTIGATION COMMISSION, NEW ZEALAND

> The rear half of the tail boom, which had been struck at about 45° on its left side by a main rotor blade and had separated from the forward half of the boom.

> The tail rotor gearbox with complete tail rotor hub assembly attached beside the main wreckage.

The main rotor blades were attached to the hub and mast, and the mast was intact. The transmission turned freely, the free wheeling unit worked correctly and both flexible couplings were intact, the tail rotor gearbox rotated freely and did not show any evidence of pre-impact failure.

The empennage attachment fitting and the tail rotor blades and their root fittings were taken to a laboratory for analysis. It was found that the empennage attachment fitting had failed in overload. However, it was established that the separated tail rotor blade, serial number 5666C, had failed as a result of debonding at the adhesive to metal airfoil interface. Once the debonding occurred, the load was transferred to eight non-loadbearing rivets. Despite the centrifugal and torsional loads induced during tail rotor rotation, the non-loadbearing rivets evidently held the airfoil to the root fitting for a period of time, but the airfoil eventually cracked from the rivets as a result of environmentally assisted (corrosion) fatigue. Final overload of the remaining sound metal ligament occurred and the airfoil separated from its root fitting.

Examination of the other tail rotor blade, serial number 5638C, showed that a debonding of its adhesive had also occurred, in a similar manner to the failed blade (see Fig 13.7). The adhesive down the trailing edge of both blades had also debonded and could be peeled off easily. This suggested that the trailing edges may have opened some time before the accident.

Consultation with the manufacturer and safety authorities determined that this type of failure had not occurred previously with this helicopter type.

The first and most obvious question was: Why did this debonding occur? Detailed analysis, in consultation with the manufacturer, established that at some stage both tail rotors had been disassembled and rebuilt. Such rebuilds are not authorized. Whoever rebuilt the blades did not prepare the surface (for application of adhesive) correctly, used no knit-carrier to allow adhesive to sufficiently bond to the surface and used a different adhesive than that which the manufacturer uses. Effectively, more than a rebuild, the tail rotor blades had been remanufactured.

Fig 13.7   TRANSPORT ACCIDENT
INVESTIGATION COMMISSION, NEW ZEALAND

Fig 13.8   TRANSPORT ACCIDENT
INVESTIGATION COMMISSION, NEW ZEALAND

With the high cost of a new tail rotor blade, whoever carried out the rebuild was obviously 'saving' money. Where did blades 5666C and 5638C come from originally? Who remanufactured the blades? Where did the operator purchase the blades? How did the documentation bear scrutiny? Why did the licensed aircraft maintenance engineer (LAME) fit the blades to ZK-HKM? Why were no defects found during maintenance checks?

Obviously, the accident investigator was intensely interested in the answers to such questions. The following subparagraphs from the accident report, while lengthy, are worthy of note:

> 1.32 . . . tail rotor blades, serial numbers 5638C and 5666C. These blades were previously installed on Robinson R22 serial number 1987, registered as N192KC. They were manufactured in September 1991 and

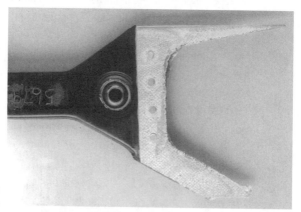

Fig 13.9   TRANSPORT ACCIDENT
INVESTIGATION COMMISSION, NEW ZEALAND

Fig 13.10   TRANSPORT ACCIDENT
INVESTIGATION COMMISSION, NEW ZEALAND

were fitted new to N192KC. They had a total recorded time of 1388 hours when they were fitted to ZK-HKM.

1.35 From photographs and a video taken of the wreckage of N192KC at the accident site, it was established that the helicopter's tail rotor blades had been extensively damaged in that accident.

1.36 In April 1994 the owner of ZK-HKM traveled to the United States on business and while in California he purchased a number of components from N192KC including its tail rotor blades. He said that the main purpose of his trip to the United States was not to purchase R22 components, but he became aware of them through reading *Trade A Plane* magazine. He said that an artificial horizon included in the components for sale had primarily attracted him to look and as a result he purchased other components as well. He could not recall from whom he purchased the tail rotor blades and other R22 components. He remembered that he paid cash for them but did not receive a receipt for the transaction. He said the components and the blades were at the same location.

1.37 He said he had to wait several days for N192KC's logbook to arrive before he could examine it. The logbook did not have any entries that showed the tail rotor blades had been damaged, removed, replaced, repaired, modified or overhauled. He said he knew the helicopter had been involved in an accident as it was quite obvious. He inspected the blades and, after comparing their serial numbers against entries in the logbook, believed they were serviceable. No form of release note or serviceability tag was attached to the blades.

1.38 The components, including the tail rotor blades and logbook, were shipped to New Zealand in a container along with other items he had purchased. The owner said he personally loaded the container in preparation for shipping. The container arrived in Auckland on May 22, 1994 and the tail rotor blades and other components were transported directly to Whakatane. He said he received the tail rotor blades in the same condition as when he purchased them.

1.39 The tail rotor blades and logbook were given directly to a LAME who was responsible for the maintenance of ZK-HKM. The LAME said he received the tail rotor blades in good condition from the owner of ZK-HKM and believed they were genuine. After inspecting the blades and examining N192KC's logbook the LAME said he assessed the blades as serviceable and locked them away in a cupboard in his hangar.

1.40 The owner said he never told the LAME specifically that N192KC had been involved in an accident. However, it was the owner's custom to only buy parts from wrecked helicopters, and he said the LAME was aware of that. In addition the logbook for N192KC did not accompany a complete helicopter, only parts of one.

1.41 On August 20, 1994 the LAME again inspected the blades in accordance with the manufacturer's R22 maintenance manual, and examined the logbook entries. He said he believed the blades were serviceable and had no reason to suspect they had been repaired, and issued a 'Serviceable' tag . . . for each of them. He subsequently fitted and balanced the blades to ZK-HKM on August 26, 1994 without encountering any difficulty. He did not carry out any repair to the blades.

3.6 The identity of the person who carried out the unauthorized repair was not established.

3.14 The helicopter had flown some 415 hours with the tail rotor blades fitted before one of them failed.

3.15 A progressive debonding within the blades occurred from the time they were fitted to ZK-HKM.

3.16 Periodic inspections of the tail rotor blades by the LAME and various pilots did not alert them to any problem.

Source: Transport Accident Investigation Commission, New Zealand

Several adverse comments could be made about individuals involved in this saga, especially the rogue who *repaired* the blades. Sadly, the two people who cannot be criticized are the deceased parties. When considering the previous owner (the man who purchased the blades) of ZK-HKM, however, bear in mind that both he and his son flew a considerable number of hours in the aircraft with the defective blades fitted. Still, this sign of good faith was not sufficient to prevent him being charged with and convicted of manslaughter. He was sentenced to a period of imprisonment.

Avoidance

If you are responsible for buying parts for helicopters, ensure that the total history of any part is revealed and can be proven to be correct. If any doubt at all exists, ask questions. The life you save may be your own or that of one of your workmates or a loved one. Similarly, if you are an aircraft maintainer, it is your job to know that a part is what it is documented to be. Can your conscience cope if you don't remove all doubt and the worst happens?

Use all available resources, including a thorough internet search, on the registration of all aircraft in which the part has supposedly been fitted. Are they still flying? Are they in any accident database? Generally, if a deal sounds too good to be true, it most likely is!

Recovery?

If you are the pilot, survive the accident, and find you are working for a company that buys bogus parts, sue and leave, not necessarily in that order. And then have a very big mouth because a bad reputation may just put them out of business — make sure you have all your facts right, though.

14

Fuel

There are three things that can go wrong with the supply of fuel:

➤ Contamination — when water or other matter gets in.

➤ Starvation — when the pumping or flow mechanisms falter or block.

➤ Exhaustion — when the fuel reservoir is empty.

At the very least, the helicopter loses substantial power as its engine runs inefficiently; at worst the engine stops completely.

CONTAMINATION

Case study 14:1

Accident Report SEA98LA064

Date: April 19, 1998

Location: Sedro Wooley, Washington, United States

Aircraft: Bell 47G-3B-1

Injuries: nil

During initial climb immediately after takeoff the helicopter lost power. The pilot attempted to maneuver to a clear area, but struck trees during the descent. Post-crash examination revealed no mechanical malfunction with either the engine or carburetor, although the carburetor had been subjected to a post-crash fire and its fuel metering orifices could not be bench tested. Fuel samples taken from the helicopter's fuel tank prior to the accident and from the fueling hose from the portable fuel truck were examined and found to contain 1,2-Benzenedicarboxylic acid, bis(2-ethylhexyl) ester.

The pre-accident aircraft fuel sample, which contained a submersed glob of clear substance similar to that observed by the pilot seeping from the cut ends of the fueling hose, was tested. The test revealed that the 'soluble gum' consisted of 1,2-Benzenedicarboxylic acid, bis(2-ethylhexyl) ester and that nearly 100 percent of the gum was soluble in the fuel. The operator had been advised by the local distributor of the 'Versicon' hose that the hose was acceptable for use with aviation fuels. Specifications provided by the manufacturer of the hose, HBD Industries, provided a note which stated, 'Not recommended for the variety of unleaded gas existing presently.'

The National Transportation Safety Board determined the probable cause(s) of this accident as:

Fuel contamination and subsequent carburetor fuel flow restriction initiated by the use of an improper material (incorrect fuel hose). A factor was the trees.

Source: National Transportation Safety Board, United States

Case study 14:2

Accident Report FTW01LA195

Date: August 24, 2001

Location: Cameron, Louisiana, United States

Aircraft: Bell 206L-3

Injuries: 2 serious

The helicopter was in cruise flight over the Gulf of Mexico when it began to vibrate and shudder. The pilot lowered the collective control to initiate an autorotation and the engine lost power. During the ensuing autorotation, the helicopter's floats were deployed. The pilot attempted to decelerate the helicopter; however, the controls became stiff, and subsequently, the helicopter hit the water hard.

Examination of the helicopter's fuel system revealed that the fuel nozzle inlet screen was collapsed and the screen was 80 to 90 percent contaminated with a brown material with a polymeric-like to varnish-like appearance. The screen was examined by a laboratory, and it was determined that DIEGME, a fuel additive used as an icing inhibitor, was present on the screen. The helicopter's operating environment is such that salt water could have been introduced into the fuel system. The presence of water and fuel would allow bacteria to grow. The combination of bacterial growth, DIEGME, and water resulted in the formation of an 'apple-jelly' type material, which then adhered to the fuel system components (fuel nozzle screen). The blockage and collapse of the fuel nozzle screen resulted in an interruption of fuel flow, and the subsequent loss of engine power.

The National Transportation Safety Board determined the probable cause(s) of this accident as:

Fuel contamination due to the combination of DIEGME, water, and bacterial growth, which resulted in formation of an apple-jelly type material that blocked the fuel nozzle screen and led to a loss of engine power.

Source: **National Transportation Safety Board, United States**

DIEGME

'Apple Jelly' is a substance composed of Fuel System Icing Inhibitor (FSII, chemical name Diethylene Glycol Monomethyl Ether) and water, in roughly equal amounts; and up to two percent of other minor constituents. The substance can range in color from light amber to very dark brown. The consistency variations can range from a gummy/gelatinous form at one extreme, through a syrupy/taffy-like form at mid-range, to a thin watery consistency at the other extreme.

The substance is found in various locations in the fuel distribution system after the injection of FSII. Most often it occurs at the base activity level, involving, for example, filter/coalescer elements and sumps, storage tanks, refuelers and aircraft fuel tanks. The primary detrimental effect of the substance is its contribution to the rapid disarmament of filter coalescer elements, requiring more frequent changes of the elements. This obviously increases maintenance costs.

There are many forms of contamination, but water is a common one, often a result of refueling from drums and not using correct techniques.

Water in aviation fuel occurs in three forms:

➤ Dissolved water, which is similar to the humidity in the atmosphere. Like humidity, dissolved water will condense into water droplets as the temperature of the fuel drops.

➤ Suspended water, which is droplets of water in the fuel that have not yet settled out by sinking to the bottom of the fuel container. Suspended water in large amounts will cause the fuel to have a hazy or cloudy appearance.

➤ Solid bodies of water, which form at the low points of the fuel container and are created from the settling of suspended water or by the introduction of quantities of water either by such devices as leaking filler neck seals or by transfer when the container is being filled from another with such problems.

There is no way of preventing the accumulation of water formed through condensation in fuel tanks, although this amount can be minimized if the tanks are kept topped up. The accumulation is certain, although its rate will vary.

Frequent checks of aircraft fuel drains are necessary. At a minimum a check for water in the fuel should always be made:

➤ Before the first flight of the day.

➤ After every refuel.

The check before the first flight will normally cover overnight (or prolonged storage) accumulation of condensation, while the post-refuel check will confirm that water has not been introduced from the fuel dispensing or storage facilities. Obviously there may be other times when water contamination may be suspected — being parked out in heavy rain may affect some fuel systems, particularly those with poorly fitting filler caps — and a check is obviously prudent in such cases.

A fuel drain will not confirm the presence or absence of water unless it is carried out properly.

Minimum settling time

Following a refuel, much of the water will be suspended droplets, and unless they are present in sufficiently large amounts to give the fuel a cloudy appearance, a fuel drain check will not disclose the water. After a refuel, the minimum settling times to produce accurate checks for water are:

➤ for AvGas, 15 minutes per foot depth of fuel

➤ for turbine fuel, 60 minutes per foot depth of fuel

Of course, the effort to do all of the above correctly can be negated if the receptacle into which the fuel is drained is not clean.

Knowing when and how to check the aircraft fuel system and contents for water is important. What if you find water? The answer is very simple — remove it immediately. Usually this will simply be a matter of continual drain checks until a nil result is obtained. Major contamination will necessitate major draining efforts, and this will usually be in the preserve of a licensed aircraft maintenance engineer, who will ensure that the total system is purged.

A short while back, a repeat winner of the NVP (Not Very Professional) Award filled his aircraft with a substance that vaguely resembled fuel and went flying — it was a very short trip. The takeoff was normal, as was the initial cruise portion of the trip (120 sec). The unscheduled approach to the trees was non-standard and the landing was very hard on the equipment.

The pilot removed himself from the bent metal and, as he waited for rescue, tried to piece together what went wrong. It took the investigators only minutes to determine that the probable causes were:

➤ The barrels used for refueling had been stored upright and pooled water on the top of the barrels had worked its way into the barrels past the bungs. The fuel was not only contaminated with water, it was also contaminated with rust as the insides of the barrels had rusted.

➤ The barrels were rolled to the aircraft just prior to refueling, thus ensuring that all the contents (fuel, water, rust, etc.) were thoroughly mixed into a blend that was only marginally combustible.

➤ No filter of any sort was used.

➤ The barrels were tipped in order to extract all of the fuel, water, rust, etc.

➤ An extension was placed on the pump stand pipe so that absolutely all of the contents of the barrels could be pumped into the aircraft.

Not only did this occurrence receive the NVP of the month award, but it was also first runner-up for the NVP of the year award.

It should also be noted that when the contents of the fuel system were analyzed the report stated that it was a wonder the aircraft started, let alone flew.

SOURCE: 'JUST BECAUSE IT SMELLS LIKE FUEL DOESN'T MEAN IT WILL BURN LIKE FUEL', *VORTEX*, 5/99

Fuel drum etiquette

Prior to the first takeoff, make sure that the aircraft tool box contains rubber gloves, a bung wrench, filters, a standpipe and collar, a diaphragm and nylon valve repair kit, grounding cables, and enough tools to do the job. Make sure that you know how to use them.

Okay. You have just landed at a fuel cache, perhaps one that is not familiar to you. All things being equal, the fuel cell in your aircraft is presently uncontaminated, and the trick is to refuel, without incident, while maintaining this uncontaminated state!

The basics

1. Ensure that the drum you are using contains the proper fuel, regardless of what is printed on the outside! Also note: different oil companies have different colors for drums, but a drum's color is not a foolproof indicator. Confirm by the appearance and odor of the fuel each time.

2. Be suspicious of any drum that seems light or heavy: water weighs 20 percent more, and AvGas 10 percent less, than Jet B. Whatever is printed on the drum cannot be trusted if the original seal is broken or missing.

3. Somewhere on the drum is a fill date. Most oil companies discourage using fuel that is more than two years old. One reason is that a nasty fungus (Cladosporium resinae) can thrive in small amounts of water in jet fuel, and will clog fuel lines. Older fuel can be used safely with caution. Check for any strange odor, or a dark or cloudy condition. If you have any doubt, do not use it.

4. Check all unsealed drums for an 'X' marked on the end. This is the accepted marking for contamination. However, the lack of an 'X' is no guarantee of quality! Many pilots who use a part drum will mark the date, aircraft registration, and approximate amount used, near the bung. (If you have any doubt, don't use it!)

5. Store the drum in the proper manner, and be suspicious of any drum that is not, especially if you have reason to doubt whether it has been well resealed (bung or vent loose; gaskets torn, missing or twisted). Even when properly resealed and stored, a part drum is more likely to contain moisture because of the increased 'breathing' (more air content equals greater compressibility).

6. All fuel drums should be stored on their side, with bungs and vents at the 3 o'clock and 9 o'clock positions. Make sure that the top of the drum (with the openings) is lower than the bottom. This will minimize breathing (air and moisture exchange from the outside).

    When opening a drum, observe the following:

    ➤ Stand the drum on end and block it with the high side at 12 o'clock, the bung at 3 o'clock, and the vent at 9 o'clock. This prevents water or dirty fuel from reaching the openings.

    ➤ Ensure that the standpipe cannot reach the lowest point in the drum. Thus, any small amount of water or dirt will remain in the drum. You should not need the last gallon badly enough to risk using it.

    ➤ If possible, stand up your drums prior to their usage (up to two days, if dry conditions can be assured) to allow contaminants time to settle out. Avoid agitating the drums when refueling.

7. If you have a helicopter and you must hot-refuel, avoid putting loose items such as bungs and wrenches on top of the drum.

    Note: Hot-refueling from drums should be done only during an emergency, or under very controlled conditions in compliance with . . . the approved company operations manual (lots of groundcrew, no passengers on board, pilot at the controls, and a developed refueling procedure complete with individual duties and signals). The potential for disaster normally outweighs the potential for time saved.

8. Upon emptying the drum, locate it (with bung and vent reinstalled) so that it will not become a rolling or flying hazard to yourself or others using the fuel dump.

9. Proper grounding is critical, especially during winter operations. Dry winter air and blowing snow transform the rotors into powerful static generators. Moreover, snow insulates, and static may not dissipate on touchdown. Avoid wearing nylon clothing or wiping Plexiglas when refueling. Dusty or sandy conditions are also conducive to static buildup. Check the condition of the ground cables, and replace any doubtful connections.

    Note: The proper sequence for grounding is: drum to ground (anchor post), drum to pump, pump to aircraft, nozzle to aircraft, then open cap. When finished, reverse order.

10. Fuel caches should be located clear of sandy, dusty, or debris-strewn areas. They should be organized to expedite refueling, with a good approach/departure path. (Remember: you will be heavier leaving, than arriving, unless you arrived with a load of uranium.)

11. Always carry and use water finding paste, such as Kolor Kut. A tube will fit unobtrusively in your map case and last for a long time. A dab on the end of the standpipe will give a positive indication of water.

12. Ensure that the pump is equipped with a clean and serviceable go-no-go filter and particle filter in series, with intact o-rings. The go-no-go is designed to bind up and prevent flow in the presence of water. Increased pressure usually means blockage or contamination. Observe the sight glass for dirt or water in the sediment filter.

13. Squirt the first pump strokes into a container before putting the nozzle into the aircraft. Any dirt downstream of the filters will be flushed out of the hose, and can thus be examined.

14. Dispose of plastic caps, metal rings, and date tags from your used drums carefully to prevent the risk of foreign object damage (FOD) in the refueling area.

15. Don't forget that the first pre-flight of the day should include a draining and catching of the aircraft's sump/airframe fuel-filter contents. Do this before disturbing the aircraft.

Vortex, 3/2002

Of course, many things can contaminate your fuel. In the days of venison recovery in New Zealand a Hughes 500 had an engine failure due to the filter being clogged with deer hair! There is even one case where, mysteriously enough, the contaminant was the remains of a muffin (the cake variety)!

The glaringly obvious advice is to go to any lengths to make certain the fuel that goes into your helicopter is clear, clean and the right sort. Then ensure that once it is in there it remains in that state until it is efficiently burnt in the combustion chamber.

FUEL STARVATION

Case study 14:3

Accident Report MIA01LA085

Date: February 24, 2001

Location: Whitesburg, Georgia, United States

Aircraft: Hiller UH-12B

Injuries: nil

The purpose of the flight was to photograph the parent's house of the adult passenger; the flight was returning when the accident occurred. While descending for approach to a ball field about 250 to 300 feet above ground level at an indicated airspeed between 60–65 knots, the helicopter yawed. The pilot corrected the yaw with anti-torque pedal and noted that the engine rpm and main rotor rpm needles were 'split'; the main rotor rpm was in the green arc and the engine rpm was indicating zero. He rolled the throttle to the right towards the fuel-on position, and then lowered full collective and maintained 40–45 knots while descending for an autorotative landing to the down-sloping road. He cleared a bridge ahead and at approximately 15 feet above ground level, he applied aft cyclic to decelerate 'somewhat', and pulled full collective. He touched down at an indicated airspeed between 20 and 30 knots on the road that curved to the left and later reported, 'We didn't hit real easy'. He lowered full collective after touchdown and while sliding, the right skid contacted grass on the side of the road causing the right skid to collapse. The helicopter then rolled onto its right side and slid approximately 4–5 feet before coming to rest. There was no post-crash fire or fuel leakage.

Examination of the helicopter by an Federal Aviation Agency airworthiness inspector revealed 8–9 gallons of fuel on board; the carburetor bowl was found approximately 1/2 full with a slight amount of contamination. The engine-driven fuel pump would not produce 2 psi (pounds per square inch) flow and would not produce any flow with a minor restriction. Additionally, the auxiliary fuel pump strainer was 'so clogged and corroded it had to be removed with a metal hook and screwdriver'. The helicopter was inspected last in accordance with a 100-hour inspection that was signed off on January 5, 2001; the helicopter had accumulated approximately 24 hours since the inspection at the time of the accident.

Review of the helicopter flight manual pertaining to engine failure when flying above 325 feet above ground level revealed that it was best to maintain 50–60 mph glide speed and to pull cyclic stick steadily back at approximately 75 feet above ground level. Additionally, 'when 10–15 feet above the landing surface, level the helicopter with forward cyclic stick and then apply collective pitch, as necessary, to cushion the landing impact'. The flight manual also states, 'To reduce the rate of descent and forward speed during an autorotation landing, the helicopter's flight path should be flared or leveled out, prior to ground contact'.

The National Transportation Safety Board determined the probable cause(s) of this accident as:

The improper emergency procedure performed by the pilot due to his electing to perform a run-on landing resulting in excessive airspeed at touchdown and subsequent roll-over following ground contact with grass. A contributing factor was the inadequate 100-hour inspection performed last to the helicopter resulting in the total loss of engine power due to fuel starvation.

Source: **National Transportation Safety Board, United States**

The aircraft has sufficient fuel on board but it is just not getting to the engine in the right quantity. There are several mechanisms that can be faulty or fail in both piston and turbine engines: carburetors, injection systems, fuel control units, fuel pumps, fuel lines, valves — an aircraft maintainer could likely add many more. Most of these components cannot be accessed by the pilot or flight crew from the cockpit and, even if they were, there is rarely a lot of warning given.

Some problems can be anticipated and controlled from the cockpit. Piston engine helicopters are susceptible to carburetor icing and this leads to fuel starvation when ice build-up limits the flow of the fuel-air mixture. Many aircraft have thermometers to indicate the temperature in the carburetor and a colored arc (usually yellow) to indicate the range in which icing is likely to occur. The most common method of control is for the pilot to select 'carburetor heat — On' and warm air is directed into the carburetor, melting any ice or raising the temperature. Some power is lost with the increased temperature of the mixture but, if too much ice accumulates, all the power will be lost so it's a very small sacrifice.

Carburetor ice can be a problem in temperatures as high as 90°F (32°C) with a high relative humidity and rarely occurs below 14°F (–10°C) as most of the moisture is already in ice form. The most severe range that ice is most likely to form in the carburetor is between 41° and 50°F (5° to 10°C).

There are three ways this ice forms:

- Impaction – super-cooled water droplets that freeze on contacting a cold surface.

- Refrigeration – a temperature drop as heat energy is used to vaporize fuel.

- Throttle icing – a restriction, such as a partially closed butterfly, that causes an area of low pressure and so a reduction in temperature.

A drop in power and rough running of the engine can occur when carburetor icing forms. If unchecked, the engine may well fail altogether. Indeed, there will come a point where carburetor heat will have insufficient opportunity to melt the ice.

FUEL EXHAUSTION

Case study 14:4

Accident Report MIA99FA158

Date: May 20, 1999

Location: Intercession, Florida, United States

Aircraft: Hughes 369E

Injuries: 1 fatal, 1 serious

The pilot was hovering out of ground effect at about 75 feet above ground level adjacent to power lines. The helicopter started to sink. He applied right anti-torque pedal and the engine noise decreased. He moved the cyclic to the right, increased collective pitch to arrest the rate of descent, and the engine-out audio activated. As soon as he cleared the power lines, he lowered the collective pitch to gain rotor rpm, and applied forward cyclic in an attempt to clear a tree to his front. He realized impact with the trees was imminent, and applied aft cyclic in an attempt to make a vertical descent. The helicopter collided with trees and terrain hard in a nose-low left-skid-low attitude.

The commercial pilot sustained serious injuries, and a lineman located on a cargo platform on the left side of the helicopter was fatally injured. The helicopter sustained substantial damage. The flight originated from a Landing Zone (LZ) located at the Intercession City substation about 1 hour 15 minutes before the accident.

The pilot stated he had been conducting operations in and out of the LZ with another helicopter, and had hot-refueled three times before the accident flight. Each time he returned to the LZ, he had about 100 pounds of fuel remaining, and would add about 220 pounds to the helicopter. He could not recall the exact time that he departed on the third flight. He returned to the work site area and was working south of the accident site. The other helicopter joined up with him before they relocated to structure No. WIC 134. He approached the structure on a heading of about 140° magnetic, came to an out of ground effect hover at about 75 feet above ground level, and moved in next to the stringing block. The other helicopter remained off his left side and was photographing their operation. Right before the emergency, he pulled off the structure so the photographer in the other machine could change film in his camera. He hovered about two to three minutes, and checked his fuel before he moved back in next to the stringing block at WIC 134. He had been hovering for about 10 minutes, moving around the structure, when the helicopter started to sink.

Examination of the main fuel tank revealed the fuel tank was empty and not ruptured. The engine was removed, transported to the engine manufacturer, and placed in an engine test stand. The engine started, and obtained takeoff power, and all cruise power point specifications.

The National Transportation Safety Board determined the probable cause(s) of this accident as:

The pilot's improper fuel management that resulted in fuel exhaustion and a total loss of engine power.

Source: National Transportation Safety Board, United States

How many pilots miss such a basic step as checking the fuel level gauge? Surprisingly, more than you would think. Hundreds of pilots run out of fuel every year. A huge percentage of helicopter accidents are caused by fuel exhaustion — access any accident database on the internet for proof.

Although it seems distinctly careless and inexcusable, pilots often plan to run very low on fuel in order to minimize weight and enhance their aircraft's lifting capability. Such a calculation is a matter of very fine judgment. Other factors — such as an undetected leak, faulty measurement of quantity, and failure of the low-fuel warning system — can be involved. But quite often having too much air in the fuel tank results from a lack of any planning whatsoever.

There was a case in the United States some years ago where the accident investigator could find no good reason for a helicopter being extensively damaged after an engine failure. There was clean fuel in the tank and all the pumps were in working order. Then a witness revealed that the pilot was seen, shortly after the accident, returning to the helicopter carrying two jerry cans. He was struggling to carry them as he approached the wreck, but toting them easily as he left. All pilots are embarrassed when they run out of fuel – except the ones that don't survive.

Fuel problems – the final decision

Whatever the causes and the symptoms, the pilot has to make a decision when problems with the fuel become evident in-flight. If the engine has stopped altogether, the decision is obvious – autorotate! But, should you try to restart the engine during autorotation? Should you try banking the helicopter to the side that favors the fuel pick-up pipe? Should you try to replace the fuel pump circuit-breaker? And, if you get it restarted or if you have starvation or contamination and the engine is faltering or fuel pressure is dropping rapidly, should you try to make the next airport, road, patch of civilization (or sometimes the next island or land mass)? Should you land immediately? So many things to consider and so little time.

Make the best decision you can, follow the plan that your decision leads to and, when everything stops moving and the excitement is all over, defend your decision no matter what. If people come up with a logical argument that shows that you could have made a better decision, accept that and learn from it.

15

Fire

While fire is one of the greatest in-flight fears for a pilot, it is, thankfully, surprisingly rare in the normal course of events. As the case studies will show, there is frequently no warning of an impending fire; fires don't happen at any particular time, although obviously refueling is a riskier occasion; and, most importantly, with appropriate plans, procedures and crew reactions, they are not always tragic.

Case study 15:1

Report 3/90

Date: 1988

Location: North Sea

Aircraft: Sikorsky S-61N

Injuries: minor/nil

The helicopter departed the Safe Felicia, a North Sea semi-submersible rig, at 1345 hours with two pilots and a full passenger load of 19 for a one-hour charter return flight over the North Sea to Sumburgh. No cabin attendant was required by regulations for the flight and none was carried. At 1423 hours, when 40 nautical miles from Sumburgh and 15 nautical miles from the Shetland Island coast, cruising at 1500 feet altitude in Instrument Meteorological Conditions (IMC), G-BEID (ID) established radio contact with Sumburgh Approach Air Traffic Control (ATC), which identified the aircraft on radar and passed an inbound Visual Flight Rules (VFR) clearance.

The co-pilot, who was the handling pilot, reported that 5 minutes later he heard a muffled crack or 'bang'. This was also heard by a number of passengers. In particular the occupant of seat No. 4B heard it as a loud bang from above, from the area of the No. 2 engine drive train. The noise was not heard by the commander, but both crew felt that there was possibly a slight change in the vibratory 'feel' of the aircraft. About 6 seconds later, while the crew were discussing this, the No. 2 engine fire warning lights illuminated brightly. Engine instrument checks revealed nothing unusual and no signs of smoke or fire were observed by the crew. Passengers heard a number of abnormal noises after the bang and the passenger in seat No. 4B described a grinding mechanical noise, again from almost directly above.

After the fire warning the co-pilot started a descent and, in accordance with the drill − 'engine fire in flight, step A: suspect fire' − the commander retarded the No. 2 engine speed select lever to ground idle and started a stopwatch. He transmitted a mayday distress call at 1428:44 hours on the Sumburgh ATC Approach frequency, informing ATC that the aircraft had experienced an engine fire, in error identifying it as on No. 1, and that the intention was to descend to 500 feet and continue VFR. About 48 seconds after the bang, the No. 2 engine was shut down by operation of the fire emergency shut-off selector handle and the main fire extinguisher fired in accordance with step B of the drill.

The No. 2 engine fire warning lights remained illuminated and 90 seconds after the initial noise, while waiting the specified 30-second period before firing the reserve fire extinguisher, the No. 1 engine fire warning lights also illuminated. The other engine indications remained normal. At about this time passengers saw oil coming from the cabin ceiling, streaking down trim panels and the inside of the left window adjacent to seat row No. 4. The oil dripped almost continuously onto the occupants of seats No. 4A and 5A. Oil covered the outside of the right windows at rows No. 4 and 6. Smoke issued from a joint in the ceiling panels in the central part of the cabin.

The co-pilot increased the rate of descent, breaking cloud at 800 feet above mean sea level, and leveled out at low height, assisted by the 'check height' voice alerting audio warnings at 250 and 100 feet radar altimeter height, from the automatic low altitude warning system. The commander, still without visual confirmation of fire, informed ATC that ID was 'now showing fire on both engines continuing VFR towards the coast'. He briefed the passengers via the cabin public address (PA) system to prepare for an emergency ditching, and took the aircraft controls. On the commander's instructions, the co-pilot deployed the floats and made a 'ditching' RT call. The co-pilot put his head out of the cockpit window to check float deployment, saw a great deal of smoke behind and reported that the aircraft was on fire. A gentle power-on ditching was made 11 nautical miles off the Shetland Island coast at 1431 hours, some 30 seconds after the visual confirmation of the fire and three minutes after the initial abnormal noise.

After ditching, the co-pilot streamed the sea anchor and the aircraft floated satisfactorily with its flotation bags correctly deployed. The seawater-activated switch which should have stopped the Cockpit Voice Recorder (CVR) on ditching did not operate and the CVR continued to record for some 5 minutes after touchdown. The Automatically Deployable Emergency Location Transmitter (ADELT) was not manually activated by the crew and did not automatically deploy.

The commander shut down No. 1 engine, instructed the co-pilot to open the cabin door and attempted to apply the main rotor brake but found that the normal resistance to brake lever operation was absent and the brake ineffective. A passenger in seat row No. 4 saw the brake lever being pulled down and simultaneously heard an abnormal metallic grinding noise from an area corresponding to the rotor brake location. While the rotors were running down, the co-pilot entered the cabin, which was filling with wispy gray acrid smoke that was rapidly becoming more dense. He donned his immersion suit, opened the cargo door at the forward right side of the cabin and launched the forward life raft. The commander did not don his immersion suit.

The commander remained in his seat while the rotors were running down, which took some two minutes, and during this time ordered the launching of the rear life raft and the evacuation of the passengers. He then entered the cabin and found that he was unable to see through the smoke beyond seat row No. 3. The crew became very concerned for the passengers in the rear seat rows, but the smoke had become so thick and nauseous that they were unable to penetrate to the back of the cabin. They were unable to find a crew smoke-hood or oxygen equipment, since none was stowed on the flight deck of this particular aircraft. There was oxygen equipment on board, but this was stowed under Row 8, near the aft end of the cabin. Meanwhile six passengers had gathered in the rear of the cabin and were attempting to manually jettison the life-hatch, which had not been remotely unlatched by cockpit action. They reportedly experienced considerable difficulty and delay in unlatching the life-hatch and removed an adjacent push-out window in an attempt to reach the external release handle. The passengers finally managed to jettison the life-hatch and launched and boarded the associated life raft.

Some four minutes after ditching the passengers in the forward life raft, which was still attached to the aircraft by its painter, who had initially seen small flames in the area around the No. 2 engine exhaust, could now see signs of a growing fire at the right side of the main gearbox housing. They exhorted the crew to leave the aircraft, which they promptly did. The two life rafts were paddled with difficulty away from the burning aircraft, linked together and the passengers redistributed to equalize the loads.

Source: Air Accidents Investigation Branch, United Kingdom

Case study 15:2

Aviation Investigation Report Number A99W0061

Date: April 28, 1999

Location: Fairview, Alberta, Canada

Aircraft: Aerospatiale AS355 F1 Twinstar

Injuries: nil

The helicopter had completed a routine gas pipeline patrol and was returning to Fairview, Alberta, with the pilot and one passenger on board. During a shallow cruise descent into Fairview, at about 800 feet above ground, the red battery temperature light illuminated on the warning caution advisory panel. The pilot observed that the voltmeter and ammeter indications were normal and turned off the battery. About three minutes later, at approximately 500 feet above ground, as the pilot was contemplating a precautionary landing, the helicopter lost all electrical power and the cabin and cockpit began to fill with smoke and fumes. The pilot and passenger opened the side windows to ventilate the cabin, and the pilot accomplished an emergency landing at once on an available farm field.

After landing, the pilot shut down the engines and both occupants evacuated the helicopter without further incident or injury. Flames were observed to be emanating from the vicinity of the right baggage compartment, and the helicopter was subsequently destroyed by an intense ground fire.

The pilot was certified and qualified for the flight in accordance with existing regulations. He had accumulated 15,000 hours of flight experience, including a total of 8500 hours on rotary wing aircraft and 3500 hours on AS 355 Twinstar helicopters. The pilot completed the landing without a loss of engine power or flight control authority. He estimated that he landed about six minutes after the battery temperature light illuminated.

The pilot and passenger quickly moved away from the helicopter after the landing and did not attempt to combat the fire due to the intensity of the flames. The pilot had experienced an in-flight battery temperature warning in the past and had landed without incident. The aircraft flight manual (AFM) states that the pilot is to turn off the battery master switch and land as soon as possible if the battery temperature warning light illuminates. The AFM interprets 'Land as soon as possible' as 'land at the nearest site at which a safe landing can be made'.

The company operated two AS355 F1 and two AS355 F2 Twinstar helicopters which were fitted with a dual-battery, cold-weather start kit.

The right baggage compartment is located immediately forward of the battery compartment on AS355 helicopters. The compartments are separated by a 0.050-

inch-thick aluminum bulkhead. The company used the right baggage compartment in each of the four company helicopters for storage of the required survival and emergency equipment. The survival and emergency equipment included a five-person survival shelter and a survival kit that contained emergency flares. The bags that housed the survival shelter and the emergency kit were made of flammable nylon, and the survival shelter was also packaged in a waxed cardboard box. The bags were not required by regulation to be flame-resistant, and during testing, the packaging materials ignited quickly, melted, dripped and were totally destroyed by fire. The survival kits contained two hand-held, marine-type parachute flares and four day/night smoke flares.

The pilot reported that he had removed the battery compartment side access panel on April 25, 1999 to visually examine the batteries and had not noticed any discrepancies within the battery compartment.

Findings as to causes and contributing factors included:

1. The auxiliary battery paralleling cable was not attached to the positive post of the main battery during routine maintenance.
2. The in-flight fire occurred when the unattached battery cable arced through the battery compartment forward bulkhead in flight and ignited the flammable nylon survival gear bags in the adjacent baggage compartment.

Source: Transport Safety Board of Canada

Case study 15:3

Transport Safety Board of Canada Accident Report

Date: January 28, 1989

Location: Buttonville Airport, Toronto, Canada

Aircraft: MBB BK117

Injuries: nil

The Medevac helicopter landed on a wooden dolly at Toronto/Buttonville Airport after completing a flight. The pilot and co-pilot, the sole occupants of the helicopter, reported that a minute or so after a normal shutdown, a 'bang' was heard and flames were observed coming from the vicinity of the passenger/medical compartment. The helicopter rapidly became engulfed in flames and eventually burnt to the ground. Neither crew member received any injury.

At the time of the incident, the fuel quantity was relatively low. Because of the system design, most of this fuel would have been contained in the supply tank; the main tanks would have been essentially empty. This is deemed significant, as the analysis suggests that the fire started through electrostatic ignition of the fuel-air vapor at a fuel vent and the subsequent flashback of the flame front into the fuel tanks. The severity of the fuel-air vapor explosion would be enhanced by a near-empty fuel cell.

During shutdown, unused fuel, heated by proximity to the engines, drained back into the tanks. The return line was integrated with a fuel vent line, allowing fuel vapor to vent to atmosphere. The insulated wooden dolly or poor bonding between the aircraft structure and the skid gear may have hindered dissipation of the precipitation static that normally builds up during helicopter operations.

As the fuel vapor vented, the differential charge between the fuel vent and the airframe provided a current path. This would supply the necessary ignition device and fuel to cause the initial 'bang' heard by the pilots. Due to the robust nature of the aircraft structure, most of the explosive damage was directed downward, away from the cabin, preventing any personal injury to the crew.

This accident was the second to occur worldwide to a BK117 aircraft: the first occurred in August 1985 under very similar circumstances. A West German Medevac-configured machine with low fuel remaining landed on a wooden dolly. Following shutdown, the crew were completing their duties when a loud 'bang' was heard in the cabin. The subsequent fire destroyed the helicopter.

A third reported incident involved a BK117 being refueled. During the operation, the person refueling the helicopter suddenly noticed a flame coming out of the left-side vent pipe, and extinguished it with his gloved hand. Preliminary investigation revealed that the helicopter was parked on asphalt, which acts as an insulator.

It should be noted that static electricity continues to claim aircraft year after year. Fixed floats, fiberglass skis, wooden dollies and poor bonding effectively insulate helicopters, preventing them from dissipating their electrostatic charge easily after landing. Add the proper concentration of fuel vapor and you have the makings of a bomb.

Source: *Vortex*, 10/90

Case study 15:4

Transport Safety Board of Canada Accident Report

Date: unknown

Location: Salmo-Creston Highway, British Columbia, Canada

Aircraft: Bell 206L-3

Injuries: nil

The helicopter was substantially damaged when fire broke out during a refueling operation. The fire started as the fuel nozzle touched the fuel tank filler neck. An examination of the fuel pump disclosed a faulty 'ON/OFF' switch that had probably caused a spark, igniting the fuel. There were no injuries, but the 206 was heavily damaged.

Source: *Vortex*, 3/94

Every make of helicopter has in-flight drills for smoke or fire, and pilots should know these procedures well. A situation involving fire can quickly become an injury accident or worse. Knowing what to do the instant such an emergency arises is vital.

Civilian operators don't seem to have trouble with fires when hot-refueling as often as their military counterparts. The military are often (especially in battlefield simulations or situations) hot-refueling with pressurized fuel. It may be compared to the difference between the number of fires everyday motorists have when refueling at the local garage and the number of fires when racing cars are in for a pit stop.

Having said that:

➤ Always use a grounding wire when refueling.

➤ Know where fire extinguishers, hoses and first aid equipment are stored.

➤ Be prepared for the unexpected.

Regularly acknowledging the possibility of a fire (giving it more than just a passing thought), with due consideration to what you would do, will often assist your instincts if the worst ever happens.

Maintain situational awareness and remember that the greater the distance you are from any unexpected fire, the better. Unless someone's life is at risk, think carefully about what may happen next before you go rushing toward a burning helicopter with a view to fighting the fire. Those machines are insured after all, and they are still being made. Danger comes not only from the heat of the flames, but also the harmful toxins of the smoke.

16

Ditching

For a variety of reasons it is not uncommon for competent and confident helicopter pilots to find themselves dropping into water. While the ordeal is never pleasant, pilots who have trained properly and prepared for a ditching have every chance of surviving to fly another day.

Case study 16.1 has been supplied by Jim Wilson, chief pilot for Helicopters New Zealand Ltd. Jim has thousands of hours of helicopter experience and is a pilot who now takes all possible precautions when flying over water. His account is an inspiration for all helicopter pilots.

Other case studies in this book include some very good examples of how things generally happen during ditching in larger helicopters. See, for example, case studies 2.2, 13.3, 15.1, 19.5, 20.10 and 21.7.

Fig 16.1 Case study 19.5 looks at how this aircraft came to ditch in the sea
AIR ACCIDENTS INVESTIGATION BRANCH, UNITED KINGDOM

Case study 16:1

A pilot's account

Date: April 1982

Location: Cook Strait, New Zealand

Aircraft: Hughes 500D

Injuries: 1 minor

One afternoon in April of 1982 I left Masterton, in the North Island of New Zealand, to fly to Omaka Aerodrome near Blenheim in the South Island. As this involved about 20 nautical miles of flight across water (Cook Strait) I took the precaution of putting on an inflatable life jacket prior to takeoff in the Hughes 500D.

As required, I'd filed a flight plan and I knew where and when to report; this was just another short cross-country and I'd done hundreds. I had planned to fly at 3000 feet but there was an overlying northwesterly wind that generated lee-side turbulence and the ride was choppy. Two-thirds of the way across the strait I traded off a couple of thousand feet and the ride was smooth at 1200 feet.

About 10 to 12 miles out from the Marlborough Coast I'd changed radio frequency from Wellington Information to Woodbourne Tower when a chip light came on. I have an engineering background and I was complacent about the caution light. Chip lights were common enough, the Allison C20Bs were very prone to $2\frac{1}{2}$ gearing failure in the accessory gear box. I intended to carry on to Omaka and there remove the magnetic plug for inspection. This complacency meant that I didn't make a precautionary call to Woodbourne to tell them of the event. I monitored the temperatures and pressures; all were in the green, and carried on — no problem.

Two or three minutes after the chip light came on, right around the time when the nose yawed and the engine surged, a domestic airline Fokker Friendship called Woodbourne Tower for taxi/takeoff clearance. The tower promptly responded and the aircraft read back the instructions. This radio traffic took roughly one minute, during which the 500's engine stopped completely with seizure indications (pronounced yaw with low rotor rpm horn sounding).

I entered an autorotation at the minimum sink rate. I'd just completed five type ratings* for civil aviation officers and company pilots and I was confident about my abilities with regard to a successful auto and engine-off landing. Indeed, the recency of experience was one of tremendous assistance to maintain minimum sink rate speeds, turns and positioning to be flying along the swells. One must

---

* A type rating is when a pilot instructor tests the knowledge and skill of another pilot in a particular model of helicopter. The flight lasts between one and two hours, with much of that time spent covering emergency procedures.

get aligned along the swells and this gave me a little crosswind (on the port forward quarter) that was neither here nor there.

It was a busy minute; I took the time to put my sunglasses on. I reasoned that this would help me with regard to water definition but I think the truth was that I didn't want to lose them. I locked the shoulder harness and loose articles, opened the doors. In the flare, the radio was finally silent and I made my mayday call. I gave my position as 'four miles off the Wairau Bar.' This referred to the sand bar at the mouth of the Wairau River.

The surface of the sea was fairly calm with a 6-foot swell. Water entry was gentle with almost no forward speed. It stayed gentle for a very short time; when the tail rotor struck the water all hell let loose. The main rotor blades made water contact and it was obvious that one was going to strike the cockpit. I released my shoulder harness and crouched forward in front of the instrument panel. It was the closest study I've ever made of a clock and it was exactly ten past one. The rotor struck the cockpit roof Plexiglas and the blades then rapidly stopped rotating. Immediately the helicopter rolled right (with the swell) on its way to being fully inverted.

Few would argue that opening the doors prior to water entry was a good move; unfortunately they both slammed shut upon water contact. Loose articles crowded me in the confined cockpit — maps over my face, my hat over my eyes, seat cushions floating about. A major problem was the Dave Clark curly cord around my neck. There isn't much room in the main cabin bulkhead of the Hughes 500. I don't remember precisely how many seat squabs there were, but there were too many! I was trying to push all this debris away to get to the door handle. Of course, there was no door jettison mechanism or quick release on the doors. With the helicopter lying on its starboard side, trying to get a foothold over the width of the cockpit to get leverage to open the (now) top-side door (which is closed in the horizontal position) and battling with all the debris while shoulder deep in cold water (aircraft now 75 percent submerged) is not, I can assure you, a fun way to spend an afternoon.

Finally I got the door open, managed to climb out as the helicopter rolled fully inverted. I attempted to climb onto the belly but that bloody curly cord was still around my neck. I tell you, no matter how hard I pulled that cord, with the plug on an angle to the jack connection, it just wouldn't come loose. If I could have tied that cord to a tree, the helicopter would never have sunk! I got free of it by taking off the headset.

I actually ended up standing on the cargo hook, believe it or not, trying to prolong the delay to the inevitable — I was in that 'this is not happening to me' frame of mind. Of course it 'definitely was happening'; I was about to be swimming so I began to inflate my life jacket. By the time the aircraft had become a submarine my jacket was fully inflated.

Bad luck rather than bad management had given me another challenge — I was in the middle of a kelp bed. I stayed calm and slowly got through the tangle into clear water. Staying calm is a big factor: get yourself orientated, know where the sun is, where the coast is and from where the swell is coming. On the tops of the swells I could see White Bluffs and even catch glimpses of the beach. I realized the current was taking me away from the coast; I was dog-paddling to South America.

Even in a 6-foot swell, the troughs make for a pretty lonely and depressing place — you just can't see anything. Get your back to the swell and avoid taking any mouthfuls of salt water because this will start a coughing fit that will distract you from your orientation. One must stay in time and sympathy with the swells and getting distracted from this is very counterproductive.

Swimming is a waste of time. All one manages to achieve is to tire at a more rapid pace. The water was cold, and this, in itself, was energy sapping enough. Stay afloat, keep the water out of the lungs and do whatever you can to be seen. As I found when talking to my rescuers, I was a very small object in a very big sea. All I could do was to, where possible, keep the bright color of my life jacket facing the sun to generate maximum reflection and 'spotability'.

Now, this legal requirement to file a flight plan, which I'd done, to report and enable 'flight following' and all those good things was all well and good. Yes, with my position report, Woodbourne Tower had a pretty good idea where I was, what my predicament was and the fact that I needed rescuing. The flight plan thing had worked very well indeed and it was unfortunate that, with all this planning and contingencies, when I actually did fall into the drink there was a massive panic: 'Well, what do we do now? Who's got the boat?' Of course, if I'd made the precautionary call when the chip light came on they would have had a bit more time. Obviously they had to find a boat but they never did! This caused quite a stink in official circles: 'Yes, you must file a flight plan to cross Cook Strait; you've got to report here and there, and follow these procedures'. Past that the procedures went out the window; they just hoped you'd get yourself picked up by a ferry or a fishing boat or something.

The point was, though, they probably could have done something if I'd given more warning. One of the guys in the tower knew someone with a jet boat that could have been launched off the Wairau Bar quite quickly. But of course they didn't know anything about my problems because I never said anything — not until it all went quiet. With the untimely radio traffic I gave the mayday in the flare and this is not particularly desirable; hardly a textbook example.

I did get rescued — luck was not taking sides after all it seems. Aside from the Friendship taxiing at Woodbourne, there was another domestic airliner just airborne from Wellington. This was being flown by Captain Don McGregor, who heard my mayday and immediately spotted my helicopter. He actually flew behind me as I was descending into the water. He had a load of passengers on board and did a circuit around me. The helicopter didn't take long to disappear and Don stayed on station for 40 minutes, at which time an Air Force Cessna 421 arrived to take over.

I was supposed to meet a chap called Phil Mellsor at Omaka, and he was the one who came out to pick me up in another Hughes 500. He said that, if the Cessna hadn't been there to direct him onto me, one head in the water would have been very difficult to spot.

Phil didn't bring a scoop net with him and I was to climb onto the skids. I didn't think I was all that tired but I soon found out that one tends to overestimate one's own capabilities in these situations. I tried to hook myself over the skid but found that very hard with an inflated jacket on. Then I thought that if I could get onto the inside of the skid I could hook a leg over it and use the step to haul myself onto it. The helicopter came down in the hover, another swell came through and all that happened was that I got mashed underneath the machine. Not pleasant to have an underwater close-up look at a fuel drain. I got over the skid on my third attempt.

I think my body compensated for what was happening because it wasn't until I was being transported to safety that I realized how cold and tired I actually was.

I had been through various water survival courses in the ocean and in lakes. I've also experienced the HUET (Helicopter Underwater Escape Trainer) and I think that this could be made a little more realistic by actually configuring the aircraft the way it is flown. Put in the seat cushions, the maps and charts, and wear a headset. The other thing you don't get is the noise of the blades smashing into the cockpit; that can be quite frightening, wondering when it's all going to stop.

For all that, the dunker trainer definitely helps with orientation underwater and this is important because helicopters, with or without emergency floats, can turn turtle very quickly. Also, have a think about surviving the cockpit injuries from the loose articles flying around and the cuts/abrasions from airborne Plexiglas when the rotors hit the cockpit. There was a lot of that. And no training will ever prepare you for how fast it happens from 1200 feet. I went from flying straight and level to swimming in 2-2 $\frac{1}{2}$ minutes!

Source: Jim Wilson

As a result of his experience, Jim offers the following advice.

Life jackets

There are two types of pilot/operator when it comes to life jackets:

1. Those who suffer from 'carry it under the seat' syndrome.
   If you're going to put it under the seat, leave it behind because it's of no use to you. You must wear the life jacket — above anything, that's most important! And, if you're going to wear it, put it on properly. If you leave the strap around the back loose when you get into the water, the buoyancy of the jacket will tip your head back and/or the jacket will push up into your face.

2. The 'I don't want to spend the money' brigade.
   Most personnel flying across the water much of the time, like those servicing off-shore oil rigs, have all the gears: survival suit, life jacket, HUET training and all that sort of stuff. But for the average guy/girl jumping into a helicopter to fly across a lake or the sea the RN Beaufort jackets are expensive and probably an overkill. But something of that ilk with a couple of pockets in which to put water dye marker and mini-flares would be a good investment. If you ever find yourself bobbing around awaiting rescue, those locating devices will be your most prized possessions.

Precautionary calls

Precautionary calls cost nothing, except perhaps a smidgen of pride if nothing further goes wrong. The day I went swimming with my clothes on was the last time I was ever complacent about a chip light.

Training

Keep practicing those autorotations and engine-off landings. If you get the opportunity to do some water survival or HUET training, take it.

In the water

➤ Stay as calm as possible but expect to be apprehensive — then get over it.

➤ Get into sync with the swells, keeping your mouth clear of water and, above all, don't try to swim an unrealistic distance. Remember, you will tend to overestimate your capabilities.

Staying prepared

Make sure you have your life jacket professionally inspected every once in a while — its condition deteriorates over time. Check the manufacturer's recommendations.

Study the flight manual for the type of aircraft you are flying. There will be a section on ditching into the sea. If you have pop-floats and a rotor brake, the procedure is common sense. If not, many manufacturers recommend that, after the flare of an autorotation, the pilot use the cyclic to bank the aircraft so that the main rotors contact the water early, virtually acting as an instant rotor brake. The thinking here is that, once forward airspeed and rate of descent are arrested and the machine is 4–8 feet above the water, the biggest hazard (other than drowning) is most likely the main rotors. There will be no danger of them injuring a passenger who gets out of the helicopter too quickly or of them smashing through the cockpit if they have been stopped by early contact with the water. The downside of this is that 'banking in' from the hover will cause a degree of disorientation but, as the fuselage will likely roll soon after immersion, perhaps this is not too bad a trade-off.

Make sure that you read the flight manual, however, as it may specify the side on which to roll the aircraft. In some helicopters a sudden stopping of the rotors can cause the transmission to break free of its mountings and move a considerable distance, destroying anything in its path. The direction in which it moves is affected by the side to which the aircraft has rolled — the manual will tell you which side will prevent it will from entering the cockpit.

Cold water survival

Several of the responses to the shock which results from sudden immersion in cold water can kill:

➤ Heart failure is possible in those with weak circulatory systems, particularly the elderly.

➤ Hyperventilation increases the chances of swallowing water.

➤ Cold limbs are weaker and uncoordinated, making swimming difficult.

➤ Breath-holding ability is severely curtailed, to about 20 seconds, reducing the chances of successful escape from a submerged aircraft.

Because panic can magnify these responses, it is important to remain calm and methodical if faced with a cold-water emergency. Consciously control your breathing as much as possible.

A life jacket can prevent you from drowning by keeping you afloat and keeping your head out of the water — but most life jackets do not give any significant protection against hypothermia.

Hypothermia means lowered 'deep-body' temperature. In cold water, the skin and peripheral tissues become cooled very rapidly. But it takes 10 to 15 minutes before the temperatures of the heart and brain begin to cool. Intense shivering occurs in a futile attempt to increase the body's heat production and counteract the large heat loss.

Survival times for persons in cold water will vary greatly. Fig 16.2 shows potential average predicted survival time of normal adults in water of different temperatures. Children cool much faster than adults. However, unconsciousness and drowning is more often the cause of death, reducing these predicted survival times considerably.

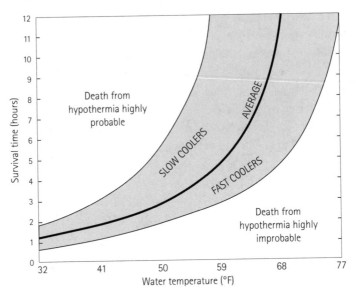

Fig 16.2 Cold water survival duration NEW ZEALAND FLIGHT SAFETY

Broadly speaking the more clothes that are worn on entry into cold water, the longer will be the survival period. This will vary significantly depending on the type of clothing and the amount being worn. Wet wool retains 50 percent of its insulating properties; wet cotton 10 percent.

Don't swim in an effort to keep warm — the heat generated will be lost to the cold water due to more blood circulation to the arms, legs and skin and increased water circulation through the clothing.

Your first consideration must be to reduce heat loss. The most critical areas of the body for heat loss are the head, sides of the chest and the groin region. Two techniques can be employed to reduce heat loss from these critical areas. Both techniques can increase the predicted survival time by about 40 percent.

Lone survivor: hold the inner sides of your arms in contact with the side of the chest. Hold your thighs together and raised slightly to close off the groin region.

Fig 16.3 NEW ZEALAND WATER SAFETY COUNCIL

Group of survivors: huddle together with the sides of your chests and lower body areas pressed together. If there are children present, sandwich them in the middle of the group for extra protection.

Fig 16.4 NEW ZEALAND WATER SAFETY COUNCIL

In both cases the principle is to reduce the amount of body surface exposed to the cold water, with the chest having first priority.

17

Loading issues

Winching up, tying down, throwing in, hooking on, attaching to, shutting in – no matter what is being carried it is all pretty basic stuff. But only the unwise would consider it easy and straightforward. There are a multitude of things that can go wrong, sometimes with tragic results. All pilots should ensure that every member of their groundcrew reads and follows the advice in this chapter.

SLING LOADING

Case study 17:1	Date: unknown
	Location: Sept-îles, Quebec, Canada
Transport Safety Board of Canada Accident Report	Aircraft: Bell 206B
	Injuries: 2 fatal

During a very short slinging operation, the locally designed and constructed cargo box became unstable. The box swung violently forward, striking the main rotor blade. It then swung rearward, and debris from the box hit the tail rotor. The main rotor flapped back, cutting the tail boom.

The Bell 206B JetRanger crashed about four miles east of the airport at Sept-îles, Quebec. The tower controller saw the fireball, heard the explosion and sounded the alarm. Both the pilot and his passenger died on impact.

The helicopter cargo hook, located 200 feet from the fuselage, showed strike marks on the outer casing, which indicated violent oscillation of the hook in all directions, particularly to the front and right. The hook was found in the closed position.

The box was designed and built specifically for helicopter operations but no flight crews were asked for input. A week prior to the accident it was loaded and slung below a Eurocopter AStar helicopter for testing. It swayed and oscillated up to 55 knots, at which point it would stabilize somewhat. At speeds over 55 knots, the box became unstable, forcing an immediate speed reduction. Several flights were conducted with differing loadings but it was agreed that the box should be sent back to the manufacturer for modification. The AStar pilot suggested to the chief pilot that the box should be transported there by truck. Shortly before the accident, the chief pilot went on holiday and was replaced by the incident pilot.

The sling cables were too short, much shorter than the box length, and the angle formed at the apex of the four cables was considerably greater than the 45 degree maximum recommended. With just one 3-foot choker, the box was suspended much closer to the helicopter than the suggested one rotor diameter.

Source: *Vortex*, 1/94

The following brief descriptions of eleven sling loading accidents that have happened in Canada serve to remind us why sling loading, like all helicopter operations, is a matter for careful planning, procedure and execution.

1. A locally designed box came apart in flight, permitting the unweighted sling rope to contact the tail rotor.

2. A Hughes 500 lost its tail rotor assembly when the sling gear, a 17-foot braided nylon strap and three-point swivel hook, came in contact with it. The pilot had been transiting for a load at 90 knots.

3. An 18-foot-long braided nylon strap was substituted for a steel cable that was starting to come apart. During transit for a load, the unweighted strap took off the Hughes 500's tail rotor.

4. A log was released when it became apparent that the helicopter could not be stopped safely in its downwind approach. Unfortunately, before the pilot could fully recover, the long-line snagged and the machine crashed.

5. An AStar was slinging two Bell 206 rotor blades when the blades flew up and took off the tail rotor.

6. A load slid off the pallet, and the unweighted pallet flew into the tail boom. Fortunately the sling gear was short enough to keep the pallet from the tail rotor.

7. A twin-engine helicopter was on final approach with a sling load when an engine failed. There were people on the approach path, so the pilot elected not to release the load. The helicopter crashed.

8. A 206 crashed and was destroyed when its water bucket snagged on a power line.

9. A pilot lost control and his helicopter crashed during a slinging operation to a factory rooftop. The pilot either didn't or couldn't release the load.

10. An unweighted 20-foot choker/lanyard combination took off the tail rotor assembly as a 206 descended down a mountainside at a descent rate of 1500 to 2000 feet per minute.

11. A 16-foot lanyard/cable combination came in contact with the tail rotor. Witnesses two miles from the crash site saw the MD369 break up in flight and heard the explosion.

Principles of sling loading

The equipment

Pilots should always know where the load is and what it is doing. A mirror that allows for both a short-slung load and the hook to be seen is therefore essential. For long-sling operations, the safest method is to look out of the door at the load during takeoff and landing. This may be difficult in helicopters which place the pilot in the right-hand seat.

Some Bell models have a special door fitted to overcome this problem. For other models, if a lot of sling work is anticipated, it may be advisable to install controls for the left seat; also engine instruments can be installed in the door or floor in the pilot's line of sight. Legal requirements may require certification, and ballast may also be required if the helicopter is designed to be flown from the right side only.

All equipment used in the slinging operation should be in good condition. Electrical and manual hook releases should be checked and the cargo mirror adjusted.

Metal to metal contact, that is, a D-ring with a clevis pin or a metal eyelet, should always be used between the hook and the sling. Synthetic fiber rope can twist up on the hook and make it impossible to release the load.

The type of sling used will vary, depending on pilot preference and the type of load to be carried. A woven polyester rope sling stows easily, will not come apart if it twists and will not rebound upwards should it break.

Fig 17.1 The slinging rope was of a material that allowed stretch and 'loading'. When the rope broke during a lift it recoiled and contacted the main and tail rotors of a Bell UH-1A

Steel cable is the strongest sling material and it is necessary to use this with large helicopters and large loads. However, it is heavy, may be too bulky to carry in small helicopters when operating in the field and must be replaced when it becomes kinked. Use of a leader will mean replacement of only a short length rather than the entire sling when kinks occur.

Chains work well for small helicopters and they store in a smaller area than cables. With loads over 1000 pounds, however, chains are not a good idea as they can bind where they cross and not tighten up.

Long-lines are single-line slings that may remain attached to the helicopter during the task (except in the case of an emergency jettison of the load). They may have a hook on the lower end and the load sling is attached to this rather than to the helicopter. The hook should have a safety catch. The length of the line should be either shorter or else much longer than the distance between the helicopter hook and the tail rotor. Even with this precaution against the line tangling with the tail rotor it is essential that the hook weigh at least 10 pounds to allow safe flight without a load on the end.

For maneuvering at the hover, a long sling is preferable. The pilot can see the load and its relationship to parts of the helicopter that are in his or her line of sight and also judge the angle the line makes with the ground. It is thus easier to position the load or to stop unwanted movement.

On the ground

In the takeoff, loading, unloading and landing areas, the safety of people on the ground must be taken into account. These areas should be clear of any debris or objects that could be disturbed by rotor-wash and strike either personnel or the helicopter.

Ground personnel may be used to advise and guide pilots but unless their capabilities are well known, never rely on them completely. The pilot is ultimately responsible if something goes wrong.

Ground personnel should be thoroughly trained and briefed on rigging, hook-up signals and safety procedures before any operation is started. They should wear goggles, gloves and ear protection. They should never stand under a load, or between the load and any immovable object. While they may have to check the security and alignment of the rigging as it is pulled taut by the helicopter, they must not place their hands in an area where they could be caught or pinched by the rigging.

Handling techniques

Some loads are difficult to fly because they will spin or oscillate. Unfortunately, there is no certain method of predicting how any particular load will fly. If it does not fly well, all that can be done is to re-rig it and try again.

Once in flight, however, the pilot may be committed, with no option other than jettisoning if the load shifts. A shifting load can make life interesting at best and can damage the helicopter, even bring it down, at worst. The key, therefore, is to make sure the load is adequately rigged and settled before leaving the loading zone. For loads most prone to shifting, that is, pipes, fence posts, poles, logs, and so on, the sling should be initially rigged over-center so that the first pull will tighten it.

It may also be a good idea to settle the load further by using the collective to 'bounce' it against the ground before departing the loading zone.

A speed that offers the most control of the load should be flown. No matter how well a load flies, however, there is not much sense in exceeding the manufacturer's recommended speed for external load carrying.

Some loads require accurate flying to avoid swinging. If a swing develops, concentrate on maintaining steady straight and level flight (it may be necessary to reduce airspeed slightly). If the damping is insufficient, position the helicopter over the load.

If, during the approach to the unloading spot, the pilot monitors the speed of the helicopter over the ground, he may misjudge the inertia of the load and it may continue past the spot. To overcome this, look at the speed of the load over the ground and imagine you are sitting on top of it, then 'fly' the load as you would the helicopter. The same applies when maneuvering at the hover. A steady speed reduction to zero — rather than 'throwing the anchors out' — will be the quickest way to stabilize the load over the spot so that it is ready for release.

Rigging

The pilot is concerned, during sling loading operations, with safety and efficiency. Proper rigging of the load is the key. If the rigging is incorrect, the load will be difficult to fly and could be dangerous. Following are both general tips and some more specific ones.

First, know how to tie knots that don't come undone and don't tighten up under a load, making them difficult or even impossible to untie.

Single-point slings

Except for items such as logs, a single-point sling is not the best method of attaching a load; if the load was to spin, it could unravel the rope and the eye could loosen and come apart. If the load demands a single-point lift, doubling the sling through the attachment point is better but the best technique is to use a sling with a swivel incorporated.

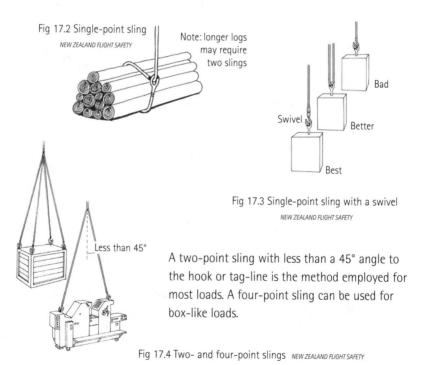

Fig 17.2 Single-point sling
NEW ZEALAND FLIGHT SAFETY

Note: longer logs may require two slings

Swivel

Bad

Better

Best

Fig 17.3 Single-point sling with a swivel
NEW ZEALAND FLIGHT SAFETY

Less than 45°

A two-point sling with less than a 45° angle to the hook or tag-line is the method employed for most loads. A four-point sling can be used for box-like loads.

Fig 17.4 Two- and four-point slings *NEW ZEALAND FLIGHT SAFETY*

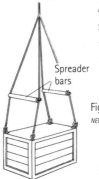

A spreader bar is useful for stabilizing a load, or where the sling may catch or damage the load if attached conventionally. Four cables with two bars are used for a four-point hook-up.

Spreader bars

Fig 17.5 Using a spreader bar
NEW ZEALAND FLIGHT SAFETY

Stabilizing a load

Box-like loads such as small huts usually fly very poorly as they tend to spin. A drogue chute can be used to stabilize the load. Attaching a pole to the load can give the drogue more leverage. A windsock-like chute will fly better than a parachute which must oscillate to spill out the air being forced into it.

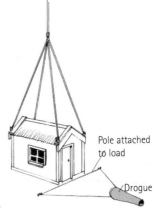

Pole attached to load

Drogue

Fig 17.6 Stabilizing a load with a drogue
NEW ZEALAND FLIGHT SAFETY

Logs or cut timber usually fly very poorly unless a tail is installed. A piece of plywood or a tree bough nailed to the load will help stabilize it.

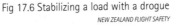

Fig 17.7 Stabilizing logs and cut timber
NEW ZEALAND FLIGHT SAFETY

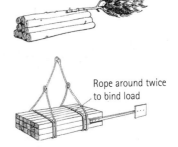

Rope around twice to bind load

Wire spool rigging

Simply threading the sling through the center of the spool is not the correct method of carrying this load as the spool will spin. The sling should be passed through the spool center, around a substantial piece of scrap wood and back up through the spool.

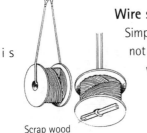

Scrap wood

Fig 17.8 Rigging a wire spool *NEW ZEALAND FLIGHT SAFETY*

Pallet slings

Pallet slings allow for quick turnarounds as the pallets can be preloaded and the sling later attached or detached rapidly. The disadvantage is in the necessity to secure the load on the pallet.

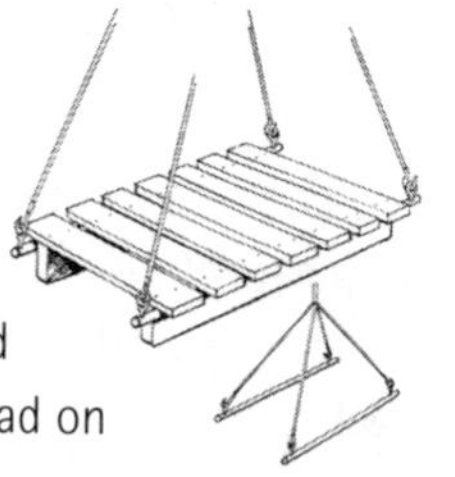

Fig 17.9 A pallet sling *NEW ZEALAND FLIGHT SAFETY*

Some pallets have no overhang. For a pallet sling to work with these it is necessary both to have a method of detaching the sling bars from the rope/chains in order to insert the bars under the pallet and also to have a chain or similar item that can go over the load to hold the bars outwards, that is, to prevent them from slipping towards the center of the pallet.

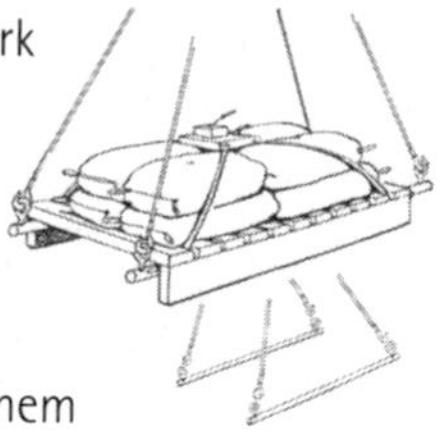

Fig 17.10 A pallet without an overhang
NEW ZEALAND FLIGHT SAFETY

Nets

Just about anything can be carried in a net. The weight must be centered, however, and the load made as symmetrical as possible. A net with a tarpaulin spread inside is useful for carrying many small items that could slip out. A steel cable net should be used for heavy loads or for items with sharp edges that could cut nylon net.

Single poles, logs

When carrying a single pole or log, wrap the rope or chain around the end of the pole twice. Logging operations use a cable choker where a ball on the end clips into a sliding catch further up the cable so that the cable chokes down on the log when it is under tension.

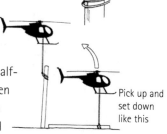

For placing a pole vertically, a clove-hitch (two half-hitches back to back) is used at the bottom of the pole; then the rope is run up to the top of the pole where a half-hitch is made. When the load has been placed the sling will loosen and can be easily removed by groundcrew. A remote hook can be useful for releasing chokers or when the long-line is to be retained.

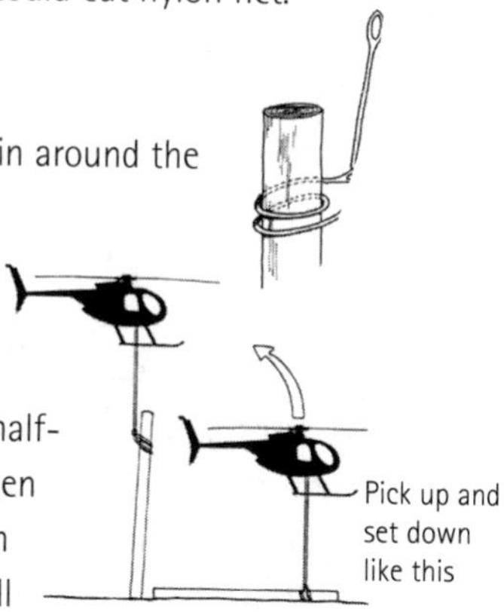

Pick up and set down like this

Fig 17.11 Carrying single poles
NEW ZEALAND FLIGHT SAFETY

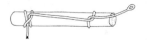

Fig 17.12 Closed clove hitch *NEW ZEALAND FLIGHT SAFETY*

Pipe shackles, baskets

Pipe shackles usually comprise two cables or chains — each longer than the pipe to be carried — which attach to the helicopter hook at one end and to short replaceable chains at the other. When the hooked end is passed around the inside (towards the center of the pipe lengths) of the lifting end and the hook engaged in a pipe end, the sling will tighten itself as the lift is applied (make certain, though, that each hook is engaged in the opposite ends of the same pipe!).

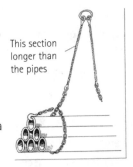

This section longer than the pipes

Fig 17.13 Carrying pipes
NEW ZEALAND FLIGHT SAFETY

The pipe basket is a good method for a light helicopter to carry drill rod or pipe and it allows for a quick turnaround.

Plywood

Plywood is one of the worst types of load to fly because of its shape. It should always be tied in a bundle or the top few sheets nailed together to prevent it coming apart like a pack of cards. A four-rope plywood string is easy to rig.

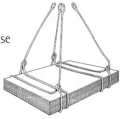

Fig 17.14 Carrying plywood
NEW ZEALAND FLIGHT SAFETY

Alternatively, a specialized plywood sling can be constructed from pipe, welded through flat stock steel with chains of about 30 inches long welded to the pipe about 12 inches in from the ends. The end pieces are supported by two 10-foot chains each.

Fig 17.15 A specialized plywood sling
NEW ZEALAND FLIGHT SAFETY

Drum hooks

Drum hooks comprise a pair of hooks designed to engage the rim of the drum. The pull on the sling provides the clamping force against the drum ends, and for unloading at unattended sites the pilot need only slacken the sling for the hooks to disengage.

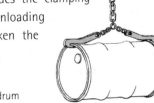

Fig 17.16 Carrying a drum
NEW ZEALAND FLIGHT SAFETY

On the other hand, this self-unloading feature means that the pilot cannot uplift the load at the takeoff point without assistance from the ground. The hooks can have a bungee cord between them to provide the clamping force for such situations but unless the bungee has a release catch incorporated, self-unloading will not be possible.

Rope drum sling

A rope drum sling can be used for one or three drums at a time; two slings at once can be used to carry a load of six drums. Lay the rope out on the ground with the ends staggered. Place the drum or drums over the rope, take the long end (A) over the top of the drum(s) and pass it through the bend (B). Loop it back and pass it through the eye of the short end (C).

Alternatively, wrap the rope twice around the drum(s) and then pass the long end through the eye of the short end, as shown in Fig 17.18.

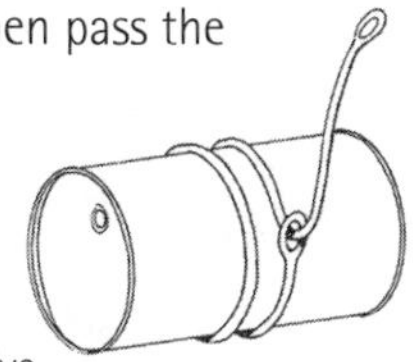

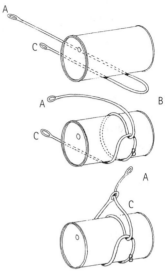

Fig 17.17 Tying a rope drum sling
NEW ZEALAND FLIGHT SAFETY

Fig 17.18 An alternative method for tying a rope drum sling
NEW ZEALAND FLIGHT SAFETY

Marshalling signals

It is essential that the pilot and any person giving marshalling assistance agree on what signals to use and what they mean. Use of a reliable radio microphone is even more advantageous.

Beware aerodynamics

Prior to any operation involving the transportation of large slung loads, helicopter pilots new to the role should carefully assess whether the type and shape of material to be carried could develop an undesirable 'aerodynamic' behavior in flight. Typical examples of the types of loads that may cause such problems are:

➤ Sheets of building materials – plywood, roofing iron, and so on.
➤ Pipes and tubing.

- Farm gates.
- Huts and small farm buildings.
- Small boats.
- Motor vehicles.
- Damaged aircraft.
- Fiberglass swimming pools.

If 'suspect' loads have to be moved some distance, consideration could be given to firmly securing some form of drogue to act as a stabilizer or 'tail' whilst in forward flight.

Gentle lift-offs

Handling technique instructions for initially taking the weight of the load should include that lift-off should be very gentle. The pilot is often unable to see the entire strop(s) and there is always the possibility that one has hooked over a skid. Sudden increases in height may initiate a dynamic rollover if this is the case. Any time during an operation when the sling gains slack (as when filling a water bucket from a body of water such as a lake), the pilot must be aware of the possibility that the sling could foul on a skid.

INADVERTENT CARGO HOOK RELEASES

Even the simplest of mechanisms has the potential to fail. Cargo hooks on helicopters are far from foolproof — if the hook itself doesn't fail, the hook release mechanism, be it electric or manual back-up, can be accidentally triggered. Some possible culprits are: the wrong combination of shackles and/or karabiners; lifting angles; foreign objects contacting the mechanism.

Case study 17:2

Aircraft Accident Report Number 95–021

Date: December 9, 1995

Location: Westland, South Island, New Zealand

Aircraft: Robinson R22

Injuries: 1 fatal

The 400-hour pilot and his crewman (shooter) departed in ZK-HUH on a deer-hunting sortie. They were in a remote area of South Westland on the west coast of

Fig 17.19 ZK-HUH TRANSPORT ACCIDENT INVESTIGATION COMMISSION, NEW ZEALAND

New Zealand's South Island. At about 2000 hours, having had no luck searching the grassy flats of the Jackson River or the tops surrounding the upper reaches of the Cascade River, the deer hunters decided to follow the Gorge River out to the coast.

On reaching the mouth of the Gorge River they turned to the northwest and flew along the beach. Shortly afterward they observed a deer halfway up a steep slip adjacent to the beach, some 200 yards from the river mouth. The pilot executed a 180° turn to head into the light southerly breeze and positioned the helicopter for the shooter to obtain a successful shot. The slain deer subsequently rolled down the slope, ending up amongst trees at the base of the slip.

The pilot thought about hovering with one skid against the slope to enable the shooter to retrieve the deer, but control of the helicopter would be sensitive with the weight transfer involved in the shooter disembarking from the skid, and the pilot considered the area was too 'tight' to allow adequate margins for safe maneuvering. The shooter suggested to the pilot that, as an alternative, he could be lifted into the site on the strop. The pilot agreed that, in the circumstances, he would prefer to do this.

The pilot flew ZK-HUH the short distance of about 60 yards to the beach and landed. The shooter disembarked and took the long 'twenty foot' strop from its stowage in the front of the helicopter. He had three short strops and karabiners around his waist which were normally used in recovering deer. The pilot lifted off and held the helicopter in a hover of about five feet to allow the shooter to reach in under ZK-HUH and attach the long strop to the cargo hook.

The pilot recalled that, during the hover, he made a visual check of the position of his hand while holding the cyclic control. He was wearing a pair of light leather gloves and verified that his right thumb was resting against the edge of the hook release bracket, clear of the hook release button.

The pilot maintained the hover, facing into the light wind blowing along the beach. He considered that rather than risk dragging the shooter through the nearby trees by proceeding directly to the slope, it would be best to first gain some height by flying to the south. Accordingly he slowly increased the height in the hover, looking out of the door and watching the shooter until the slack in the strop had been taken up. The shooter gave him a 'thumbs up' signal indicating that all was okay.

The pilot continued to increase height slowly, lifting the shooter off the ground and beginning to move forward through translation. Some 5 to 10 seconds later the helicopter had reached a speed of 30 to 35 knots and was about 50 feet above the beach when the pilot felt a lurch as the weight on the strop altered suddenly. He looked down immediately and saw the shooter falling, with the strop trailing him through the air. It was evident that the strop had released from the cargo hook.

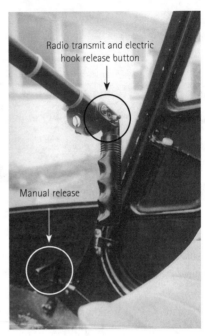

Radio transmit and electric hook release button

Manual release

Fig 17.20 TRANSPORT ACCIDENT INVESTIGATION COMMISSION, NEW ZEALAND

The pilot observed the shooter strike firm sand feet first and roll over several times in the direction the helicopter had been moving. The pilot turned the helicopter downwind without delay and landed on the beach beside the shooter. His first impression was that the shooter was badly winded and had injured his left arm and shoulder, but it soon became clear that he had received internal injuries and needed medical help urgently.

Given the remote location and the late hour, the pilot decided to fly the shooter to the nearest settlement (Jackson's Bay some 31 miles to the north) to summon assistance. He managed to load the shooter into ZK-HUH. The injured crewman appeared to be conscious but was unable to speak or to coordinate his movements. Concerned about the shooter's movements interfering with the helicopter controls, the pilot spent some time positioning him and securing his limbs. At this time, quite unexpectedly, another person arrived at the scene.

This person lived with his wife and family on the southern side of the Gorge River mouth. He had witnessed the entire accident with the exception of the shooter's actual contact with the sand and had crossed the Gorge River in a canoe to render any assistance possible. He helped the pilot secure the shooter in the helicopter.

The shooter's condition was such that he became very agitated when airborne and despite several attempts it proved impracticable for the pilot to fly him any significant distance. He eventually landed on a beach about 2.5 miles from Gorge River and left the shooter in the care of the local resident while he flew to Jackson's Bay. On arrival, shortly before 2100 hours, he advised the police at Greymouth by telephone about the accident. He returned to the beach, landing at dusk, and learned that the shooter had died about half an hour earlier.

The pilot had only recently taken employment in the deer recovery industry. This was at the request of the shooter who leased ZK-HUH. The shooter had experience in deer recovery and himself held a private helicopter pilot license. The pilot was working under the shooter's guidance in regard to deer recovery.

During the 19 hours that the pilot had flown ZK-HUH he had carried deer by strop on about 15 different occasions and had lifted the shooter on the strop once. No problems had been encountered with the cargo hook which had locked effectively and released as required.

The shooter had obtained the cargo hook from an Australian source and it had been installed on ZK-HUH by an approved aircraft maintenance organization before the pilot and shooter commenced deer hunting operations.

It was a Classic Aircraft Products CH400-1 type with a 1000-pound capability

Fig 17.21 Cargo hook in open position TRANSPORT ACCIDENT INVESTIGATION COMMISSION, NEW ZEALAND

but in the R22, for which it was an approved type, it was limited to a maximum load of 400 pounds.

The installation in ZK-HUH included a 'push to operate' electrical release mounted on a bracket attached to the pilot's cyclic control. The switch 'button' was not shrouded but the attachment of the mounting bracket to the forward site of the cyclic control tubing partially protected the switch within the elbow of the control handle.

ZK-HUH had been fitted with a cargo hook previously. The electrical circuit utilized one of the circuit breakers (CBs) in the aircraft's standard CB panel, forward of the left-seat base, as protection against current overload. The CB could be pulled manually, if required, to isolate the hook release switch from the aircraft's electrical power supply.

The CB wired for this purpose was one of several spares located among the double row of CBs in the panel. It was rated at 25 amps and was placarded 'HOOK'. The pilot was unaware that the electrical hook release could be disarmed by this means. The extent of the shooter's knowledge of the cargo release electrical system is not known.

A foot-operated cable release connected directly to the cargo hook was fitted on the lower right side of ZK-HUH's cockpit. This could be used by the pilot to open the hook mechanically if the electrical release failed to function satisfactorily.

The cargo hook itself had a manual release arm, normally in a vertical position on the right side of the outer case which, when rotated clockwise some 15° (rearwards movement of the arm), opened the hook mechanically.

The cargo hook incorporated a 'keeper' or safety latch at the end of the hook load beam. This acted as a one-way latch, allowing a karabiner or looped strop to be slid onto the load beam with the hook reset in the closed position. Thus it was retained securely unless the hook was released electrically or mechanically from inside the helicopter or externally by movement of the manual release arm.

Normal operation of the cargo hook required only a momentary 12-volt DC electrical supply to release the load. Once the load beam had been released it would remain in the unlocked, hook open, position until it was reset manually by ground personnel.

When ZK-HUH was examined following the accident the cargo hook was open. There was no marking or other evidence to indicate that the cargo hook assembly or its load beam had been subjected to unusual loading by the strop being incorrectly positioned or misaligned in some manner when it was attached.

Functional tests, with no load on the hook, showed that the hook reset normally and the electrical and mechanical release systems operated satisfactorily.

Detailed inspection of the cargo hook wiring disclosed no evidence of electrical shorts, chafes or other faults. The helicopter was suspended by its lifting fixture and a load of 160 pounds attached to the hook. The cargo hook wiring was pulled and flexed but no defect was revealed. The load was released only electrically by activation of the release button. It was noted that the contacts at the rear of the switch were not covered. A remote possibility existed that these could be bridged inadvertently by some conducting object or material and the switch circuit completed.

With regard to the strop and fittings, all were in new condition. No swivel was fitted. It could not be determined how the shooter attached the arrangement to the hook, however trials and tests of all methods and loading angles failed to readily simulate an undesired release.

In deer recovery operations, the shooter 'riding the sling' is a very common practise despite its inherent dangers. Sub-paragraph 1.39 of the accident report reads:

Previous accidents have demonstrated that the practise by shooters of 'riding the chain' when deer have been lifted out, or of being transported in some other manner suspended beneath a helicopter while deer hunting, holds considerable potential for a fall with the consequent likelihood of serious or fatal injury. Instances have occurred in which karabiners have twisted in a certain way and released from cargo hooks, and strop and karabiner assemblies have become unexpectedly detached, resulting in fatal falls. The occurrences have indicated that extreme care must be exercised by any individual 'hooking on' to a cargo hook, or attaching a strop and karabiner combination, to ensure that the method of attachment will be safe and dependable throughout the period of suspension.

Source: Transport Accident Investigation Commission, New Zealand

One doesn't have to be around general helicopter operations for long to hear of or even see an inadvertent hook release.

As a helicopter enthusiast before embarking on any flight training, I was frequently a groundcrew member on many types of operation. I would regularly 'ride the sling' and not really give it a second thought; at twenty-one years old, I considered myself reasonably invincible. By the time I was twenty-three, I was still quite 'gung-ho' but I didn't ride any more slings, having seen a bent and twisted seed (agricultural) bucket that had been dropped from a very great height. I also heard several first-hand accounts of mishaps, and it finally dawned on me that cargo hooks and electric release systems are to be treated with great respect.

It wasn't for another 10 years that I finally saw it happen. I was helping a friend with a contract fertilizing forests using Global Positioning System tracking (a new concept then) and a Bell JetRanger. He took off from a pad on a bush-clad hillside and had not gone very far before the full bucket dropped into the bush. I ran to the edge of the pad expecting to see him executing an autorotation into the valley below and was surprised that he was climbing and returning to the pad. When we retrieved the dropped bucket and performed a 're-enactment' we found that, when he had landed beside the bucket, with everything still connected, the D-shackle had twisted in the hook and was not secure at all. I now know that it's an all too common occurrence.

During training, pilots are told to be well aware of third parties on the ground who can be endangered by objects falling from aircraft. Whenever a sling load is carried careful thought should be given as to the best possible flight path so that, should the load itself (or something from the load) plummet from on high, the only thing that will be damaged is the load itself, not some unsuspecting member of the public or some expensive property.

OVERLOADING

The term 'maximum all-up weight' in respect to a helicopter suggests that weights above this amount cannot be lifted and that any weight below it can be. Neither conclusion is accurate. The ability of a helicopter to lift loads varies hugely and depends on factors such as: ambient temperature; atmospheric pressure and elevation of the lifting location, combining to give density altitude; and wind.

On a cold winter day at sea level in the middle of a high pressure system with a 20-knot head wind, most helicopters will lift an impressive amount more than the manufacturer recommends in the flight manual. Put the same helicopter on a 6000-foot high hill on a hot humid day with no wind and the pilot should go to the flight manual to calculate just how much can be lifted in these conditions; it is considerably less than on the cold day. Indeed, it might be just the pilot and the fuel in the helicopter!

There is a finite weight figure on any given day. Often, pilots working to the last commercial dollar have been known to pull maximum allowable power in no-wind conditions and get a sling load light on the ground but not clear of the ground to enable build up to translational lift and subsequent departure. The groundcrew then lifts the load manually (with the helicopter at power) and begins running in the takeoff direction until the helicopter has enough translational lift to support the load by itself and increase speed and lift. Loads have been 'thrown' off bluffs using this method, with the under-slung load dragging the helicopter down into the valley until airspeed and power even out the equation in the hope that the aircraft comes out on top. These practises are not recommended but show how pilots can and do cheat the limitations of their machines.

So what is the consequence of overloading a helicopter? With careful handling by an experienced pilot who knows what to expect and flies within his own limitations, nine times out of ten the operation will be successful, but there is no margin for error. Many of the bad things that happen to helicopters happen far more readily to the overloaded helicopter. Vortex ring state, overpitching, recirculation, severe turbulence, retreating blade stall, pilot stress are some; and woe betide the overloaded helicopter that suffers a mechanical failure.

Case study 17:3

Aircraft Accident Report Number 90–013T

Date: December 1990

Location: Richmond Range, South Island, New Zealand

Aircraft: Robinson R22 Beta

Injuries: 2 minor

The operator was based in the Nelson (South Island, New Zealand) area. One day just before Christmas of 1990, a customer (a hiker) wished to be dropped at Slaty Peak Hut in a remote area in the Richmond Range. He indicated the spot that he wished to be flown to on a folded map that he had brought with him. The operator noted the elevation of the site as 1450 feet. Knowing that the load; pilot, fuel, passenger and hiking pack, would be a full load for the Robinson, the operator was pleased that the flight was to be made in the cool of the early morning.

It was the operator's practise to use commercial helicopter pilots on a part-time basis. He chose one he had trained himself — a 46-year-old man with 325 hours' experience, of which 17 hours were in the last 90 days, all on R22s.

The pilot assessed the weights to be carried as follows:

Pilot	224 pounds	Hiker	168 pounds
Pack	60 pounds	Fuel (60%)	108 pounds

The pilot considered this load to be within the aircraft's capability. He was not unduly concerned because the elevation of the destination landing site was, after all, only 1450 feet. The pilot had the hiker stow some boxes of food from his pack beneath the seats and the pack was placed at the hiker's feet.

A hover check (in ground effect) at the base, which was near enough to sea level, required an intake manifold pressure of 24 inches of mercury. The ambient pressure (QNH) was 1017 hPa and the temperature was 55°F (+13°C).

The hiker guided the pilot to Slaty Peak Hut. On arrival it was found that the hut was only a short distance from the peak. The elevation was in fact 1450 meters above mean sea level, that is, 4750 feet. The hiker's map was a metric conversion!

Fig 17.22 TRANSPORT ACCIDENT INVESTIGATION COMMISSION, NEW ZEALAND

There was no helipad at the hut, so the pilot elected to land on a nearby tussock covered saddle. He assessed the wind as calm or light southerly so planned an approach towards the south, oblique to the ridge, which would permit him to break off the approach or to overshoot should problems arise. A power check at 50 knots showed 19 inches MP (Manifold Pressure), which the pilot considered sufficient for a zero-speed landing.

The approach was made at a relatively shallow angle at a power setting of about 22 inches MP. At about 15 feet above the landing point the pilot noticed that the dual tachometer (displaying both engine and main rotor rpm) was indicating 97 percent. (The normal operating range, or 'green arc', extended from 97–104 percent). He opened the throttle until it was fully open but this had no effect on the rpm.

By this stage the pilot considered that he was committed to the landing and controlled the flight path by increasing collective pitch. He was unable to arrest the forward motion completely and touched down heavily at a low groundspeed.

The aircraft pitched slowly onto its nose then fell onto its right side, extensively damaged. The pilot and passenger, both restrained effectively by their lap and diagonal harnesses, opened the upper door and climbed out. They repositioned the emergency locator beacon and then descended to the hut to await rescue.

The maximum all-up weight permitted for the R22 was 1370 pounds. With an empty weight of 840 pounds, using the pilot's estimates, the takeoff weight at Nelson was 1400 pounds. The aircraft was overloaded. To hover the aircraft in ground effect required 24 inches MP; at 4750 feet the available MP was 23.1 inches, so insufficient power was available.

Even if the entire load from the pack had been stowed under the seats, the center of gravity would have been about one inch ahead of the forward limit; in practise it must have been further forward. As fuel was consumed the center of gravity moved even further forward since the tanks were toward the rear of the aircraft.

The power check performed by the pilot at 50 knots prior to landing was a military technique which had been demonstrated by the flight testing officer at the time of the flight test for the commercial pilot license. The intention was to estimate

Fig 17.23 TRANSPORT ACCIDENT INVESTIGATION COMMISSION, NEW ZEALAND

the power required to hover; IGE by adding 5 inches MP, or OGE by adding 6 inches MP. It was then necessary to open the throttle briefly to confirm that this manifold pressure was available. Had the pilot made this calculation he would have found that 24 inches MP was required but only 23 inches was available.

Alternatively, the pilot could have attempted an OGE hover (with sufficient room to lose height to get through translation) close to adjacent terrain to get a visual vertical reference and he would have found that he was well above OGE hover ceiling.

At the point on short final (15 feet above ground level) when the pilot noticed that rotor and engine rpm were down to 97 percent (bottom of the green), the already limited power of the R22 was reduced by some 7 percent. Opening the throttle would have no effect as the aircraft was already at full throttle height. Attempting to control the approach path with collective pitch had inevitable consequences.

The pilot should have noted at Nelson that his power required to hover (24 inches MP) was only one inch below the maximum permitted. Had he appreciated the significance of this, a number of steps were open to him: he could have de-fuelled to the minimum required (reducing weight by about 104 pounds) but this would not have been sufficient in itself; he could have decided to make two trips, one with the hiker and one with the pack; or he could have referred the problem to the operator. All of these options of course entailed commercial penalties.

If the center of gravity problem had been noticed by the pilot in the hover, a far more practical solution would have been to sling load the pack and contents on the belly hook. If the pilot had recognized he was going to have insufficient power for the high-altitude landing he could have easily taken the pack to a lower elevation and dropped it. After taking the hiker to the hut, he could have gone back for the pack. Even if he didn't recognize the power problem, he could have jettisoned the pack on short final.

Source: Transport Accident Investigation Commission, New Zealand

Principles of overloading

A helicopter is designed, approved and safe to fly in accordance with the performance section of the flight manual for that helicopter. A pilot who allows the all-up weight to exceed the weight indicated by the performance section on a given day is overloading the helicopter and accepting the ensuing risks. It is really that simple.

Once in a while, the pilot of a helicopter can be tricked by people he trusts. I have heard of a pilot of a Bell 212 on a heli-skiing operation who suffered a heavy landing from a descent he could not arrest. The skid gear was bent dramatically. A head count revealed that the operator had loaded an extra two passengers without telling the pilot, meaning the machine was overloaded by 250 pounds. Customers have even been known to alter the weight notices on cargo.

For a commercial pilot, particularly a new one eager to impress, pressure from a customer or employer can be keenly felt. It can be very difficult to say no, but this demanding person can kill you and get away with it! Don't be pushed beyond your better judgment. When you need to say enough, say it with commitment.

If a demanding customer is to be a passenger, a practical demonstration may be a good approach. Allow him or her to load the 5000 pounds onto your light helicopter, then sit the customer in the passenger seat and explain what you expect to see happen with all the dials as you take off. When you are pulling 103 percent torque and the helicopter hasn't moved an inch, back the power off and explain the consequences of overloading and how easy it would be for one or both of you to perish. He or she might be more realistic next time.

More information relevant to overloading and required power can be found in Chapter 7 Overpitching.

CENTER OF GRAVITY

While the gross weight of the helicopter is important, the pilot must ensure that the load is positioned so that the center of gravity (C of G) is within the range specified in the weight and balance section of the helicopter flight manual. The C of G is the point at which all the weight of the system is considered to be concentrated. If the helicopter was suspended by a cable attached to the C of G point, the helicopter would hang level, parallel with the ground. The allowable range usually extends a short distance fore and aft of the main rotor mast; the range varies with each model but it is small, commonly within a few inches.

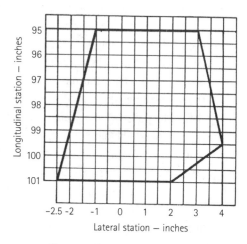

Fig 17.24 Center of gravity envelope
SCHWEIZER AIRCRAFT CORP

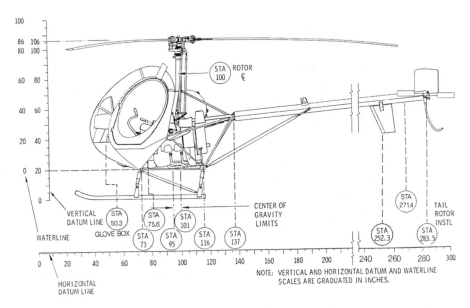

Fig 17.25 Center of gravity limits SCHWEIZER AIRCRAFT CORP

Case study 17:4

Accident Report FTW84FA102

Date: December 24, 1983

Location: Hughes, Arizona, United States

Aircraft: Bell 47G-2A-1

Injuries: 1 fatal, 2 serious

The pilot and observer were on a flight to rescue two duck hunters stranded in a boat that had become stuck in ice 500 to 600 feet from shore. After removing the right door, they planned to pick up one hunter at a time on the right skid in gusty winds. The pilot was able to hover at the bow of the boat as the first hunter climbed onto the right skid, but when the helicopter banked to the right it entered a spin to the right. The pilot regained control and continued to the landing area while remaining in a constant right turn. As he entered a 5-foot hover to land, the aircraft pitched forward. The pilot pulled the collective to gain altitude but the aircraft entered a right roll and turn, and subsequently struck several trees and crashed. An investigation revealed that the C of G exceeded the forward limit by .33 inches and the right C of G limit by 5.2 inches. The aircraft was not equipped for lateral loads and no lateral C of G limits were listed in the flight manual. Those equipped for litters (lateral loads) were rigged with additional cyclic control.

Source: National Transport Safety Board, United States

Principles of weight and balance

Helicopters are designed to have the C of G directly below the lifting point, that is, the rotor mast. Rigging of the cargo hook is an exacting science; ideally it should be directly below the rotor mast. Therefore the C of G datum (that ideal point where the C of G should be for any flight) usually lies between the rotor mast and the cargo hook.

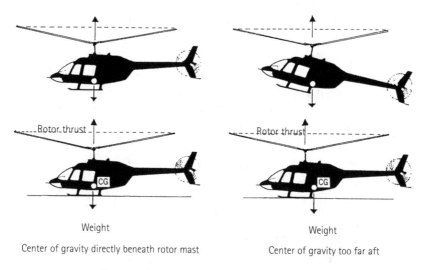

Weight	Weight
Center of gravity directly beneath rotor mast	Center of gravity too far aft

Fig 17.26 Centre of gravity concerns *NEW ZEALAND FLIGHT SAFETY*

A C of G that is forward or aft of this point has a direct effect on the position of the cyclic control for a no wind hover. If the C of G is well forward of the datum, the cyclic will have to be aft to maintain the hover; if aft of the datum, the opposite applies. The cyclic control can move only so far before it reaches its stops. If the C of G is too far forward or aft there is a point where a stable hover is not possible. Also, if the C of G is too far forward of datum and the helicopter is airborne, it will be 'nose low' and, when the pilot wishes to slow the machine by raising the nose, he or she may not have sufficient cyclic control to do so. The opposite applies if the C of G is too far aft.

The C of G also changes in flight. An insecure load may shift and suspended loads may swing. During winching operations, the winch arm introduces a lateral C of G factor. Fuel tanks are rarely located on the balance datum so burn-off will move the C of G. A passenger may move within the aircraft. More commonly a person causes problems by boarding or alighting abruptly (giving the pilot insufficient opportunity to counter with cyclic) or a person moves to the extreme front or rear of a skid, putting the helicopter hopelessly out of C of G limits. One of the catastrophic elements of losing a tail rotor gearbox is the massive forward movement of the aircraft's C of G.

Lateral C of G is less of a problem but light helicopters specify which seat the aircraft should be flown from as the balance is set up to enable the pilot to fly alone without ballast. Ballast is necessary if the pilot is to be the sole occupant and wishes to fly from the other seat.

Many case studies in this book mention C of G as a factor. See, for example case studies 5.1, 6.1, 12.1, 17.3, 22.2.

Few light helicopter pilots complete a weight and balance calculation prior to flight. Experience and knowledge will generally trigger concern where C of G may be approaching range limits. Large helicopters regularly flying commercial transport operations operate more like an airline, and weight and balance is calculated and documented as a matter of procedure.

Pilots can get a reasonable estimate on the C of G position in the hover. Lift the helicopter just inches from the ground, pause and take note of the position of the cyclic stick. Also, if the helicopter is incorrectly loaded it will want to roll, pitch or drift from the hover. In the worst case scenario the front or rear skids, depending on which limit has been exceeded, won't lift off the ground! A low initial hover allows the pilot to set the machine down safely and reload.

Many of the hazards mentioned in this book are aggravated by an out of limit C of G. Even an allowable C of G that is approaching the limits can turn a controllable situation into an accident.

For most pilots, C of G is something they are naturally aware of, even subconsciously. Like many other things that come with experience, considering weight and balance becomes habitual.

LOOSE OBJECTS IN THE COCKPIT

Controls can become immobilized in many ways. Most are foreseeable and preventable; a few are not. The following case study shows why a clean and tidy cockpit, with everything secure and in its rightful place, is a necessity.

Case study 17:5

Report Number 2001-007

Date: September 20, 2000

Location: Casement military airbase, Ireland

Aircraft: Eurocopter EC-120

Injuries: nil

The helicopter, which was operating in the private category, departed Cork routing for Hook Head, County Wexford, planning to then continue to Galway. As the helicopter approached the Waterford area, the pilot found that he could not move the yaw (rudder) pedal right of center. At this stage he suspected that something was jamming the pedal, and he further suspected it might be his mobile phone. As he had $2\frac{1}{2}$ hours fuel on board, he decided to contact the Air Corps Helicopter Detachment at Waterford Airport to ask for their advice. Following discussion with the Air Corps Detachment, he decided to divert to the military airbase at Casement, which gave better wind/runway options and a larger airfield to effect a landing without yaw control.

He duly arrived at Casement and practiced several approaches, expending his excess fuel in the process. There was some discussion with helicopter pilots at Casement as to the selection of a concrete runway or a grass area for the final run-on landing. The pilot decided to perform the landing on the concrete runway and the Air Corps Crash Rescue Service laid a blanket of foam. When the unrequired fuel was expended the pilot performed a run-on landing on the foam

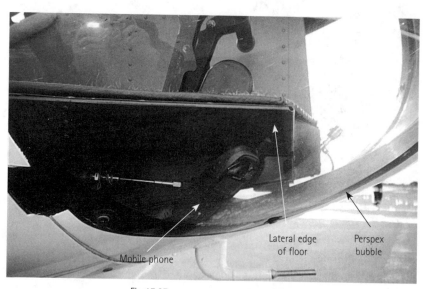

Mobile phone   Lateral edge of floor   Perspex bubble

Fig 17.27   AIR ACCIDENT INVESTIGATION UNIT, IRELAND

Fig 17.28 AIR ACCIDENT INVESTIGATION UNIT, IRELAND

Fig 17.29 AIR ACCIDENT INVESTIGATION UNIT, IRELAND

blanketed runway. He was able to hold the required heading after touchdown, until the aircraft decelerated to a slow forward speed. At this point the helicopter turned uncontrollably to the left, through about 110°. The helicopter then came to a standstill on the runway but off the foam blanket, without damage.

The final landing was observed by an AAIU inspector who arrived on the scene as the pilot was exhausting the surplus fuel. On inspection of the helicopter after landing, a mobile phone was found to be wedged between the horn of the yaw control and the Perspex bubble window at the front of the helicopter, on the starboard side. The phone was firmly wedged, and the maximum degree of movement to the right was a pedal position a little under half an inch right of center.

During the investigation, one loose item, a pencil, was found on the cabin floor, under the rear seat.

The pilot recollected that he probably left his mobile phone sitting on the rear seat of the helicopter, behind his own seat. When the control became jammed, he could not locate the phone, and he suspected that it had fallen into the under-floor area.

Source: Air Accident Investigation Unit, Ireland

If, like the pilot in case study 17.5, you are alone in the helicopter, a 'stuck pedal' situation is almost impossible to resolve. Attempting unusual attitudes to try to relocate the item is not a good idea as excessive maneuvering without full control is a very risky option. A passenger with the dexterity of a contortionist may be of some assistance. Videotaping this solution would also be of enormous entertainment value at a later date.

The pilot in the case study made some very good decisions after his one initial unfortunate oversight. He asked for assistance, he didn't panic and he acted in accordance with the advice that he was given. On this occasion, traveling to an airfield with good emergency services was a particularly prudent step.

Every pilot should be briefed and asked to demonstrate 'stuck pedal' procedure. This involves varying the power input (and therefore engine torque) to control yaw for a run-on landing at the lowest speed that directional control can be maintained. At the end of the day, however, always take care that there is nothing loose in the cockpit that might give rise to such problems. Loose objects in the cockpit can be even more of a problem if a door is open or if doors have been removed. Many accidents have resulted when something has flown from the cockpit (or a boot locker) and struck the tail rotor with enough force to cause failure.

Of course, the pedals are not the only control that can become jammed. While case study 17.6 demonstrates a stuck collective resulting from a maintenance error, the collective is equally susceptible to loose object interference; fortunately, it is generally quite an accessible area.

Case study 17:6

Accident Report SEA02IA096

Date: May 20, 2002

Location: Cocolalla, Idaho, United States

Aircraft: Kaman K-600

Injuries: nil

At approximately 1745 Pacific daylight time, a Kaman K-600 being flown by a commercial pilot, sustained a jammed collective condition while entering a hover at a logging site. The pilot was uninjured and the aircraft sustained no damage during the pilot's use of alternate controls to effect an emergency landing. Visual meteorological conditions prevailed and no flight plan had been filed. The flight was engaged in logging operations and originated from a local site nearby.

The pilot reported that while flaring the helicopter in anticipation of a log loading cycle, the collective became jammed at a high power setting and the pilot was unable to stop the helicopter from climbing. The pilot utilized throttle and a reduction of rotor rpm to execute a safe landing.

Post-event examination revealed a bolt that had been installed 180° from its normal orientation. The reversed installation allowed the threaded end of the bolt to interfere with the clevis on the throttle cable preventing the reduction of collective.

Source: National Transportation Safety Board, United States

Where the collective lever is 'stuck', the pilot is faced with a torque setting that cannot be changed. More often than not the pilot is left holding onto a handful of helicopter with more power set than is needed for a descent to the hover. It can be a daunting prospect and requires a cool head and a bit of thought. But the pilot needs to get it down in a reasonable time frame because weight is reducing with the burning of fuel and adding to the problem.

Without the use of the collective lever, the pilot's controls of choice include cyclic, yaw pedals and the rpm control — either the N2 governor (turbine helicopters) or the throttles.

Let's first look at one option that should not be considered: the engine-off landing. A critical element in establishing and maintaining autorotation is to lower the lever. Forget it — it's stuck! Another critical element in the engine-off landing process is the judicious use of collective lever at the end of the flare (to cushion the aircraft to the ground). You can't — it's stuck!

The first action to consider is to maneuver the aircraft to a position from where a landing is possible, that is, descend. A combination of rpm reduction, perhaps to the bottom of the green range, and cyclic can be used. Whether the cyclic is used to descend at high indicated airspeed (IAS) or low IAS depends on your geographic location. Just remember that a high-speed descent will lose altitude for you, but at the end of it you will have to reduce speed to land and that speed reduction may give you an unwanted gain in altitude. A low speed descent, below minimum power speed will help and is more manageable. However, beware of speeds so low that vortex ring state becomes possible. Remember that an angle of bank will assist the aircraft to descend. An initial spiral descent at a healthy angle of bank and low IAS is an option.

As you get to the final approach area, with luck a long flat runway, a combination of slow speed descent at reduced rpm — out of the governed range and under throttle if necessary — will have you able to produce a run-on landing at a safe speed.

This unusual and, thankfully, rare emergency is one to think about and plan in the shower or lying horizontal in the rest mode where all good ideas come from. It is not one to think about for the first time when it actually happens.

18

Winching

Winches on helicopters are primarily used as lifesaving equipment. Many people who have been hoisted from areas of jungle, mountain or sea owe their lives to the skill and teamwork of the helicopter crews, and the ingenuity of winch manufacturers. Some operators use winches to insert and extract harbor pilots from ships. Other operators, particularly military, use the winch to place personnel into and extract them from areas where a landing is not possible.

Whatever the task, the helicopter crew must work as a team to use the winch effectively, efficiently and, above all, safely. The winch was designed for saving lives so the equipment must be serviceable and the crew capable of performing the task.

Case study 18:1

Aircraft Accident Report Number 93-017

Date: November 14, 1993

Location: Tasman Sea, North Island, New Zealand

Aircraft: Bell 212

Injuries: 2 fatal

On November 12, 1993 a winching exercise was held which involved five company employees as winchmen. A winchman is the person who is suspended on the hook of the winch cable to assist the people being lifted or lowered. The exercise lasted three hours and was completed uneventfully. As the winching was done over land it was referred to as 'dry' winching.

At 0800 hours on November 14, company participants assembled at the base for a briefing in preparation for a period of deck winching to be conducted from the company's offshore support vessel. The briefing was conducted by the winch crew supervisor, and included the viewing of a safety video, a tour of the helicopter and a discussion of winching. Neither of the two pilots, one of whom was to be winch

Fig 18.1 TRANSPORT ACCIDENT INVESTIGATION COMMISSION, NEW ZEALAND

operator on the flight, attended this preliminary briefing. The senior pilot was aware of the intended program and familiar with the topics discussed at the briefing. The pilots stated that they attended while the participants were briefed on the detailed procedures for the winching operations, but those involved could not recall their presence at the time. This briefing concluded at 0910 hours and the pilots prepared the helicopter for takeoff.

While the pilots were preparing the helicopter the other participants checked their equipment, after which the winch crew supervisor briefed the pilots and participants on the proposed program for the deck winching.

The aircraft departed from base at 0925 hours. The first three sorties of approximately 30 minutes each involved a total of 24 winch cycles which were completed without any difficulties.

The fourth sortie involved an experienced winchman and a substitute 'survivor' as the designated person was not available. As the aircraft left base it established communications with the support vessel which was stationed some 2 nautical miles off the coast steaming at 2 knots with the wind from 120° at 30 knots on its port bow. The sea state was estimated by the first mate as a 5- to 7-foot swell with a 5-foot sea. The helicopter pilot estimated the swell as 13 to 16 feet.

On board the helicopter were the pilot flying the aircraft, the winch operator (who was also a Bell 212 pilot), the winchman, the survivor and an observer (who was also a trained winchman). The aircraft's intercom was live throughout the winching operations. This allowed the winch operator to use both hands for the winching, and all stations on the aircraft to hear the winch patter.

The first maneuver was to lower the winchman and survivor to the middle of the deck, some 33 feet from the ship's stern. Here the survivor was unhooked from the winch and remained on the deck while the winchman was winched in and the pilot flew a circuit about the ship.

The next winch operation was to return the winchman to the ship's deck and to winch him in with an (empty) stretcher. The helicopter remained immediately above the ship during this winch cycle. The pilot in command declined to move clear of the ship when told he was clear to do so by the winch operator.

When the winchman arrived at the helicopter door with the stretcher he was seen to converse briefly with the winch operator. The winch operator stated that the winchman asked him to relay a request to the pilot that he remain 'on station' while he went down to uplift the survivor. The winch operator conveyed this exchange to the pilot as a request 'to remain over the ship' and the pilot agreed.

Fig 18.2 Bell 212 above the ship's deck TRANSPORT ACCIDENT INVESTIGATION COMMISSION, NEW ZEALAND

The winchman was then lowered to the deck of the ship. After he landed on the deck at least 49 feet of excess cable ran out after him and coiled on the deck. This amount of cable was in excess of the normal amount winched out on previous occasions to compensate for the rise and fall of the ship as it responded to the local swell. The winch operator stated that he winched out some excess cable 'to prevent the winchman and survivor being whipped off the deck', when he lost sight of the winchman briefly. He believed that less cable run out was involved, but sufficient for him to advise the pilot, 'I may have a runaway', or words to that effect.

The operator's offshore operations manual specified that, if the winch operator was unable to control the cable, he was to call 'runaway cable in/out'. The winch operator was overheard advising the pilot that a 'runaway out' had occurred. In response the pilot switched off the hoist hover switch (the master switch for the winch). This was in accordance with the operator's instructions. The offshore operations manual didn't indicate the procedure to be followed after isolating the power to the winch.

Fig 18.3 The winch pendant
TRANSPORT ACCIDENT INVESTIGATION COMMISSION, NEW ZEALAND

Subsequently, both the operator's manager and the winch manufacturer were adamant that no further winching should be attempted. In the event, the pilot restored power to the winch after obtaining agreement from the winch operator. The winch operator stated that as soon as the power was restored, he checked that his winch pendant was controlling the winch correctly.

As the winch operator winched in the excess cable, the winchman flicked the cable in what seamen on the ship believed was an attempt to uncoil some half loops which had formed as a result of the random coiling of the excess cable on the deck. Meantime, the survivor had hooked onto the winch hook and the winchman signaled to the winch operator with a 'thumbs up' that they were ready to be winched in. The pair were lifted clear of the deck by the aircraft and the winching in commenced. The winch operator did not include the 'clearance to move left' in his patter to the pilot during the winching in because the pilot had agreed to his request to remain over the ship. The normal procedure is for the pilot to climb the aircraft vertically until the load is just clear of the deck; that way, the pilot can put the load back on the same surface if an adverse lateral center of gravity situation occurs. When satisfied with power requirements and lateral center of gravity, he clears the winch operator to start winching in. The observer in the aircraft overheard the normal patter for this sequence of events.

The pilot maintained the aircraft's hover some 50 feet above the ship's deck during the winching in. As the men on the hook neared the helicopter's skids a pause, normally made by the winch operator to ensure the personnel on the hook were clear of the skids and facing the door, did not occur. The winch operator said this was because he could see they were lined up satisfactorily. He continued winching in waiting for a 'closed fist' signal from the winchman to indicate the winching in should be stopped. He could not recall receiving such a signal and said that the cable ran to the end of its travel when he endeavored to stop winching in on his own initiative. The closed fist signal was not a recognized signal in the operator's winching procedure and it was incumbent on the winch operator in any event to stop winching in before the hook's buffer hit the up-limit switch.

As the hook reached the end of its travel it hit the up-limit switch. This was followed by a 'mechanical screech' and, one or two seconds later, by a cable break which resulted in the two winched personnel falling clear of the aircraft and onto the starboard quarter of the deck, sustaining fatal injuries.

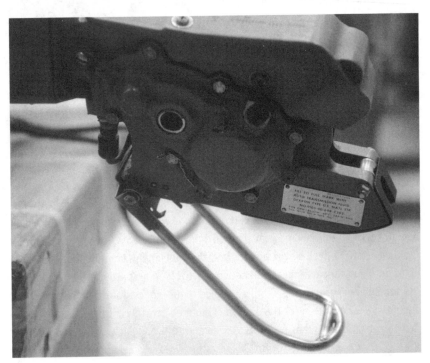

Fig 18.4 Up-limit bar TRANSPORT ACCIDENT INVESTIGATION COMMISSION, NEW ZEALAND

The subsequent accident investigation identified a number of relevant safety issues:

➤ The dependence of the hoist design on the integrity of the winch cable.

➤ The adequacy of inspection techniques to establish cable serviceability.

➤ The design of the hoist cable circuitry.

➤ The adequacy of existing instructions on winch operation.

➤ The crew's compliance with instructions.

The winch cable was built up by twisting seven stainless steel wires into strands and then twisting 19 of these strands in such a way that the seven central (inner) core strands were twisted in the opposite direction to the exterior 12. This ensured that the wires lay flat and each took an approximately uniform load without any tendency to twist. The investigators in this accident found that the cable strength had been degraded by internal abrasion and was not of the required strength in the area of the failure at the time of the accident. The wear is obvious on a cable during inspection by areas where small kinks occur. The localized section was 12 inches from the hook end of the cable.

The winch had a number of safety features designed into its manufacture. Normal winch speed is 150 feet per minute. To winch in at full speed up to the limit switch, the sudden stop, especially with weight applied, places stress on the

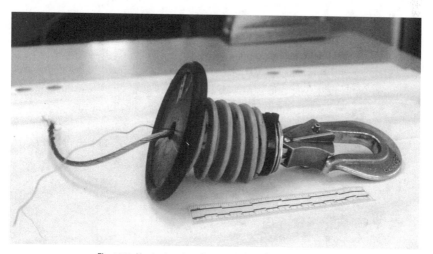

Fig 18.5 Hook showing the position of cable separation

cable which eventually degrades its strength. The manufacturers designed a mechanism into the winch whereby the speed of winching in was reduced automatically. At 20 feet from fully in (hook about 15 feet below the skid), the winch speed was reduced to 75 feet per minute (50 percent of full speed), and at two feet from fully in, the speed was further automatically reduced to 15 feet per minute.

When the hook met the up-limit bar, it triggered two microswitches in sequence, each of which was designed to stop the winch winding in by interrupting the power to the winch drum motor. If, in spite of these features, the winch continued to wind in at maximum speed, a clutch system restricted the tension on the cable to less than a third of its ultimate strength and, if the current drawn by the motor at any stage exceeded 160 amps, a cut-out protection was provided.

In this case, investigation revealed that the winch's circuitry had a short circuit caused by the failure of two transistors in the control box circuit's power switch/ heat sink. This fault rendered the winch operator's pendant control and the pilot's independent override facility ineffective. The circuit fault also deactivated the two speed reductions and the two up-limit microswitches. In this situation, the only way to stop the winch was for the pilot to switch off the power supply to the winch at the hoist power switch.

The inquiry found that the operator's operations manual included appropriate instructions for avoiding the serious consequences of a cable failure during deck winching. Excerpts quoted in the accident report include:

Winching in

With the aircraft accurately positioned over the 'survivor' the winch operator calls, 'Up gently to take the weight of one/two'. The pilot climbs the aircraft vertically until the load is clear of the ground. At this point, if the pilot is happy with the power and center of gravity, he'll say, 'Clear to winch'. The winch operator starts winching in. As he does so, he calls, 'Clear forward and down', whereupon the pilot will, provided the area is clear, ease the aircraft into a gentle forward descent. This avoids the survivor being suspended so far above the surface that injury would be sustained if:

➤ The cable broke.
➤ Winch suffered a runaway out.
➤ Survivor slipped from strop.
➤ The cable was cut due to aircraft malfunction.

Deck winching

The procedure is standard, except that once the crewman/survivor is lifted clear of the deck and obstacles, the winch operator should direct the pilot to move the aircraft left, clear of the vessel. Once clear, a normal lift is completed with the aircraft moving forward and down over the water while winching in.

Under the special instructions and emergencies section of the operations manual, mention was made of the consequences of allowing the cable to run up against the limit stop at full speed. 'All winching is to be carried out without reliance on the up limit switch', and warned of the possibility of overstressing the cable if it is allowed to run up against the limit stop at full speed.

Also from the manual, 'where personnel are to be winched for training purposes they are not normally to be lifted higher than 15 feet above the surface'.

The investigation concluded that, had the operator's instructions been complied with, the consequences of the winch cable failure would have been limited to the personnel on the hook falling into the sea from a height of some 15 feet.

Source: Transport Accident Investigation Commission, New Zealand

Potential pitfalls

There are a number of potential pitfalls associated with winching operations. By the very nature of rescue work, rescue winching is seldom a straightforward operation. Mountaineers rarely fall to the top of a mountain, and rescues are rarely performed on nice sunny days. Also, with between 35 and 175 feet of cable winched out, a winching helicopter is sitting well inside the height versus airspeed 'avoid' curve. An aircraft emergency during that time will usually have adverse consequences for those suspended on the wire, as well as the crew. Therefore, prudence dictates that the aircraft remains in that vulnerable position for as short a time as practicable.

Ways of avoiding, or at least lessening the potential for disaster, fall into three areas:

1. Care and maintenance of the equipment

Care of the winch cable is a big issue. A winch cable may look frighteningly thin when you are suspended from it, but it is remarkably strong and stays that way if looked after. Most rescue winches have a manufacturer's limit of two people or 600 pounds. However, the cable is designed with a tensile strength three times the maximum that could be applied. Cause of deterioration in strength include kinks, sudden jerking stops with a large weight attached, and running the cable up to the top at full speed. If the cable near that end has deteriorated, the stress will be marked at that point. Modern winches have variable speeds which automatically slow the winch as it nears the aircraft. Manufacturers specify the nature and frequency of cable inspections to ensure integrity. Buy your cable from an approved supplier, and certainly know the history of your cable.

I have heard of only one other instance of a cable breaking. On a particularly stormy day off the South Island of New Zealand, a crewman (with a very healthy build) was winched down to the deck of a yacht to tend to a sick yachtie. The time came to winch the sailor and crewman back to the aircraft. With a rolling, pitching vessel, and the mast threatening to interfere with the good intentions of the hovering Iroquois, the crewman's ankle became caught in the deck safety cable. There were three possible outcomes: leg pulled off, untangle the leg from the cable, or activate the cable cut switch. However, a fourth outcome happened — the winch cable broke. The yachtie had a sailing companion for a while.

2. Standardization of procedures and patter

Standard procedures should exist and be followed. On final approach to a winch site, the winch operator counts the aircraft down in both distance and height with relation to the survivor (in the sea) or winch area on land. The pilot will concentrate on the line or track to the survivor so that the aircraft becomes positioned directly over the survivor. However, the pilot will lose sight of the winch site with some distance still to go. For the next few minutes, the pilot will place the winchman/medic on the ground, and execute a precise pickup of the winchman and survivor all through the eyes and voice of the winch operator. The winch operator may already be winching the winchman/medic out and down while the aircraft is still transitioning to the hover, and be 'positioning' the aircraft, via the pilot, so that it remains clear of obstacles — rock faces, trees, radio masts, ships gantries and the like.

Both the pilot and winch operator have their individual tasks, but each must think and act as one to do the job efficiently and effectively. The words, phrases and units of distance used by the winch operator must mean the same thing to all crew members — they must be standard, to avoid confusion. For example, never say 'winch down' or 'winch up' because, if the first word is at all muffled, the pilot hears only the 'down' and 'up' and will move the aircraft accordingly. Similarly for acknowledgments, the winch operator will say 'yes', 'affirmative' or 'correct', rather than 'right' for obvious reasons. Confusion between crew members increases the time spent over the site and creates the possibility of further injuring people.

If there is an emergency with the winch or aircraft, standard calls will produce pre-briefed or acknowledged actions to save the situation or, at least, mitigate the damage. There is a definite need for specific procedures, patter and training.

3. Correct training

A standard training winch circuit centered on an airfield is normal fare for winching helicopter operators. The circuit was originally based on winching people from the sea, so it is fairly tight, low-level, and allows the survivor to be kept in sight throughout by at least one crew member. It allows the winch operator's tasks and patter to be practiced. Crews train using this standard circuit, and then modify it to the different situations — jungle pad, deck or other confined area — during the real rescue.

Most operators carry limitations in their operations manuals to the effect that 'during training, winchman should not be lifted higher than 15 feet above the surface'. Some would say that you must train the same way you operate, and there is validity in that argument. There are ways that crews can practice the art of hovering at 100 feet above ground level in adverse conditions while a winch operator patters the pilot through a 'simulated' winching cycle without having a live body on the wire. Although training and realism are essential, there is little point in harming people unnecessarily in training.

19

Weather

Weather hazards have an incredible record of triggering helicopter tragedies. But there are books, videos and training materials available so that pilots can be fully informed of the dangers. These resources are listed under Further Reading at the back of this book. The discussions of hazards in this chapter will be dealt with individually where possible, although in reality several different dangers may be present at the same time.

Flying into adverse weather is unwise and can be terrifying. Most of the time what you actually experience is worse than anything you envisaged. The danger is then quite evident and very real — far more real than reading about it in some book or talking about it with other pilots. Your life is on the line. Why are you flying in such atrocious conditions? Do you know what to expect next?

Case study 19.1 should give readers some idea of just how many things a pilot can experience in adverse weather conditions. Disorientation through limited visibility in cloud and heavy rain make for a navigational nightmare. Then there is the worry about airframe and mechanical integrity in severe turbulence over rugged terrain. And just when you are almost safe . . .

 Case study 19:1

Aircraft Accident Report Number 96–011

Date: March 31, 1996

Location: Gisborne, North Island, New Zealand

Aircraft: Robinson R44

Injuries: 1 fatal

The pilot, an overseas visitor to New Zealand, had hired privately owned ZK-HJD for a North Island tour. A condition of the hire was a dual check under the auspices of an Ardmore training establishment, and he arranged to meet the instructor at the helicopter's Albany base on March 26, 1996. They flew the helicopter to Ardmore where the dual check flight took place and, on completion, the pilot proceeded on tour with a traveling companion as passenger.

The arrangement was for the pilot to return the helicopter to North Shore Aerodrome on Monday April 1 as he was booked on an overseas flight either that night or the following morning. On the morning of Sunday March 31, he visited the Hawkes Bay and East Coast Aero Club at Hastings Aerodrome. There he discussed the weather and possible flight routes with an instructor. The instructor obtained for him a faxed weather briefing from Napier Tower and he suggested that the pilot not attempt to go anywhere because of the conditions; he himself had just turned down an ambulance flight to Wairoa because the weather was below IFR [instrument flight rules] minima. The pilot left the club and returned to the nearby private property where he had parked the helicopter. He took off from there just after midday and proceeded to Napier Airport where he refueled the helicopter. Fuel company records indicated that 16.78 US gallons of 100/130 AvGas were uplifted.

After refueling the pilot filed an abbreviated flight plan by radio with Napier Tower, took off at 1228 hours and vacated the control zone to the north, terminating his flight plan when clear of the zone. He did not indicate to the Napier controller what his intentions were after this point.

Between 1255 and 1305 hours a helicopter fitting the description of ZK-HJD was seen at Waikokopu, 25 nautical miles from Gisborne, on the western side of the isthmus connecting Mahia Peninsula to the mainland. The witness who reported the sighting described the helicopter as hugging the coast about 150 feet above sea level and 'pounding along'. The weather conditions were described as 'misty rain' with poor visibility out to sea.

At 1306 hours the pilot of ZK-HJD called Gisborne Tower advising that he was 20 miles south and requesting a clearance into the control zone. The tower controller cleared him into the zone Special VFR [visual flight rules] on a QNH [air pressure at mean sea level] of 995 millibars and reported the actual weather conditions at Gisborne airport as:

surface wind 110° (M), 26 knots gusting 33 knots, visibility 1500 meters [approximately 1 statute mile], moderate rain, scattered (cloud) at 500 feet, broken at 800 feet, temperature 17(°C) [63°F], forecast 2000 foot wind 140 (degrees) magnetic, 40 knots.

The tower controller had not been previously aware of the pilot's intended arrival at Gisborne.

About 1315 hours a resident at Muriwai, 5.5 nautical miles to the south of Gisborne airport, heard a helicopter fly over 'so low it shook the house', and judged the speed to be high as the noise came and went very quickly. He ran outside immediately and climbed onto his patio table to see if he could see the helicopter, but by this time it had 'disappeared into the murk' and he heard the sound fade away. He described the weather as 'blowing a gale; drizzly rain' and said that he could barely make out the shape of a nearby ridge, some 550 yards from his house. The sound indicated that the helicopter was headed directly for Muriwai Beach, half a nautical mile to the northeast.

Nothing further was heard from the helicopter and, after it failed to arrive at Gisborne, search and rescue action was initiated. Two locally based helicopters conducted a search during the afternoon of March 31 but no sign of ZK-HJD was found. Next morning, the crew of one of the search helicopters located a body and some helicopter wreckage on Muriwai Beach. The wreckage was identified as ZK-HJD and the body as that of the pilot.

Beach searches at each low tide over the next three days and nights yielded a quantity of wreckage and other items from ZK-HJD. Most of the wreckage comprised small fragments of the airframe, including some heavy items such as a portion of the lower left tubular frame. Other items included one life jacket which had not been inflated and showed no sign of having been worn, and the pilot's video camera from which serviceable videotape was recovered. However, later investigation found that nothing had been recorded since March 29.

The entire main rotor and head washed ashore on the night of April 4 and this was the last significant item of wreckage to do so.

Several attempts to locate the remaining wreckage by boat were unsuccessful.

The 49-year-old male pilot was a United States citizen who held a US airline transport pilot certificate endorsed for airplanes and helicopters. He held flight instructor certification on both categories. He was an airline pilot in current flying practise on DC-8 aircraft. He held a current Federal Aviation Administration Class 1 medical certificate and also a New Zealand private helicopter pilot license.

Helicopter type ratings included the Robinson R22, Hughes 369, Bell 206 and Bell 212. In February 1996 he had completed type training on the Robinson R44 at the Robinson factory, and had also attended three safety awareness courses at the factory since 1988.

The approximate total flying time of the pilot was established to be 27,000 hours airplane and 700 hours helicopter, including 400 hours on the Robinson R22 and, at the time of the accident, 15 hours on the R44.

Post-mortem examination of the pilot revealed that he had died of multiple injuries consistent with a severe deceleration, with fractures to the right hip, right forearm, lower left leg, ribs and neck. No evidence of a pre-existing condition which could have affected his ability to control the helicopter or of sudden incapacitation was found and toxicological tests disclosed no evidence of alcohol or other drugs.

With regard to the aircraft, ZK-HJD had accrued approximately 150 hours of flight time since new and maintenance had been in accordance with the manufacturer's specifications. There are very relevant excerpts in the approved flight manual for the R44 that need to be noted. A Civil Aviation Authority addition to the limitations section reads:

a) The following limitations (1–3) are to be observed when the pilot manipulating the controls has not taken the awareness training specified in AIC/GEN A113/95, and has logged less than 200 hours of helicopter flight time and less than 50 hours of flight time in the RHC Model R44 helicopter.
   1) Flight when surface winds exceed 25 knots, including gusts, is prohibited.
   2) Flight when surface wind gust spreads exceed 15 knots is prohibited.
   3) Continued flight in severe or extreme turbulence is prohibited.
   4) Adjust forward airspeed to between 60 knots indicated airspeed (KIAS) and 0.7 Vne^2 [velocity never exceed] but no lower than 60 KIAS upon inadvertently encountering moderate, severe or extreme turbulence.

Note: Moderate turbulence is turbulence that causes:
1) Changes in altitude or attitude;
2) Variations in indicated airspeed; and
3) Aircraft occupants to feel definite strains against seat belts.

Estimated weight and balance calculations at the time of the accident were adjudged to be within allowable limits.

In addition to altimeter, airspeed indicator and vertical speed indicator, ZK-HJD was also equipped with an attitude indicator and directional gyro. Apart from a VHF transceiver and a transponder, there were no other radio aids to navigation.

At the time of the accident, ex-tropical cyclone Beti lay to the east of East Cape, moving slowly southeastwards. A strong moist southeasterly airflow prevailed over the Gisborne region. Gale force winds and heavy rain caused damage, extensive surface flooding and road closures in the area over the weekend of March 30/31.

The beach from where the helicopter wreckage was recovered was, in southeasterly conditions, directly in the lee of the ridge delineating the southern boundary of Poverty Bay, and the seaward extremity of which is Young Nicks Head. The ridge is a 'razorback' feature, 1.5 miles in length and 1100 yards at maximum width, tapering to a point at each end. Its height exceeds 400 feet over most of its length and the highest elevation is 580 feet. The northern aspect of the ridge consists of steep eroding papa [mudstone] cliffs.

Assuming a free-stream flow of 40 knots over the ridge, the magnitude of up-draughts and down-draughts could be expected to equal at least half that of the horizontal speed, that is, 20 knots or about 2000 feet per minute. With such a steep feature, the transition between up-draughts and down-draughts was likely to be sharply demarcated. The pilot of one of the search helicopters, a Hughes 369D, described the conditions in the area on the afternoon of the accident as 'wild'.

A local fisherman reported to the police that there were [13 feet] high waves, by his estimate, breaking on Muriwai Beach around the time of the accident.

The wreckage recovered from Muriwai Beach over the few weeks following the accident comprised the complete main rotor assembly and a large number of fragments of the severely disintegrated airframe. The aftermost section of the tail boom was also recovered, complete with the horizontal and vertical stabilizers and the tail rotor and gearbox. Part of one tail rotor blade was missing.

One main rotor bore evidence of having flapped to extreme, resulting in mast bumping severe enough to cause mast failure. The blades themselves showed no

Fig 19.1 Recovered wreckage of the R44 TRANSPORT ACCIDENT INVESTIGATION COMMISSION, NEW ZEALAND

sign of having struck any part of the airframe but one blade had a significant sweepback over the outer 6.5 feet of its length together with compression wrinkles on the trailing edge consistent with a water strike. It was the root end of this blade which had struck the rotor mast and caused the mast failure. The droop stop 'tusk' at the root of the other blade had been driven with great force into the upper portion of the rotor head itself, as that blade flapped down in relation to the other, before the tusk failed. The tusks normally engaged static stops on the rotor mast, preventing mast contact by the rotor head at low or zero rpm.

Many of the airframe fragments were unidentifiable at first, but some were found to have part numbers stamped on them. These did not appear in the parts catalogue, but submission of the list of part numbers to the manufacturer resulted in the positive identification of most fragments. This confirmed that the airframe break-up had been more severe than first thought. One crumpled panel, thought initially to be from the aft fuselage, was identified as part of the second section of the tail boom.

The pilot had successfully followed the Hawke Bay coast as far as the base of the Mahia Peninsula and it was probably soon after the helicopter was sighted there that the pilot made his initial call to Gisborne Tower. Tracking north from that location by any route other than coastal was unlikely due to high inland terrain and the weather conditions.

Arriving over the shoreline at Muriwai Beach, the pilot would have encountered severe turbulence directly downwind of Young Nicks Head as well as poor visibility due to rain and sea spray. Additionally, there was likely to have been poor definition due to lack of contrast between sea and sky. His altitude is unlikely to have been more than about 400 feet in order to maintain visual contact with the surface. The pilot's probable intention was to follow the beach north to the airport.

The distribution of the wreckage suggested that the accident occurred close to the beach, probably just outside the breaker line. The extent of structural disintegration of the helicopter and the injuries to the pilot indicated that the helicopter had struck the sea surface while in forward flight. Possible reasons for striking the sea include:

➤ Flying into a down-draught exceeding the helicopter's climb capability.
➤ Loss of visual reference.
➤ Loss of control in turbulence.
➤ Malfunction, such as an engine failure.

In the turbulence in the accident area, the pilot would have found the control workload demanding, although, given the nature of the damage to the main rotor, it is unlikely that control was lost sufficiently to result in mast bumping in flight. The main rotor blades had not struck the airframe at all, whereas airframe strikes were the rule rather than the exception in cases of mast bumping and rotor separation in flight (in teetering rotor systems such as on the R44).

Despite theorizing on several scenarios, the exact sequence and mechanism of ZK-HJD's collision with the sea remain unknown.

The weather at Napier was suitable for the pilot to have attempted the flight but deteriorated as the flight progressed. The surface wind was only 19 knots from the south at Napier, but gusting to 33 at Gisborne at the time of initial radio contact. The pilot had not discussed weather conditions directly with either tower controllers prior to leaving Napier but would have been aware of the conditions likely to be encountered at and en route to Gisborne. Neither tower controller had been aware of the pilot's intention to proceed to Gisborne.

The three most noteworthy subparagraphs in the findings section of the accident report were:

- ➤ Some or all of the weather components were probable contributing factors.
- ➤ No definite cause was established as to why the helicopter collided with the sea.
- ➤ Main rotor separation prior to colliding with the sea probably did not occur, but could not be entirely ruled out.

Source: Transport Accident Investigation Commission, New Zealand

The pilot's next of kin sent a representative to New Zealand and he consulted Bernie Lewis. Bernie reported:

I read and noted the contents of the accident report and arranged for a local operator to fly us around the area in a Hughes 500. By the times of the sightings and reporting points we calculated that the ground speed of the R44 was at least 120 knots in pretty turbulent conditions with low cloud and probably not very good visibility all the way up the coast from Hastings. We flew around the area of the Mahia Peninsula, which is where the last position report was given. We could only assume that this was a correct position report. Going across the peninsula were two or three little narrow gullies. The rotor head was found on Muriwai Beach, which is just on the other side of the Mahia Peninsula. I surmise that what the pilot has done, given the low cloud and general conditions, is to have flown up one of these gullies with about a 30-knot tail-wind component and possibly 40 or 50 knots of airspeed. So he would have shot up that gully at about 70 or 80 knots or more . . . on the north side of the peninsula it is very, very precipitous. Coming out of the top of the gully and over the cliff at quite a high speed, the pilot would be looking for Gisborne in the distance and suddenly not seeing anything. With the bad visibility in low cloud and rain he would have seen very little, everything gray in front and the sea down below. Being a fixed-wing pilot, his natural reaction, in shooting out over this cliff edge, would be to shove the cyclic forward . . . a classic way of getting mast bumping. The rotor mast that was attached to the recovered rotor head was squashed in and this is typical of mast bumping failure.

While we will never know for sure if the pilot in case study 19.1 did indeed induce the mast bumping, be in no doubt whatsoever that moderate to severe turbulence alone can cause a negative gravity situation significant enough to cause mast bumping in a teetering rotor system.

WIND

It is vital that a helicopter pilot know the direction of the wind in any flight regime. So much so that the simplest maneuver can turn into a disaster if the wind is coming from an unexpected direction.

A great many of the hazards and solutions in this book have the horizontal movement of air (wind) as a primary element. Translational lift is present for many machines in the hover with a 15-knot head wind; but with a 15-knot tail wind you might need 30 knots of indicated airspeed to reach translation. An autorotation is doomed if the flare is executed with a tail wind. A power-on approach to landing will risk vortex ring state with a tail wind. A turn with low airspeed will suffer greatly if you are turning downwind. The list goes on and on.

Except for severe turbulence, it is often better to have a stronger rather than a lighter wind, for the simple reason that it is easier to pick wind direction. One of the trickiest times to be working a helicopter hard is when the wind is light and variable, say less than 8 knots. Without continuous concentration, a complete 180° wind shift can go almost unnoticed, until it's too late.

Very experienced helicopter pilots know wind direction without even consciously thinking about it. They know the signs to look for — ripples on a lake; movement of branches, leaves and blades of grass; dust or smoke drift; or the luxury of an actual windsock; to name just a few. And remember, lateral drift (having to offset the nose to keep the track) is another good indicator of wind direction. Always err on the side of caution. For example, when on a landing approach at some remote location, overshoot and reassess the wind if the airspeed indicator is reading 0–10 knots but the trees and landscape are zipping by far faster than that.

Pilots of fixed-wing aircraft mostly land at airstrips where there is a windsock or at airports where someone tells them wind speed and direction on the radio. But a huge percentage of helicopter pilots are operating miles from the nearest windsock or controlled airspace. It is little wonder that good helicopter pilots can immediately tell you wind direction, to within 10° or so, without giving it any thought whatsoever.

If you are in the early days of your training, practice working out wind directions. Read the forecast every day so that, if you are at a reasonably high altitude and the engine fails, or if there is no smoke or dust, at least you have some idea of the direction from which the forecast wind was supposed to be coming. Even when you are not flying, get to know an area by noting the wind and weather. It was forecast to be an easterly, so why is it a westerly? Is what you are experiencing a sea breeze, land breeze, anabatic or katabatic wind? Does this happen here regularly? If so, under what conditions? Such information could be useful next time you are flying in the area. Perhaps it was none of those things; just a weather system moving faster than predicted.

Keep paying attention to the wind and the weather. Discuss it with other pilots. Chatting about the weather is not just passing time and making polite conversation when you fly a helicopter. Pilots find it truly interesting, and with good reason.

There are many factors influencing wind direction — for example, pressure gradients, cumuloform clouds (discussed later in this chapter) and localized phenomena — but none of these change the fact that knowledge of the direction (and to some extent the strength) is right up there with fuel on a list of priorities for helicopter pilots.

VISIBILITY FACTORS

A great many accidents are blamed on 'bad' weather. When applied to Visual Flight Rules flying, bad weather is often simply a visibility issue. Spatial disorientation is discussed in Chapter 21 Human factors. Aside from darkness, the conditions that limit visibility are generally cloud, fog, snow, mist and rain. In most cases impending limited visibility can be readily seen and avoided during daylight hours, with the exception of a cloud burst or sudden blinding snow. If fog is closing in, if heavy rain or snow is approaching, if the cloud base is lowering and the terrain ahead is rising, land the helicopter and wait! If that is not possible — for example, if you are at sea and a distance from land or over terrain with no landing sites evident — you may have to accept instrument meteorological conditions and invoke your plan to avoid spatial disorientation.

A pilot's decision to fly into a no- or low-visibility situation and risk spatial disorientation is more realistically the cause of an accident than the weather conditions at the time. Indeed, with the amount of information given to trainee pilots and the mountain of available articles on the dangers of weather, 'accident' may be the wrong term to use.

Some individuals even resort to blaming the forecaster for getting it wrong. Experienced pilots discovered long ago that the weather actually takes no notice of meteorologists; it does just exactly as it pleases!

Logan Sharplin, now a veteran helicopter pilot with over 9000 hours of flying, recounts the times he flew into whiteouts that he could not see or predict:

I'm a so-called 'experienced chopper pilot'. In my career, I've exposed myself to a lot of different situations that have resulted in substantial hair loss, blood pressure and deep wrinkles! Among various work tasks, I've undertaken a lot of agricultural work, line stringing, poles, skiers, shooting, bucket and so on.

My first whiteout experience came on a beautiful clear-sky day in North Canterbury (South Island of New Zealand) in mid-spring (October). There is no fog or low cloud in sight as I raise the collective of the JetRanger and climb to approximately 400 to 500 feet. I glance right to evaluate the head of a long valley for fog. Directly ahead of me are crisp clear skies for around 15 miles to fog in the upper valley. I turn to look out front and, in less than two seconds, all my screens (including the floor) are 'whited out'. Almost immediately, the aircraft begins a right-hand orbit toward the ground. About 70 feet from the ground, in a relatively high rate of descent, I slide my door window open and focus on a hedge. I 'sink' in the landing and realize I'm totally 'uncoordinated' in my control of the aircraft and actually land while in a slow right-hand orbit.

The topography I had been flying in was undulating hills to 500 feet. I was heading to the summit of a near hill. The area below me was shaped like a large 'amphitheatre', forming one side of the valley where the base was situated. The hour was 5.45 a.m.

In February of the following year, I travel to a forestry block and land in a clear-felled area amidst mature pine trees. The block is rectangular shaped and the trees completely encompass the 25-acre area to a height of around 40 feet. The forest is in a very big flat basin. To the eastern end fog is laying just above and among the tree tops, at least two miles away. On the approach and landing, there is no sign of fog or screen build up. My crew loads my spray tank and I pull away from the load site. At approximately 30 feet, in less than three seconds, all my screens whiteout. There is no apparent sign of fog. This time, I land the JetRanger on a skid site using my open side window to focus. The degree of whiteout is less severe than my previous experience, although a passenger (a forest worker who was to point out the areas to be sprayed) is very distressed and appears to reach for the door handle.

I can assure you, if you were a 'rookie' or hadn't experienced this phen-omenon before, you (like me) could very easily lose control of your ship with regrettable consequences whatever the outcome! Remember, at this stage I had been flying for over 10 years on rotary wing alone, involved in fairly intensive operations.

After this occurrence, I thought, I'm going to give this locality a damn good lookover and see if I can actually 'see' something. It was early morning. The block was on an east-west access. I flew low-level to the south of the block and then slowly toward the east at varying altitudes, looking into the block. What I saw absolutely amazed me. There, in a very thin layer was sheet fog, which was invisible from above or below. You could see it only when almost level with it and with sun low on the horizon.

It became obvious that, when traveling through it at speed and with such a large surface area, it would 'lay-up' on any metal surface. Obviously, a Perspex screen at slightly higher temperature than ambient temperature would concentrate the whiteout effect.

This incident caught me by surprise. I had flown in fog-prone areas on numerous occasions and in all sorts of weather (usually very cold). I was, however, surprised to hear many of my colleagues had experienced similar microclimactic conditions and results; one in particular resulted in a right-hand spiral dive and fatality.

The common thread here is: early morning flying in skies that look clear (but with distant fog conditions) combine with topographical or artificial features that have the potential to cause a bowl or physical depression, where a thin layer of fog can be trapped and remain all but invisible until evaporated by an increasing temperature.

What can you do to prevent the heart race of a whiteout in clear skies? If you have the luxury of a heater, use it. If you don't and you have windows, open them prior to liftoff; the short-term inconvenience of a draft is a small price to pay. In the event you don't have windows and you encounter this condition, get the door open smartly and get a visual reference quickly.

TURBULENCE

Turbulence can be caused by numerous things: unstable air in certain weather conditions; significant wind disturbed by such obstructions as trees, topography and buildings; aircraft wake; and, from the passenger's perspective anyway, rough handling of the controls by the pilot. Although moderate and noticeable, turbulence is often of little consequence to the experienced helicopter pilot, but it can be a source of acute discomfort to inexperienced passengers. Severe turbulence is another matter. Modifications to flight, depending on the make and model of helicopter, must be made to ensure that the airframe and running gear handle whatever is encountered. Turbulence cannot generally be seen unless there is smoke or wisps of cloud or mist present.

Every helicopter pilot has a story about turbulence. I recall one occasion when pilots experienced it for an entire day.

One winter afternoon in the early 1980s, it was reported that a light aircraft had gone missing near a 10,000-foot mountain in the rugged country of the central North Island of New Zealand. I immediately went up (as a passenger) in a JetRanger for a brief search before we were overtaken by darkness.

At the time I held only a private helicopter license and wasn't required to do any flying myself; but I was involved in the search as I was a member of the emergency services. When I arrived at the search base (a small airfield closest to the search area) the following morning I was very aware of the wind strength even sitting inside a parked vehicle. Inside the base I learned that the wind was a steady 60 knots, gusting to 70 knots. Visibility was unimpaired; the cloud base was at least 500 feet. Six helicopters were each assigned an area to search. Because of the remote location and the hasty organization of the search, there were few observers available. I was to go in a Hughes 500 with a very experienced and very competent pilot I knew well, and I managed to locate two members of the news media who were readily talked into volunteering to sit in the back of the helicopter. I forgot to mention to them that it might be a little bumpy. The pilot of one Bell 206 actually taped an airsick bag to the front of her helmet prior to takeoff.

I thought I knew what to expect that morning, but I didn't. It is difficult to describe with any hope of capturing just how rough and uncomfortable it was for the six hours I sat in that machine that day. I think the scariest thing was feeling the longitudinal whip through the airframe as the vertical stabilizer adjacent to the tail rotor caught significant gusts. I tried not to envisage what this might be doing to the tail boom and tail rotor drive shaft.

The odd sideways glance and almost imperceptible shake of the head from the normally unflappable pilot was extremely unsettling.

The only thing the two media personnel actually observed was their breakfast returning with gusto and they chose not to go up for the second sortie. We never did find anyone to sit in the back that could actually look outside for any length of time that day.

The search went on for almost a week without success. Some weeks later, the wreckage was spotted miles from the search area by pilots on a cross-country.

Principles of turbulence

Airflow over gentle rolling terrain, except for localized and low-level disturbances around trees and small buildings, tends to follow a streamline flow in a fairly predictable way around the contours of the land. Rarely will this cause unforeseen problems for a pilot even with relatively little experience.

As wind speed increases, however, its kinetic energy also increases (relative to the earth) and at a rate that nears the square of the speed increase— that is, a doubling of the speed increases the energy by four times.

Tall buildings, oblique cliffs and bluff obstacles

Unlike the great majority of frequent low-level fixed-wing maneuvers, it is quite easy to visualize everyday helicopter operations that take place in or near these kinds of situations: the beach rescue at the base of a cliff, the landing adjacent to a multi-story building such as a hospital. In Fig 19.2, note the recirculation system on the windward side of the obstacle. Not knowing that is likely to be present and not predicting the resulting turbulence can lead to frights we can well do without. The wind has to be significant and the feature very steep, if not vertical, for turbulence to be a problem. Lazy, gentle winds tend to meander around obstacles; any unforeseen turbulence or flow is inconsequential and may spur control inputs that are largely subliminal.

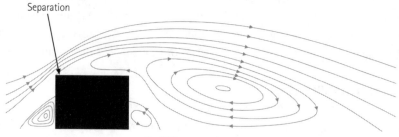

Separation

Fig 19.2 *NEW ZEALAND FLIGHT SAFETY*

Rugged terrain

Pilots will be familiar with 'separation' as applied to an airfoil. Beyond a certain angle of attack the air molecules cannot 'turn the corner' and break completely away from the upper wing (or rotor) surface contour.

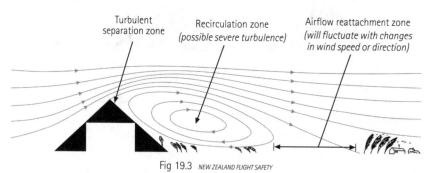

Turbulent separation zone

Recirculation zone *(possible severe turbulence)*

Airflow reattachment zone *(will fluctuate with changes in wind speed or direction)*

Fig 19.3 *NEW ZEALAND FLIGHT SAFETY*

A similar situation occurs when strong winds flow over hill country. Fig 19.3 shows the flow pattern for a stand-alone ridge. Separation occurs just beyond the crest (or point of maximum camber) and circulation is still obvious but to a lesser degree.

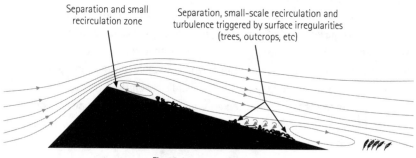

Separation and small recirculation zone

Separation, small-scale recirculation and turbulence triggered by surface irregularities (trees, outcrops, etc)

Fig 19.4 *NEW ZEALAND FLIGHT SAFETY*

In Fig 19.4, with a modification to the leeside of the hill, the airfoil analogy becomes very obvious; the general nature of the airflow is much more streamline. The ground surface roughness in the form of rock outcrops and small trees breaks up the smooth streamline flow and creates localized pockets of eddies and low-level turbulence.

Ridges and valleys

The airflow on the windward slope can vary, depending on whether the wind originates from a plain or has already been influenced by a preceding ridge. The flow on the initial ridge is virtually all upward, except for a small pocket of circulation near the base. At the crest the flow separates and an area of recirculation forms on the lee side.

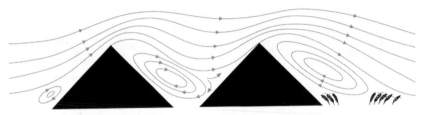

Fig 19.5 *NEW ZEALAND FLIGHT SAFETY*

But, when the free-flowing air above nears the next slope there is a change. While the bulk of the air is forced to flow upwards again, some is actually influenced to flow down the slope, particularly if it is very steep. This creates a region of descending turbulent air where a pilot might perhaps have expected all the airflow on the slope to be ascending.

Right at the bottom of the valley there may exist a region of relatively calm air, depending on the width of the valley floor and any airflow along the valley. The illustration is a simple two-dimensional flow across terrain. It pays to bear in mind that air does not just flow over objects; it also goes around them, modifying the basic patterns shown here.

There are many variables: wind more aligned to the ridge and only blowing across at an angle will tend to corkscrew in much the same way as wing-tip vortices (see 'Wake turbulence' opposite) but, in general, there are strong down-draughts and turbulence on the lee side of ridges. Where possible, operate on the windward side of any hill. While there may be some upward flow on the lee side, it will tend to be turbulent, unpredictable and often a little too close to the hillside for comfort.

The best place to operate in Fig 19.5 is the downwind side of the valley, as high up the slope as possible (cloud permitting) in order to avoid the area of potential turbulence and circulation that may be present lower down.

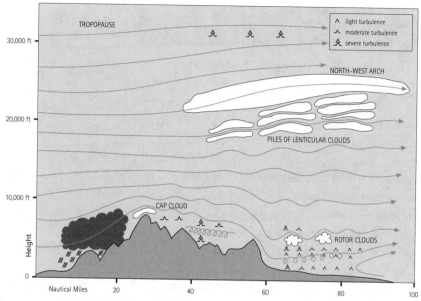

Fig 19.6 Cloud formation can indicate turbulence *NEW ZEALAND FLIGHT SAFETY*

WAKE TURBULENCE

As we saw in Chapter 2 Vortex ring state, helicopters, like fixed-wing aircraft, produce vortices from their rotor tips (see Fig 19.7).

Lift is generated by the creation of a pressure differential over the wing surface, with lower pressures being above the airfoil. This pressure difference induces a rolling motion in the airflow behind the aircraft, producing swirling air masses which trail behind the wingtips. The size and power of these spiraling, counter-rotating airflows, which have been described as horizontal tornadoes, is proportional to the weight of the aircraft (lift generated) and is usually inversely proportional to the airspeed of the generating aircraft. Trials conducted with Boeing 747 and 727 aircraft have revealed tangential velocities of up to 166 knots (see Fig 19.8).

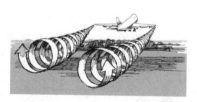

Fig 19.7 *NEW ZEALAND FLIGHT SAFETY*

Fig 19.8 *NEW ZEALAND FLIGHT SAFETY*

Case study 19:2

Accident Report NYC99FA032

Date: December 3, 1998

Location: Newark International Airport, New Jersey, United States

Aircraft: Eurocopter EC-135

Injuries: 2 minor

At about 1736 Eastern Standard Time, a helicopter on a newsgathering mission was cleared into Class B airspace to cover a story near downtown Newark, New Jersey.

The planned flight path took the helicopter west, across the final approach course for Runways 22L/R of Newark International Airport. At 1738:02, prior to the helicopter crossing the final approach course, the Newark local controller advised the pilot of traffic, four miles away, a MD-80 airliner, descending out of 2000 feet. The pilot was instructed to report when he had the traffic in sight.

At 1738:08, the pilot transmitted, 'chopper four has the MD-80; we'll maintain visual separation', which was acknowledged by the local controller.

At 1739:05, the pilot transmitted, 'chopper four got the next arrival behind the MD-80'.

As the helicopter proceeded west toward its destination, the pilot was advised of other helicopters in the same area, and reported that he had visual contact with them.

The onboard gyro-stabilized camera was pointed toward Newark airport, and transmitting to the parent television station. An airplane similar in lighting configuration to an MD-80 was seen descending into Newark. At 1739:53, as the helicopter neared the extended centerline of Runway 22L, the camera recorded a momentary vertical oscillation. There was a discussion between the cameraman and the pilot as to what had happened.

The pilot then inadvertently rolled the throttles to manual. There were subsequent variations in engine speeds and temperatures. One engine over-temped and experienced a turbine failure; the other was still capable of producing power. The pilot declared that he had a double-engine failure and autorotated into a river.

The probable causes of the accident were reported as:

the pilot's failure to maintain proper rotor rpm and his improper in-flight decision to enter autorotation due to his lack of knowledge of the power plant controls. Factors in the accident were the night conditions and the pilot's improper decision to fly through wake turbulence.

Source: National Transportation Safety Board, United States

Case study 19:3

Accident Report MIA00LA135

Date: April 16, 2000

Location: Athens, Georgia, United States

Aircraft: Hughes OH-6A

Injuries: nil

Three helicopters departed in the following order — UH-1, AH-1, and the OH-6A — en route for a static display at an airshow. When the three helicopters neared the destination airport, the pilot of the UH-1 contacted the tower about six to seven miles out and received landing instructions for the group. About one mile south of the field on a downwind for runway 09-27, the tower instructed all three helicopters to turn left [north] for an approach to the grass infield south of runway 09-27 then to cross and land in the area north of runway 09-27 and adjacent to the aircraft parking area.

Each aircraft turned to a northerly heading for the approach. The pilots reported that they maintained approximately 150–250 feet separation from the aircraft in front. The UH-1 followed by the AH-1 landed uneventfully ahead of the OH-6A. The pilot of the OH-6A slowed the helicopter to 30 knots and reduced the pitch. He stated that when he:

> increased the collective to maintain a normal approach profile . . . the increased collective had no apparent effect. He increased the collective again to arrest the increasing rate of decent. Again, the increased collective had no effect.

As he passed through about 50 feet above ground level, he noted that the engine/rotor rpm had 'dropped below 100 percent'. He announced, 'I have a loss of power', and rotated the throttle to flight idle and initiated collective pitch pull at 5 feet prior to touchdown. The pilot/passenger in the left seat stated:

> as we approached for landing, the AH-1 abruptly slowed. We closed within 75–100 feet behind . . . the AH-1, and [the pilot] raised the aircraft nose to maintain separation . . . and offset to its right [east] 75 to 100 feet . . . [the pilot] called out that the aircraft was settling and he had no power response. He increased collective pitch but the aircraft did not respond and settled vertically . . . I noticed no warning lights but, as the aircraft hit the ground, I saw the transmission oil pressure light come on. No other warning light came on.

The helicopter slid about 50 feet before it came to a rest upright, and the pilot announced over the intercom that 'he could not shut the engine down' so the pilot/passenger pulled the fuel shut-off valve, turned off the switches, and turned off the battery.

According to the Federal Aviation Administration inspector's statement:

based on the relatively low time and experience of the pilot and crew member who were in control of this aircraft at the time of the accident . . . the accident occurred as the result of the pilot's loss of control of the aircraft when he inadvertently flew it into the wake turbulence of the two larger helicopters which landed in front of him . . . the pilot was unable to maneuver the aircraft out of trouble and, consequently, the aircraft settled under power to a very hard landing.

The engine was test run and no discrepancies were found. No discrepancies were found with the airframe or controls.

The reported wind conditions at the airport at 1053 were from 270° at 7 knots. This wind condition had set up a crosswind condition from west to east or from the landing helicopter's left to right. The pilot had a total flight time in all aircraft of 5100 plus hours, 3253 hours in helicopters, and a total of 13.9 hours in this make and model helicopter.

Source: **National Transportation Safety Board, United States**

Principles of wake turbulence

Behind a fixed-wing aircraft of any size, the vortices are generated only when the wing is generating lift. On takeoff, the vortices begin when the aircraft rotates to become airborne.

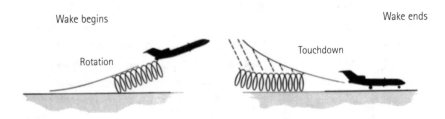

Fig 19.9 *NEW ZEALAND FLIGHT SAFETY*

Throughout all aspects of flight the aircraft will produce vortices until such time as it lands. On approach there should be no wingtip vortices forward of the touchdown point. Studies indicate that the greatest vortices will be generated when the aircraft has a 'clean wing' (no flap used) and is flying at slower speeds, like in a holding pattern.

The life of wingtip vortices, particularly near the ground, is affected significantly by local wind velocity and air temperature. Those vortices generated near the ground typically last between one and two minutes, while those generated at higher altitudes can last as long as five minutes, leaving a trail of turbulent air several miles long.

The movement of these areas of violently disturbed air is predictable in most instances, making it easy for a pilot to visualize and subsequently avoid the vortex core. The movements differ depending on whether the turbulence was generated close to the ground (during landing, takeoff or go-around) or at altitude.

Large aircraft

touch and go / low missed approach

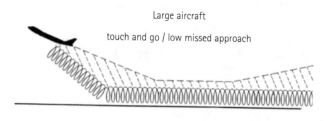

Fig 19.10 *NEW ZEALAND FLIGHT SAFETY*

On takeoff, in nil wind conditions the vortices settle to the ground and then move outwards at a speed of about five knots (see Fig 19.11). A light crosswind will arrest or impede the outward movement of the upwind vortex so that this will tend to linger almost directly behind the path of the aircraft while the downwind vortex will move outwards more quickly, assisted by the wind (see Fig 19.12).

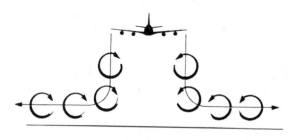

Fig 19.11 *NEW ZEALAND FLIGHT SAFETY*

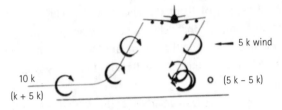

Fig 19.12 *NEW ZEALAND FLIGHT SAFETY*

Any component of tail wind may well drift the turbulence forward of the location where other pilots will expect to encounter it; that is, forward to the rotation or touchdown points.

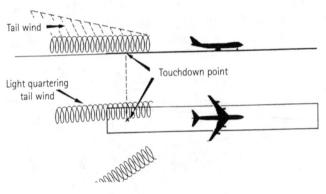

Fig 19.13 *NEW ZEALAND FLIGHT SAFETY*

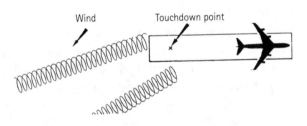

Fig 19.14 *NEW ZEALAND FLIGHT SAFETY*

Stronger winds will tend to cause vortices to break up well before their natural rate of decay. It would be a mistake to try to assume at what stage the wind will dissipate the vortex, however.

Vortices generated at altitude sink at a rate of 400–500 feet per minute initially, then gradually stabilize 900–1000 feet below the flight path of the generating aircraft. Eventually the vortices break up, the process being hastened by any atmospheric turbulence.

An aircraft entering a vortex area from the side is most affected in pitch — either nose up or nose down in successive movements. An aircraft flying parallel to the axis of the vortices, in the 'gap' between them, encounters a downdraft, but as it enters one of the vortex cores a severe rolling movement results.

Armed with this information, pilots should be able to visualize the location of the vortex trail behind large aircraft and adjust their flight path to avoid this area. Where practicable, air traffic controllers will advise pilots of the possibility of wake turbulence hazards by issuing an advisory, 'Caution — wake turbulence'. But they are neither obliged to issue the warning nor are they responsible for the accuracy of any warning.

So the cardinal rule is to keep clear of the area behind and below a generating aircraft. All precautionary techniques are based on that principle. Be particularly vigilant in calm conditions where dissipation can be expected to be slow.

CUMULONIMBUS CLOUD

Cumulonimbus cloud, which pilots usually refer to by the abbreviation Cb, is a clear example of the power of the forces of nature. It pays to have a very good understanding of these clouds as among the hazards they pose to aviation are: lightning (discussed later in this chapter), severe turbulence, microbursts and downdraughts. Knowledge of the formation, behavior and dissipation of Cb clouds is knowledge well worth having.

Case study 19:4

Accident Report FTW02LA200

Date: June 29, 2002

Location: Gulf of Mexico

Aircraft: Bell 206L3

Injuries: nil

At 1000 Central daylight time, a Bell 206L3 helicopter was substantially damaged when the main rotor blades contacted and severed the tail boom while attempting to shutdown when on an offshore platform known as Brazos A-19 in the Gulf of Mexico. The instrument-rated commercial pilot, sole occupant of the helicopter, was not injured. The helicopter was based at Galveston, Texas.

Visual meteorological conditions prevailed for the positioning flight for which a company Visual Flight Rules flight plan was filed. The positioning flight originated about five minutes prior to the accident from another nearby offshore platform known as Brazos A-19 JD.

The pilot reported that he was well aware of the convective activity approaching the offshore platform. The pilot reported that after landing at the helipad for the Brazos A-19 platform, which is located 200 feet off the surface of the ocean, he aligned the helicopter into the prevailing head wind and initiated the engine shut-down procedures. The pilot stated that when he applied the rotor brake, the main rotor blades impacted and severed the tail boom.

The winds at the platform at the time of the landing were reported as 'steady' from 135° at 20 knots. The pilot reported that the winds gusted in excess of 70 knots when the helicopter encountered the microburst.

The operator reported that the effect of the microburst spun the helicopter about 15 feet toward the center of the platform after the tail boom was severed.

Source: **National Transportation Safety Board, United States**

Principles of cumulonimbus clouds

Not all cumulonimbus clouds are thunderstorms; some do not produce lightning. There are three main stages for Cbs: growth, maturity and decay.

To begin growth there must be instability through a deep layer of the atmosphere, moist air, and a trigger action to make the air rise. This air rise might be (but is not limited to) orographic lifting (air blown up the side of a geographical feature) or convective heating from the earth's surface. Large bush fires can provide enough lift to trigger a Cb to form.

During the early part of the growing stage, all the draughts are upward (see Fig 19.15) and massive condensation occurs. These updraughts can be up to 100 knots in (vertical) wind strength. Entrainment then begins to take place, whereby air from outside the cloud is rolled into the cloud. The cauliflower edge that is so distinctive in cumuliform cloud is partly formed by air punching out from the growing cloud tower and partly from the outside air being rolled into it (Fig 19.15).

This new air is below 100 percent relative humidity so some of the cloud droplets or rain droplets begin to evaporate into it. Evaporating the liquid water uses up energy and the air cools, releasing latent heat and therefore adding to instability. But this air then becomes more dense (cooler) than the bulk of the air in the cloud so it begins to sink. Of course it sinks into higher pressure (closer to earth) and so

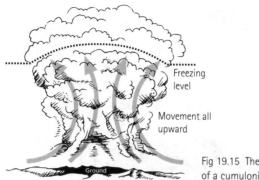

Freezing level

Movement all upward

Fig 19.15 The growing stage of a cumulonimbus cloud

its temperature rises, causing it to remain below 100 percent humidity. In this way, more moisture evaporates into the cloud and it cools further. From this point the action is like a runaway train, and strong downdraughts exist immediately adjacent to the strong updraughts. Turbulence within the cloud is vicious.

The cumulonimbus cloud is now approaching maturity; rain is freezing to hail above the freezing level and an anvil-shaped top has formed on the cloud. This top, called cirrostratus cloud, is mostly made up of ice (see Fig 19.16). The point of the anvil will indicate the direction of travel of the Cb.

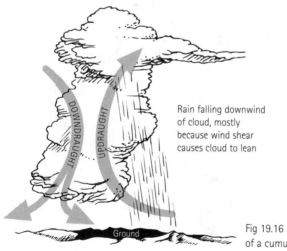

DOWNDRAUGHT

UPDRAUGHT

Rain falling downwind of cloud, mostly because wind shear causes cloud to lean

Fig 19.16 The mature stage of a cumulonimbus cloud

The amount of water and hail carried up to an enormous height — Cb tops regularly reach above 40,000 feet — is now so great that it comes rushing down. Largely due to wind-shear, the cloud leans in its direction of travel, and it is from here that most of the precipitation falls.

At the bottom of the cloud, however, the runaway downdrafts are creating their own hazard (aside from the turbulence). As the cold air falls out of the Cb, it hits the earth's surface and spreads out rapidly in all directions, forming a gust front that can travel a long way from the main cloud (see Fig 19.17). This is called a microburst, and it can rapidly change a 10-knot head wind to a 20-knot tail wind. If you are on approach to land, particularly in a confined area, this can lead to disaster.

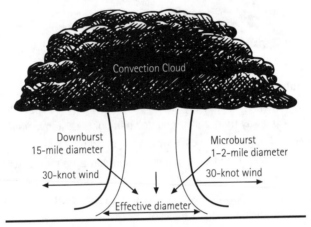

Fig 19.17 Downbursts and microbursts VORTEX

As the Cb begins to decay the airflow is all downward as the anvil-head spreads out and begins to lower. Precipitation pours from beneath the entire spread of cloud (see Fig 19.18).

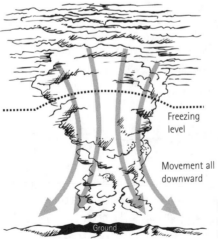

Fig 19.18 The decaying stage of a cumulonimbus cloud
NEW ZEALAND FLIGHT SAFETY

Every area of the world has its own set of circumstances where Cbs can be expected to form. Duration and strength vary but these clouds can grow very quickly, from non-existent, to be sufficient to return a very strong radar echo in as little as 15 minutes. They cannot always be visually detected as they often grow embedded in a much larger cloud sheet.

While the best way to deal with a Cb may be to avoid it completely, this is not always going to be possible. Quite often these storm clouds will form in a line, perhaps right along a cold front as it approaches (see Figs 19.19 –19.21 on pages 270–71).

If you must pass through a Cb, do so at right angles so that you are in the cloud for the least time possible. If you suspect that you are approaching clouds of this nature, secure passengers and freight, and mentally prepare yourself for a possible pounding. If the turbulence is threatening to become too much, particularly if you are in a very small machine, remember that these clouds move and individual cells don't last long when compared to other bad weather conditions. So give strong thought to how urgent your trip is: would a precautionary landing and a wait really be that much of an inconvenience? An accident is usually a huge nuisance to schedules, often for a very long time!

At night, turn cockpit lights to bright in case lightning is encountered.

What can we expect to encounter if we fly into a microburst or severe downdraft? On entering the down-flowing air the helicopter will descend with startling suddenness and speed. You will most likely be thrown upward against the seat restraints (of course, we keep them snug at all times). Unless your helicopter can out-climb the sinking air, or you are lucky enough to fly quickly out of the other side, you can expect to contact the ground with dire consequences.

The most severe conditions are produced by intense thunderstorm activity. Dangerous conditions, however, can be produced by cumulus and towering cumulus clouds. Severe downdrafts can be encountered when flying through seemingly innocent rain showers falling from cumulo-type clouds. Downdrafts can also be found in the phenomenon known as virga. Virga is most commonly found in arid regions when rain, originating in high-based clouds, evaporates before reaching the ground. You can expect such occurrences when surface temperatures exceed 77°F (25°C), surface winds are light, and the temperature dew point spread is greater than 59°F (15°C).

What can we do to minimize the effects of wind-shears? The most important rules govern the use of power and the management of airspeed.

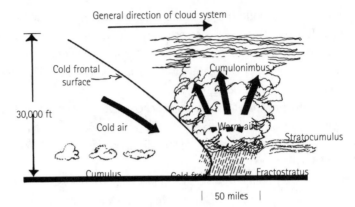

Fig 19.19 Vertical section through a cold front showing cumulonimbus clouds just ahead of the front *NEW ZEALAND FLIGHT SAFETY*

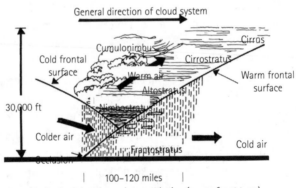

Vertical section through an occlusion (warm front type)

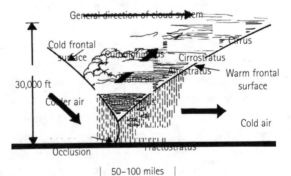

Vertical section through an occlusion (cold front type)

Fig 19.20 Cumulonimbus clouds can be expected in occluded fronts, both warm and cold *NEW ZEALAND FLIGHT SAFETY*

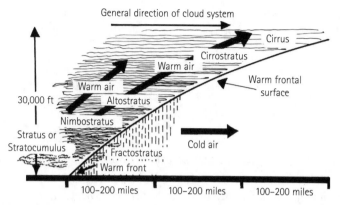

Fig 19.21 Vertical section through a warm front showing that cumulo-form cloud is minimal and cumulonimbus cloud absent *NEW ZEALAND FLIGHT SAFETY*

Fly well below Vne (Velocity Never Exceed), allowing a minimum 20 to 30 knot margin to compensate for any rapid changes in indicated airspeed and to help prevent structural damage due to turbulence. This reduced speed will also make it easier to maintain control of the helicopter. If you encounter a downdraft, apply maximum power and configure the helicopter for its best rate-of-climb airspeed. Do not allow your speed to decrease even if maximum power does not stop your descent. If impact with the ground is inevitable, maintaining your best rate-of-climb speed will provide you with the capability of initiating an effective flare to lessen ground impact.

In summary, a helicopter pilot has two alternatives when dealing with severe turbulence and wind-shear — avoidance and use of excess power. Since most helicopters do not have excess power to counteract high sink rates brought on by downdrafts, the only real alternative is 'avoidance'.

SOURCE: 'HELICOPTERS AND TURBULENCE', JOE MASHMAN, *VORTEX*, 3/94

LIGHTNING

We have all seen lightning, but have you ever tried to define it?

Lightning, basically, is a very long electrical spark which extends between one center of electrical charge in a cloud and another center of opposite polarity charge on the ground, in another cloud, or even in the same cloud.

Whilst most types of cloud are electrified, it is the cumulonimbus which acts as an immense generator and capacitor, and it's the only cloud which builds up an electrical force powerful enough to produce lightning.

SOURCE: *NEW ZEALAND FLIGHT SAFETY*, OCT/NOV 1983

The same article goes on to discuss the theories with regard to lightning and aircraft prevalent at the time:

Serious accidents caused by lightning strikes are fortunately rare. A lightning strike can, however, puncture the skin of an aircraft, damage communication and electronic equipment, or momentarily blind or disorientate a pilot, reducing his ability to fly on instruments or navigate visually. Nearby lightning can also induce permanent errors in the magnetic compass and disrupt radio communications on low and medium frequencies.

That statement is still reasonably accurate for most fixed-wing aircraft, and lightning is not a subject that invokes any great amount of trepidation. The great majority of pilots believe this to be true for all helicopters as well. Helicopter pilots with high flying hours have described how they can find a corridor between two reasonably closely located cumulonimbus clouds at sea. But there is bad news out there for today's helicopter pilot, as the following case studies and text show.

Case study 19:5

Report 2/97

Date: January 19, 1995

Location: North Sea

Aircraft: Aerospatiale AS332L Super Puma (Tiger)

Injuries: nil

The helicopter was conducting a charter flight, ferrying 16 maintenance engineers from Aberdeen to the Brae oilfield. Having just passed a position 120 nautical miles on the 062° radial from the Aberdeen VHF omnirange (VOR) radio beacon, and whilst beginning its descent from 3000 feet above mean sea level, the helicopter was struck by lightning. This resulted in severe vibration. The handling pilot was unsure as to whether the apparent directional stability of the helicopter was being maintained by the tail rotor or by the 'weathercock' effect of the airspeed, so he gently deflected the yaw pedals to see if there was a response. He had just commented to the commander that everything seemed to be in order when there was a 'crack' and the helicopter gave a violent lurch to the left, rolled right and pitched-down steeply.

Realizing that a ditching was now imminent, the first officer transmitted another mayday informing Brae of his decision to ditch and carried out the 'Tail Rotor Drive Failure' checks, which included shutting down the engines in order to contain the yaw, and then arming and inflating the floats. The ditching (in heavy seas) was executed successfully and the helicopter remained upright, enabling the passengers and crew to board a heliraft, from which they were subsequently rescued. There were no injuries sustained and the passengers and crew were later returned to Aberdeen by helicopter and ship.

Despite 20- to 23-foot high waves and a 30-knot southerly wind, the helicopter remained afloat for some three hours and thirty minutes before it was brought alongside a safety vessel. However, whilst secured to this vessel the helicopter's flotation bags punctured and it sank some two hours later.

The investigation identified the following causal factors:

1. One of the carbon composite tail rotor blades suffered a lightning strike which exceeded its lightning protection provisions, causing significant damage and mass loss.

Fig 19.22 Recovered tail rotor, gearbox and pitch servo assembly

AIR ACCIDENTS INVESTIGATION BRANCH, UNITED KINGDOM

2. The dynamic response of the gearbox/pylon boom assembly to the tail rotor system imbalance induced rapid cyclic overstressing of the gearbox attachments which was accelerated by the early failure of the upper mounting bolt locking wire, allowing consequent loosening and fatigue failure of this bolt.

Fig 19.23 (right)
Outboard side of 'White' tail rotor blade root with leading edge erosion shield and brass strip missing, and failed braided bonding strip
AIR ACCIDENTS INVESTIGATION BRANCH, UNITED KINGDOM

Fig 19.24 (below)
Sample tail rotor blade with leading edge anti-erosion shield, plastic cover overlying brass strip and braided bonding strap to root bolt
AIR ACCIDENTS INVESTIGATION BRANCH, UNITED KINGDOM

3. Complete loss of the yaw control system and a momentary pitch-down as a result of detachment of the tail rotor, gearbox and pitch servo assembly.
4. The lightning strike protection provisions on this design of carbon composite tail rotor blade were inadequate due to it having been developed from an earlier fiberglass blade which had been certificated to lightning test criteria which have since become obsolete.

Source: Air Accidents Investigation Branch, United Kingdom

Case study 19:6

Special Bulletin EW/C2002/07/04

Date: July 16, 2002

Location: 28 miles northeast of Cromer, Norfolk, United Kingdom

Aircraft: Sikorsky S76A (modified)

Injuries: 10 fatal, 1 missing

The aircraft had been scheduled to complete five multi-sector flights from Norwich on the day of the accident. The first four flights were completed without incident and the aircraft departed Norwich Airport at 1731 hours for the final scheduled flight, consisting of a series of sectors between installations in the Sole Pitt and Leman gas fields of the southern North Sea.

The first four sectors again went without incident and the aircraft departed on its penultimate planned sector between the gas production platform Clipper and the drilling rig Global Santa Fe Monarch. The purpose of this sector was to transfer one passenger between the two installations before returning the remaining eight passengers to Norwich.

The departure from the Clipper was described as normal by the helideck crew and the aircraft climbed to 1500 feet for the planned ten-minute sector to the Global Santa Fe Monarch. During the cruise, the crew spoke to Anglia Radar before establishing radio contact with the Monarch's radio operator. There was some confusion at first as the Monarch had not been expecting any further flights that evening. However, the Monarch's helideck crew was quickly assembled and the aircraft commenced its approach.

With the aircraft at a height of about 320 feet on a southeasterly heading, workers on the drilling rig heard a loud bang. No witnesses were watching the aircraft at the time but some subsequently saw the aircraft dive steeply into the sea. One witness also described seeing the main rotor head with the blades attached falling into the sea after the remainder of the aircraft had impacted the surface.

The alarm was raised by the radio operator on the Monarch and the first response was from the rig standby vessel, Putford Achilles. This vessel was holding station approximately 1.5 miles from the accident location and immediately launched its two fast rescue craft to the observed area of wreckage. They arrived at the scene seven minutes after initial notification and recovered four bodies and some light debris. Shortly afterwards they were informed that there were 11 persons on board the aircraft and the search was continued, resulting in the recovery of another body. Great Yarmouth Coastguard launched rescue helicopters and other vessels arrived on the scene that night but no survivors or further bodies were recovered from the surface search.

The floating wreckage indicated that the break-up of the helicopter was extensive and that the accident was not survivable. An underwater search for the six missing persons was commenced on July 17. Five more bodies were recovered on July 19. The underwater search for the one remaining body continued unabated but was eventually suspended on July 23 after the likely area had been thoroughly searched. A surface vessel search was maintained for two more days and an aerial search was continued until July 30 without success.

Wreckage recovery operations were hampered by strong currents which constrained diving operations to periods of approximately three hours at each tide. More than 97 per cent of the structure of the helicopter was recovered by divers working from the Diving Support Vessel Mayo.

The audio recordings revealed that the crew was not aware of any significant abnormality until the flight from the Clipper platform to the Monarch platform. About 4.5 minutes into this sector the crew discussed an increase in vibration. The non-handling pilot carried out a 'rotor track and balance' procedure. The increase in vibration did not cause the crew any immediate concern and the rotor track and balance procedure was carried out to enable the IHUMS (Integrated Health and Usage Monitoring System) to log main rotor track and balance data for subsequent analysis by the IHUMS ground station. Preliminary frequency spectrum analysis of the recorded audio information indicated an increase in the amplitude of frequencies associated with main rotor vibration towards the end of the recording. A more exhaustive analysis is in progress. The audio recording ends abruptly with three unusual, probably structure borne, sounds.

The recorded flight data did not show any anomalies during the five-hour recording. The recording ends with data time histories showing the aircraft was in level flight at about 320 feet above sea level, traveling at a speed of 100 knots on a heading of 150° (M). The recorded data became corrupted about two seconds before the end of the recording. Work is continuing to decode the corrupted data and to analyze the recorded rotor track information to determine if any trends are evident from the time histories.

Among the wreckage were two items of major significance. Firstly, three of the main rotor blades exhibited only superficial damage whereas the fourth was fractured at a position approximately 76.75 inches from the blade root. The missing outer blade section was not recovered from the main debris field. The second significant clue was the condition of the main rotor gearbox. The casing had fractured and there was visible evidence that the gearbox, together with the rotor head, had broken away from the fuselage mountings in-flight.

The fractured blade was taken to QinetiQ's materials laboratories where the fracture surface was cleaned and prepared for microscopic analysis. Clear evidence of fatigue was present, indicating that approximately half the circumference of the blade's titanium spar had failed in fatigue before the outer portion separated. Thus it was clear that the blade fracture had initiated the catastrophic event; the gearbox had separated from its fuselage mountings due to the severe imbalance created by the loss of the separated blade section.

The metallurgical examination revealed two areas that contributed to the blade separation. The fatigue initiation point of the blade's titanium spar was on the upper surface in the area of the inboard edge of the scarf joint between the two-piece titanium leading edge erosion strip. Microscopic examination of the initiation point indicated that it had suffered intense thermal damage. The area had the appearance of and discoloration similar to an electrical 'spot weld'. There was no evidence of thermal damage to the surrounding composite materials, resins or paint.

During the metallurgical examination, evidence was found of an anomaly in the scarf joint between the two titanium leading edge erosion strips. The tip of the tang on the inboard end of the outboard erosion strip was bent and folded under the outboard end of the inboard erosion strip. This resulted in a doubling of the thickness of the erosion strip material in that area which, in turn, resulted in virtual contact between the erosion strip and the blade's titanium spar especially in the areas at either end of the tang's fold line. This anomaly had occurred during the blade's manufacturing process some 21 years prior to the accident.

The initiation point of the fatigue failure of the blade's spar was at the rear point of the tang's fold line.

This rotor blade was manufactured in March 1981. In 1999 when fitted to Sikorsky S76A G-BHBF it was damaged by a lightning strike. At that time the blade had accumulated 8261 hours usage. The blade was returned to the manufacturer for assessment where, following inspection, it was repaired and returned to service. Neither the thermal damage to the spar nor the manufacturing anomaly were detected during this inspection. At the time of the accident, the blade had accumulated 9661 hours usage. The airworthiness limitation life of the blade is 28,000 hours.

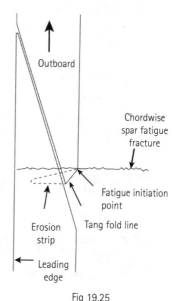

Fig 19.25

AIR ACCIDENTS INVESTIGATION BRANCH,
UNITED KINGDOM

The Air Accidents Investigation Branch and the helicopter's manufacturer are of the opinion that the electrical energy imparted by the lightning strike in 1999 exploited the anomaly that was built into the blade at manufacture and damaged the spar.

A full investigation and report is underway.

Source: Air Accidents Investigation Branch, United Kingdom

Case study 19:7

Accident Report IAD98RA090

Date: July 18, 1998

Location: Nueva Colonia, Columbia

Aircraft: Bell UH-1H

Injuries: 7 fatal

About 1840 local (equivalent of Central daylight time), a Bell UH-1H helicopter, operated by the Colombian National Police (PNC), as PNC-159, crashed in a swampy/wooded area about one mile north of the village of Nueva Colonia. The two pilots, two flight crewmembers, and three passengers were fatally injured. The passenger-transport flight from Turbo to Carepa, Colombia, had stopped at a soccer field at the Nueva Colonia. It remained with rotors turning for approximately 20 minutes, awaiting, according to witnesses, the passage of a weather system that included low visibility and thunder showers. The helicopter was observed to take off from the field, about the time of official sunset, in low visibility and low clouds. It then circled low-level to the north of the field. Ground witnesses observed a lightning flash, and shortly thereafter heard the sound of impact.

Source: National Transportation Safety Board, United States

Case study 19:8

Report 11/99

Date: December 12, 1997

Location: 10 nautical miles east-southeast of Sovereign Explorer Oil Rig, North Sea

Aircraft: Eurocopter AS 332L

Injuries: nil

The captain reported that he landed on the Sovereign Explorer in clear weather at 1226 hours. Whilst the aircraft was being refueled for the next sector, a heavy shower passed over the oil rig and moved away, but there were no indications of lightning activity. The captain further reported that the aircraft took off at 1245 hrs and climbed to 3000 feet on a southeasterly track. There were none of the usual indications of lightning activity, such as fluctuations of ADF [automatic direction finder] needles or radio interference. It was possible to see the surface intermittently, although there was a fair amount of low scud around.

At 1253 hours and without warning, there was a bright flash and a loud bang. The aircraft immediately developed a one-per-rev vibration and a loud one-per-rev noise could be heard. The handling pilot reduced power at once and the crew assessed the situation. It was obvious to them that the aircraft had suffered a lightning strike and that the main rotor blades were damaged.

In his report the commander observed:

> Control of the aircraft was still good, however, and we considered that, provided nothing else separated, the aircraft could remain airborne.

The handling pilot then executed a 180-degree turn back towards the Sovereign Explorer and the non-handling pilot made a PAN call to the Sovereign Explorer's radio operator, who was maintaining a flight watch service. He advised the passengers of the situation and instructed them to secure their survival suits in preparation for a landing.

An incident-free landing was carried out at 1301 hours. The one-per-rev vibration and corresponding noise had remained present from the incident until the landing.

Conclusions of the bulletin read:

> The aircraft suffered a lightning strike of a severity well above the level to which the aircraft was originally certificated and in the region of the more demanding present-day certification requirements. The rotor blades suffered substantial damage although this was insufficient to reduce their strength to an immediately dangerous degree. However, the blade damage reduced the stiffness to an extent that their susceptibility to blade sail on shutdown may have been increased.
>
> Fortuitously, the major damage/mass loss was symmetrically disposed between two blades situated on opposite sides of the rotor head. Had this not been the case, the vibration level would probably have inflicted rapid damage in the rotor head and structure. The crew would have been faced with a difficult decision as to whether they should remain airborne or carry

out a controlled ditching. The situation was greatly helped by their close proximity to the helideck from which they had just departed.

The damage appears to have been inflicted by an isolated strike which was not preceded by any other observed or detected activity in the immediate area.

The forecast available at the time included isolated thunder storms and embedded cumulonimbus clouds, though no lightning was observed nor was Cb activity detected close to the aircraft in the period immediately preceding the strike.

Source: Air Accidents Investigation Branch, United Kingdom

Principles of lightning

As there is no radar to 'paint' an area of potential lightning strikes it is difficult to know when your helicopter might become a target. Perhaps the best forewarning is to have an intelligent understanding of the general theory of lightning discharges from clouds. Although the following article is fixed-wing initiated, it is very relevant to helicopter operations and will give pilots the necessary information.

Since the atmosphere has a nett positive polarity, this implies a corresponding nett negative polarity on the ground for a cloud/ground strike to occur. Here, a basis for confusion exists, for in fact the electron flow is from negatively charged areas to positively charged areas which . . . may be on the ground, in other clouds or in the same cloud as the negatively charged area.

Cloud-to-ground lightning starts with a step leader, a column of negative electrons descending from the cloud to meet the welcoming discharge of positive electricity from the ground

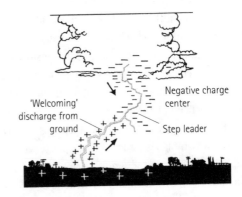

Fig 19.26 Cloud-to-ground lightning *NEW ZEALAND FLIGHT SAFETY*

The air itself serves as a natural insulation between centers of opposite polarity to the point where the negatively charged regions of the cumulonimbus cloud build up sufficiently to overcome that insulation, and electrons dart along a narrow channel of air to the positive charge center, creating lightning.

For obvious reasons more data exists on cloud-to-ground lightning strikes than any other, but all types of lightning are characteristically similar — a faintly luminous streamer, known as a 'stepper leader', passes from cloud to ground. It is 'stepped' because it advances in a series of 150 foot steps, with pauses between each step of 100 millionths of a second duration, each new step being brighter than the already established track. Its pattern is a stop-and-go, zigzag movement at a speed which varies from 130 miles per second to one-sixth the speed of light, forking and branching with every favorable accidental variation of the air.

With each step the cloud charge is effectively lowered, increasing the strength of the electrical field between the leader and the ground. This greatly intensified field causes streamers of positive ions to spring from trees or other ground projections reaching out to 'welcome' the step leader.

Lightning variations showing electron flow from negatively charged areas to positively charged centers on the ground, in other clouds, or in the same cloud

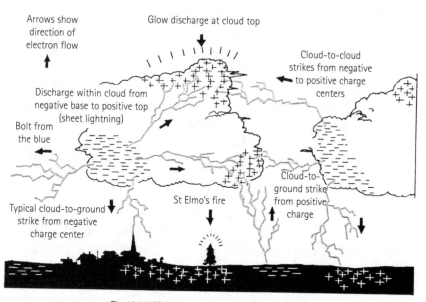

Fig 19.27 Lightning variations NEW ZEALAND FLIGHT SAFETY

Following contact between the two, a brilliant flashback known as the 'return strike' occurs. The process is repeated several times between a leader and a return strike (usually three, but as many as 42 repeat strikes have been observed on one contact) with increasing brilliance each time.

Once it reaches the ground the return strike dies out, but the charge remaining in the cloud may drain off through the conducting channel to ground, forming continuing currents. Any additional charge centered in the cloud will discharge to ground through the same channel causing a phenomenon known as a 're-strike', with the channel becoming brighter with each discharge, causing flickering.

A lightning flash just described may last up to a full second, with currents reaching as high as 200,000 amperes on occasions. The visible lightning stroke is a channel of incandescent air not more than an inch or so in diameter, while the accompanying thunder is an explosive report caused by the sudden expansion of air heated by the lightning flash.

Although height and temperature are naturally correlated, studies of lightning frequency as a function of temperature and height show a strong tendency for strikes to occur when the air temperature is within the 14° to 50°F (±10°C) range, with the heaviest concentration around freezing level. This relates to the fact that the negative charge center is also found near this temperature and altitude.

Why does lightning strike aircraft? This question has not yet been answered to everyone's satisfaction, there being several different theories. The most prevalent opinion is that aircraft are struck only when they happen to pass very close to the natural stroke path of lightning (within about a wingspan). Except for possibly large, wide-bodied aircraft, it is believed that aircraft flying in electrically saturated areas don't actually 'trigger' a discharge; they simply become a conductor through which a high charge passes. Therefore a strike probably comprises two separate events: an attachment of a high energy charge at one point on the airframe, and its exit at another.

As an aircraft flies through precipitation or an area where there is a strong electrical field (near a charge center or advancing leader) it acquires its own electrical charge. Sharp points of the aircraft structure such as the nose, wing tips, stabilizer tips and blade type antennae, act as discharge points to trickle away this electrical charge under normal circumstances. If it becomes sufficiently intense, a streamer may form and move outward from the aircraft toward a leader or charge center, causing it to advance directly toward the aircraft until streamer and leader meet.

A charge may now flow onto the aircraft but since there isn't room for very much to remain on the structure, some of it 'overflows' in the form of an intense streamer from some other aircraft extremity to progress onward. The aircraft thus becomes a link in the conducting channel from cloud to ground (or from cloud to cloud). Once within its clutches the aircraft is trapped, and when the return stroke passes through the channel those on board will experience the bright flash and loud noise associated with lightning strike. The aircraft will 'let go' only when the channel dies out naturally.

During a lightning strike a current of up to 200,000 amperes may flow through the aircraft between entry and exit points, followed at regular intervals by smaller continuing currents flowing along the same path. In most cases the only visible evidence of a strike may be a series of small evenly spaced pit marks on the airframe where the lightning flash attached.

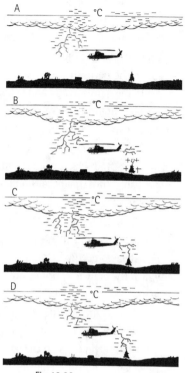

Fig 19.28 *NEW ZEALAND FLIGHT SAFETY*

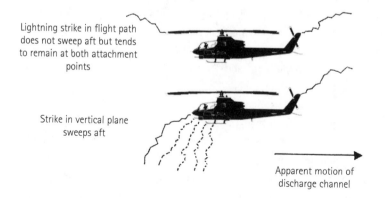

Lightning strike in flight path does not sweep aft but tends to remain at both attachment points

Strike in vertical plane sweeps aft

Apparent motion of discharge channel

Fig 19.29 Lightning discharge attachment to aircraft *NEW ZEALAND FLIGHT SAFETY*

At the trailing edges or other places where the flash can hang on longer, a hole may be melted and the fuselage skin may be crimped or deformed by the intense magnetic fields which surround lightning stroke currents. Thus, aircraft skin thickness is increased at critical areas where arc penetration cannot be permitted, such as fuel tanks.

If the lightning channel becomes drawn alongside the fuselage and a high amplitude strike occurs, the consequent overpressure which is produced can cause compressor stall or flame-out of fuselage-mounted turbine engines when the flash is swept in front of the engine inlet.

Pitot and electrical systems and flight or engine instruments also are liable to damage, although continued research and development has lessened the danger of major malfunction of aircraft systems.

The most hazardous effect on the crew from lightning strike is temporary blindness as a result of the bright flash caused. This blindness usually (but not always) occurs at night and may persist for up to 30 seconds, during which the crew may not be able to read instruments unless prior precautions are taken.

In any aircraft the electrical potential of everything inside the skin remains the same during lightning strike due to careful bonding and construction techniques, so the danger of electrical shock to the crew is minimal. However, in non-metallic aircraft such as a glider, the control cables may be the only electrical conductors and thus the pilot is placed in a direct path between lightning attachment points. It could have fatal consequences.

Another electrical phenomenon that should be mentioned is precipitation static. This is generally encountered in the vicinity of electrically charged clouds and often precedes a lightning strike. When an aircraft is flying through rain, sleet, hail or snow, the impact of these particles on the airframe will cause a charge to separate from the particle and join the aircraft, leaving a preponderance of positive or negative charge (depending on the form of precipitation) and thereby elevating the aircraft's potential with respect to its surroundings.

Since the aircraft has room only for a small amount of this charge, some of it will begin to leak off in the form of ionization at sharp extremities. This continues as long as the aircraft is flying in precipitation static charging conditions, and is visible as a bluish corona, known as 'St Elmo's fire'. Sometimes it may intensify and appear as a fluctuating column of fire snaking out from the nose of the aircraft. This ionization radiates broadband electromagnetic radiation received as interference or 'static' in aircraft communications receivers and may render the equipment temporarily unusable.

Static dischargers or sacrificial probes installed at wing tips and trailing edges reduce this interference by making it easier for the charge to leave the aircraft. They are not always 100 percent efficient, however, especially in heavy precipitation. Since the conditions that produce precipitation static may also produce lightning, a strike should be considered possible when it builds to a high level. Except for providing an easy replaceable attachment point, a static discharge affords little protection against a lightning strike.

The article concludes with several precautions that pilots can take against lightning strike:

➤ Careful study of weather forecasts and reports, and intelligent use of weather radar (if installed) can help avoid thunderstorm areas and possible lightning strike;

➤ Circumnavigate cumulonimbus clouds by at least five miles if possible when flying below 20,000 feet.... Cloud-to-ground strikes normally occur below 10,000 feet and are more severe than cloud-to-cloud discharges;

➤ Avoid flight in the vicinity of heavy thunderstorm cells when the outside air temperature is between 14°–32°F (0°–10°C). Statistics show that the heaviest concentration of lightning strikes occur in these atmospheric conditions;

➤ Precipitation static is often a prelude to lightning strike and should serve as a warning of active cells in the area. Also, the tops of the typical thunderclouds often glow like St Elmo's fire since the positively charged upper regions tend to draw in electrons from the surrounding air;

➤ Be cautious about the possible indirect effects of a lightning strike. Flight and engine instruments, and sensitive electronic equipment have occasionally malfunctioned as a result of electromagnetic fields inducing transient voltages and currents in the aircraft's electrical wiring, even though there was no direct connection between the equipment and location of the strike;

➤ Be cautious about the accuracy of the aircraft's magnetic compass following a lightning strike. The induced electrical energy may cause excessive deviation, requiring recalibration of the compass after landing;

> When the possibility of lightning discharges is indicated, cockpit lighting should be turned up fully to minimize the effects of flash blindness;

> Any aircraft that has been struck by lightning must be thoroughly inspected by a qualified engineer. Hidden damage, not visually evident from the outside, could be severe enough to effectively reduce the fatigue life of a critical component to perhaps a few hours, or even minutes!

SOURCE: *NEW ZEALAND FLIGHT SAFETY*, OCTOBER/NOVEMBER 1983

Nothing has changed with regard to lightning since that article was published. But there have been changes in aircraft design. For decades, manufacturers have been trying to design and construct composite main and tail rotors for helicopters. They succeeded in doing so and progressively these blades are gaining certification for use on more and more models of helicopter. They are considerably lighter than metal blades and therefore increase the payload of the helicopter. While the lightness may create issues with regard to blade inertia in the event of a power failure, manufacturers claim an increase in airspeeds and a greater service life for composite blades, making the helicopter cheaper to operate. All good news. The bad news is that they simply do not handle a lightning strike as well as metal blades. Do not fly anywhere near lightning if your machine has composite blades, either main or tail rotor.

HAIL AND HEAVY RAIN

It is easy to imagine how hail, which on rare occasions contains stones the size of baseballs, can damage a rotor blade, main or tail, and other aircraft components in-flight. Hail comes from cumuloform clouds, and it is a very good idea to give it a wide berth where at all possible. If you cannot avoid it and the hail stones are large, consider landing and shutting down.

Heavy rain is also, at times, thought to 'take the edge' off rotors and, if not reduce the actual serviceable life of a rotor, certainly make it have the appearance of being old before its time.

If the helicopter pilot . . . is caught in extremely heavy rain, he may be fortunate enough to escape the wind-shear but encounter yet another associated problem. A NASA study shows that a heavy build-up of water could seriously degrade the lift produced by the rotors and increase the associated drag. The study showed a 30 percent loss of lift and an increase of 20 percent in drag. Combine this with associated downdrafts and turbulence and you have a potential disaster.

SOURCE: 'HELICOPTERS AND TURBULENCE, JOE MASHMAN, *VORTEX*, 3/94

Another aspect of any precipitation is decreased visibility. This is addressed under spatial disorientation in Chapter 21 Human factors.

WINTER DANGERS — SNOW, ICE AND COLD

A harsh winter brings its own set of dangers for helicopter pilots and requires corresponding knowledge, expectations and precautions.

Case study 19:9

Aircraft Accident Report
Number 92-014

Date: July 10, 1992

Location: Glenfalloch Station, Upper Rakaia, South Island, New Zealand

Aircraft: Aerospatiale AS 350B

Injuries: nil

At about 0930 hours on the morning of July 10, 1992, the 14,000-hour pilot arrived to prepare for the day's heli-skiing operations. ZK-HWW had last been flown four days previously for an uneventful 2.2 hours on heli-skiing work. Parked in the open, the main rotor blades, tail rotor and engine intake were protected with fitted PVC covers suitably positioned and secured.

Another party had removed snow from the main rotors the day before due to a heavy overnight fall but about 12 inches remained on the exposed parts of the helicopter. The pilot noted that there was a large accumulation of snow on the rotor head and on the engine cowls and engine intake cover. He had been alerted about the snow that had earlier settled on the main rotor blades and took special care to check the rotor head assembly thoroughly. To avoid loose snow falling into the transmission cowl he did not sweep the rotor head cover at this time, believing that the pile of snow would be thrown clear once the rotors were turning.

After removing the covers from the main rotor blades and tail rotor, the pilot uncovered the engine air intake, tipping the snow off to one side carefully and checking visually that the intake duct itself was clear. Fuel samples from the main tank and engine fuel filter were tested and found free from water.

At about 0950 hours the pilot started the engine to carry out a ground run. The start cycle was normal. There was slight initial vibration due to the residue of ice and snow on the rotating blades and the pilot accelerated cautiously through the vibration range to normal ground idle, continuing to govern rpm once the vibration decreased. Governed rpm was maintained for about five minutes before power was reduced again to ground idle. All engine indications were normal and the heater was selected on full to dry out the cabin and warm the cockpit transparencies. After about 10 minutes at ground idle the engine was shut down.

Following the arrival of the heli-ski party some 35 minutes later, the pilot conducted a pre-flight inspection and used a broom to remove remaining snow from the cabin roof. It appeared that the snow had gone from the rotor head cover. He was satisfied that there was no snow lying on the transmission deck or within the engine bay. All oil levels were normal. The subsequent engine start was again uneventful and the pilot lifted ZK-HWW from the parking site and hover-taxied the 55 yards to the raised passenger loading platform. All indications were normal. After landing the pilot remained in the aircraft with the engine running while the heli-ski guide loaded the equipment, assisted the four passengers to strap into the rear seats and then seated himself in the front left seat.

ZK-HWW was fitted with a ski-pod on the left side. The 50 percent fuel load, skis and survival equipment and the six occupants brought the all-up weight close to the maximum authorized takeoff weight of 4300 pounds.

The loading process took approximately five minutes. After completing the standard pre-takeoff checks, the pilot lifted the helicopter to a low hover and confirmed, as anticipated, that 80 percent power was required to maintain a 2- to 3-foot skid height above the pad.

There was considerable loose blowing snow so the pilot carried out a towering takeoff. Power was increased to 82–83 percent and the helicopter was allowed to gain height before achieving significant forward speed. Once clear of the cloud of blowing snow the pilot lowered the nose to commence translation.

At about 200 feet above ground level with the airspeed increasing between 35 and 40 knots, the aircraft's nose unexpectedly 'twitched', the audio warning sounded and panel lights flashed. Realizing that an engine failure had occurred the pilot lowered collective, maintained forward airspeed and set up an autorotative descent straight ahead. The aircraft had been climbing above a large paddock and the unbroken expanse of snow and the flat light conditions made it difficult to judge height above the surface. The pilot had not raised collective completely to arrest the descent when the helicopter contacted the ground heavily and slid for some distance in the soft snow before coming to rest.

After checking that no one was injured the pilot switched off the electrics, stopped the main rotors with the rotor brake and told the passengers to vacate the aircraft. During the shut-down checks the pilot confirmed that, although the engine appeared to have stopped, the fuel lever was still in the normal governed position.

Ground impact occurred in a level attitude. Both skids collapsed but the helicopter remained upright, rebounding and sliding for 200 feet. Damage was confined to the skid assembly, the lower fuselage panels and frames and the tail boom. The tail rotor was damaged severely due to ground contact and the drive shaft distorted.

The engine failure occurred approximately 275 yards northwest of the takeoff pad, which was located at an elevation of about 2000 feet above mean sea level. There was little wind at the time. The local weather was clear with some cloud cover at about 6000 feet on the surrounding hills. The sun was just beginning to break through and estimated ambient temperature was 34° to 36°F (+1°to +2°C).

Post-accident investigation with regard to the engine was quite exhaustive. Test running showed no abnormalities and when stripped there was evidence of minor damage due to the heavy landing, but no pre-existing defects.

In the absence of other evidence to account for the engine failure, the lack of damage to the compressor section and the general circumstances and sequence of events supports the conclusion that, as ZK-HWW gained speed in forward flight shortly after lift-off, sufficient snow was ingested into the engine intake duct to result in flame-out.

No photographs or video records were available to confirm the pilot's suggestion that a cap or layer of snow may have remained on the rotor head cover at the time of departure. However the location of the engine intake grid aft of the rotor head and below its level rendered it feasible that some residual snow dislodged from the rotor head cover could have been ingested. This was decided to be the probable cause of the flame-out.

Source: Transport Accident Investigation Commission, New Zealand

Case study 19:10

Date: December 27, 1983

Location: Newark International Airport, New Jersey, United States

Aircraft: Aerospatiale SA-360C

Injuries: 1 serious, 7 minor

Accident Report NYC84FA061

The helicopter arrived at the airport at 1531 hours after an uneventful flight from New York with two en route stops. At 1540 hours, the pilot taxied from the passenger gate for departure. According to him, he had no control problems while taxiing. He checked the annunciator panel before starting his takeoff and saw no warning lights. Collective pitch was increased to reduce weight on the gear, followed by further pitch increase to obtain a hover. At that time, the cyclic control moved abruptly to the left and was accompanied with the right main gear becoming airborne. The pilot applied right cyclic and decreased the collective, however the helicopter continued to roll left. The helicopter then hopped on the left main gear three times and the main rotor blades hit the pavement. Subsequently, the aircraft came to rest on its left side. An examination and functional test of the flight control hydraulic systems revealed no reason for the malfunction. The temperature was below freezing at 29°F (–1.7°C), but no system icing was found.

Source: National Transportation Safety Board, United States

Case study 19:11

Date: February 4, 1985

Location: Sabine Pass, Texas, United States

Aircraft: Aerospatiale AS355 F1

Injuries: nil

Accident Report FTW85LA111

While en route at an altitude of 500 to 700 feet, the number one engine spooled down. The pilot secured the engine and turned back toward the heliport to make a single-engine landing. However, he could not maintain level flight and the rotor rpm began dropping. The pilot noted that the number two engine had high temperature (above 900°) and low torque. Subsequently, he inflated the pop-out floats and made a partial power landing in a marsh. Due to low rotor rpm, the helicopter touched down hard and was damaged.

The number one engine was removed and checked in a test cell. It operated normally. The number two engine was damaged; however, its fuel management system checked normal. The aircraft was being flown in rain showers without using the engine anti-ice. Operator personnel suspected that inlet icing caused the number one engine to flame-out and the number two engine to lose partial power. A witness estimated the surface air temperature was 33°F (0.6°C).

The National Transportation Safety Board determined the probable cause of the accident to be improper planning and decision-making by the pilot in command, with contributing factors of flight in rain with icing conditions and the pilot in command not using the anti-ice or de-ice system. Also noted were the wet and rough/uneven terrain conditions.

Source: **National Transportation Safety Board, United States**

Case study 19:12

Report Number A96W0204

Date: October 23, 1996

Location: Snare River, Northwest Territories, Canada

Aircraft: Bell 206L-1 LongRanger

Injuries: 1 fatal, 2 serious

The helicopter was chartered by Environment Canada, Water Survey Branch, to conduct hydrometric measurements at sites in the vicinity of the Snare River, 110 miles north of Yellowknife, Northwest Territories. The pilot and two technologists departed Yellowknife at 0925 Mountain daylight saving time (MDT) and arrived at the first site one hour later.

High water levels had flooded the landing site, so an alternate drop-off site was found on a small ice-covered bay nearby. The pilot maneuvered the helicopter close to shore and remained in a hover, with the skids resting lightly on the ice, while the technologists got out and checked the ice surface. They signaled to the pilot that the ice was okay. The pilot bumped the helicopter up and down twice to further test the ice strength. Both technologists walked to the left rear cabin door to unload equipment, when suddenly, without warning, the helicopter broke through the ice. Both technologists slid into the four-foot-deep water.

The pilot applied power and lifted the helicopter out of the water; however, the tail rotor had separated, and he had no tail rotor control. The helicopter rotated clockwise, rolled right and sank. The pilot was submerged, and evacuated the helicopter with assistance from one of the technologists. The second technologist was fatally injured when struck on the head by a portion of a main rotor blade after, or as, the blade fractured. The helicopter was substantially damaged.

The pilot was certified and qualified for the flight in accordance with existing regulations. He had approximately 21,000 hours total flying time, of which approximately 10,500 hours was on Bell 206 helicopters. The pilot and technologists had worked together previously, conducting water surveys. The procedure they used to check the ice had been carried out before with success.

The load-bearing capability of ice is dependent on its quality, thickness, and temperature. It is not possible to determine ice thickness from the air; however, the color of the ice will usually provide some indication of ice quality. Ice color will vary from clear blue to gray to white, with clear blue being the strongest.

The Handbook of occupational safety and health published by the Treasury Board of Canada provides a rule of thumb of 'one inch of clear blue ice for every thousand pounds' to determine the required thickness of fresh water ice. But that recommendation comes with a caution: 'Ice that is less than six inches thick should not be used for any crossing. Because of natural variations, thickness may be less than two inches in some areas.' For loads that are stationary, as would be the case for a helicopter landing and shutting down, the weight bearing capacity of the ice should be decreased by a factor of 50 percent.

Moving water under the ice can affect ice thickness and may not be apparent from the surface. A small stream, hidden from view under the groundcover foliage, fed into the bay below the ice layer. The ice was gray in color and approximately three inches thick where the helicopter broke through.

The helicopter was equipped with a watertight emergency kit with a mirror, matches, and flares secured on the outside under plastic wrap, to facilitate access to these essential items in an emergency. The survivors made use of the emergency kit after the occurrence; however, duct tape used to secure the kit was very difficult to remove with cold hands in freezing temperatures.

The helicopter, on a flight notification, was not expected to return to Yellowknife until late in the day. However, the survivors used a portable satellite phone to call the water survey office in Yellowknife. A rescue team, including Royal Canadian Mounted Police members and a nurse, arrived at the scene around 1300 MDT. Without the satellite phone to report the emergency and the prompt response by the rescue team, the survivors would likely have been stranded overnight and not been rescued until the following morning. Both individuals were suffering from hypothermia when rescued about three hours after the accident.

Source: Transportation Safety Board of Canada

Principles of winter hazards

The first thing to know is what the flight manual for your helicopter type says about cold weather operations. Know the manufacturer's recommended procedures if conditions are conducive to icing in-flight. Have an instant recollection of any emergency procedures.

Pre-flight inspections, while always important, are even more critical in winter. The pre-flight needs to be thorough even though the temperature is probably cold. If possible, do your pre-flight in the hangar. If this is not possible, dress as warmly as you can so that you don't rush the inspection because of the coldness. Wearing gloves is a good idea; they will stop you sticking to any ice on the aircraft.

If the aircraft has been parked outside, all snow, ice and frost should be completely removed. Even a small amount of ice remaining on a portion of the rotor blades could set up a vibration leading to loss of control of the helicopter. Do not scrape or chip away ice from the aircraft. The act of chipping may damage the component you are trying to clear.

During the pre-flight check, walk the main rotor blades around before start-up, listening for any dragging noise that could indicate a bearing seizure. This action also helps to relieve component stress during start-up caused by cold-thickened lubricants.

Those thickened lubricants mean that oil pressures will be high until the oil temperature gets to a normal operating range. In very cold weather this can take a long time. Some helicopter types or operator procedures include the pilot working the flight controls extensively prior to start. This is to get the hydraulics warmed up and to make sure that everything works.

When you have the helicopter started, warmed up and are satisfied that everything is 'good to go', be very cautious when lifting to the hover. The aircraft could still be stuck to the ground (or ice) and dynamic rollover is a very real danger. If this is likely, a slight back-and-forth pressure on the pedals will help to break any contact. Apply power very slowly.

Light helicopters are generally not cleared for flight in icing conditions. Flight tests show that such aircraft experience a rapid deterioration in aerodynamic characteristics and handling qualities, and a marked increase in vibration levels. Tests by the US army with a Bell 205 Iroquois aircraft found that moderate ice accumulation (approximately 0.4 inches) on inboard portions of the rotor blade was sufficient to preclude a safe autorotation in the event of an engine failure. Ice accumulates in greater amounts near the inner portions of the rotor disc, greatly affecting blade efficiency with regard to upward airflow during autorotation.

Estimating ice build-up on rotor blades poses another problem, as it cannot be observed directly as in fixed-wing operations. Pilots are cautioned not to judge the ice build-up on the main rotor by observed build-up on the windshield or other parts of the aircraft because icing occurs at an accelerated rate on the rotor blades. A more reliable method of estimating ice build-up on the blades is to monitor the power required. Research indicates that blade icing of 0.4 inches or greater will be accompanied by a five to six psi torque increase over the 'no ice' power requirement. If the pilot has anti-ice or de-ice equipment, it should be already turned on.

In-flight shedding of ice from the rotors can and does occur. It is just as likely to create a problem as it is to relieve one. Symmetrical shredding in-flight is beneficial in restoring the rotor blades to a more efficient configuration and in reducing weight. Asymmetrical shedding — that is, affecting only some of the main rotor blades — can create severe vibrations and can be extremely hazardous. Ice shedding from the main or tail rotors can also cause structural damage if the pieces that are shed strike the helicopter's fuselage, rotors or engines.

Asymmetrical shedding can be minimized by avoiding air temperatures below 23°F (-5°C). Research has shown that shedding will generally occur symmetrically above this temperature. At these temperatures, symmetrical shedding may be induced by rapidly varying the main rotor speed or by entering autorotation. At lower temperatures it is not possible for the pilot to induce shedding.

If asymmetrical shedding is occurring, another option is to climb or descend. Get out of the conditions that are wreaking havoc on you and your helicopter.

If a helicopter has picked up ice, it is quite likely that some of it will shed from the rotors during arrival and shut down. Pilots should avoid operating near ground personnel or warn them of the possible hazard of flying ice.

Many larger helicopters are approved for flight into known icing conditions. Once again, the flight manual for each type will detail the prudent use of the anti-icing and de-icing systems installed.

Turbine engines can suffer from icing. Although anti-icing systems are normally fitted, they can deal only with comparatively light icing. Piston-engined helicopters can suffer from carburetor icing (see fuel starvation, pp 183–85).

Snow ingestion can cause a flame-out with turbine-engine helicopters. A particle separator will prevent this, but it is itself subject to icing. The degree of snow ingestion can be controlled to some extent by careful flying. Hovering over loose snow should be avoided to minimize the risk of both snow ingestion and whiteout conditions (see spatial disorientation, pp 357–77).

Extreme caution should be exercised when landing on snow, particularly at an unfamiliar site. There may be doubt about the depth of snow and the condition of the underlying ground. The weight of the aircraft should be transmitted to the landing gear carefully and gradually, so that an assessment can be made of the site's ability to take all of the weight. The pilot must be ready to take off immediately if there is any doubt.

For those living in the really cold climates, landing on ice is common practise.

Every year, during initial freeze up in the fall and again during spring break up, helicopter crews fall prey to 'the ice just wasn't thick enough to support my machine'. Some are just plain lucky, while others suffer the loss of a machine and perhaps even the loss of a life.

There's no excuse for dropping through the ice on a lake or river — with preparation and planning all of these occurrences can be prevented. Here are a few bits of sage advice to assist you with your ice landing preparation:

> Ice color can be a big clue in gauging its thickness. White or blue ice is normally the first to have frozen and is therefore usually the thickest and safest to land on. On the other hand, beware of black ice. This color indicates thin ice, possible because of moving water;

> Types of landing gear dictate the thickness required for a given weight. The type of gear affects the foot print or number of square feet of landing gear placed on the ice. The bigger the foot print, the better. Floats are better than skis, which, in turn, are better than bear paw or plain skids (see table on p 296). On a Bell 206B at 3200 pounds, the foot print/weight distribution for a skid/bear-paw combination is 8 square foot and 400 pound per square foot while the same 3200-pound helicopter on floats is 24 square feet and 133 pound per square foot;

> Look for large puddles or sheets of water. These are clues to unsafe conditions;

> Avoid granular and dirty looking ice. This is normally a sign of melting;

> When landing, be smooth. Lower the collective slowly and be prepared to take off if you notice anything happening to the ice;

> Under no circumstances permit a passenger to exit the helicopter until you are satisfied with the ice condition. A few years ago, a passenger was killed because he had exited the helicopter just before it broke through the ice;

- If a crack or water appears on the ice or the ice settles, depart without delay;

- Stay away from inlets and outlets of lakes, rivers and streams;

- Remember when you're parked on an ice surface with the engine running all the vibration produced by your machine is transferred through landing gear to the ice. This could cause the ice to crack and weaken, so be prepared to depart;

- Increase the minimum required thickness by 50 percent if the helicopter is to remain on the ice overnight;

- If you are parking over night it's a good idea to land on a temporary log pad if possible. The pad will help distribute the load and it will prevent the helicopter from freezing to the ice surface;

- If the skids have, or even look like they could have, frozen to the ice, make sure they are completely free prior to departure. Remember, dynamic rollover is a probability if one gear breaks free before the other. If possible, push down or sideways on the stinger to move the landing gear;

- Gauging the ice thickness through a snow cover is impossible, so be extremely careful.

- Winter flying can be very enjoyable, but not if you drop your machine into a lake or river.

Ice thickness in inches (3200 pound gross weight)

Outside Air Temperature		Floats		Skis		Skid/Bear Paw	
°C	°F	Fresh Water	Salt Water	Fresh Water	Salt Water	Fresh Water	Salt Water
−1	30	4.0	9	6.25	14	10	17
−6.5	20	3.25	7	5.25	11	8	14
−12	10	2.75	5	4.0	9	6	12

SOURCE: 'WILL IT SUPPORT MY HELICOPTER?', W.E.C. (BILL) LOFTUS, *VORTEX*, 5/99

Perhaps the last, but definitely not the least, thing to consider when you are to fly in cold conditions is your body.

Cold weather also stresses the body. I know I sure move slower in the cold. Or maybe it's because I'm getting old. People who have had a cold injury in the past are now more susceptible to cold injuries. I personally have a standard that, if it's cold where I'm flying and the heater is not working, then the aircraft is grounded. This is because I know that I'm an integral part of the operations of that helicopter, and if I don't work properly, then the aircraft can't work properly.

I know many people who fly with jackets on as well. This is nothing to be sneered at. In the event that they must leave the aircraft in a hurry, they won't have time to look around and dig out their cold-weather gear. This is especially true of the folks that operate in remote areas.

SOURCE: 'CAN HELICOPTERS FLY AT THE NORTH POLE?', CHUCK MEAGER, *VORTEX* 6/96

20

Crew and pre-flight hazards

INADEQUATELY BRIEFED PERSONNEL

All experienced pilots know that flying with personnel who are inadequately briefed is the cause of a great many accidents. Many disasters featured in the case studies here could have been prevented had a person realized the danger, or understood some basic helicopter principle, or acted prudently around a rotary-wing aircraft.

 Case study 20:1

Report 59/82

Date: November 25, 1982

Location: Storulvån, Sweden

Aircraft: Aerospatiale AS350B
SE-HIA

Injuries: nil

By about 1100 hours the 32-year-old, 2464-hour pilot had flown a few trips transporting freight and was parked at the landing site at the mountain station with the rotors and engine running.

At this time 15 slings were loaded through the right door of the helicopter. After the slings had been loaded, it was decided that two should be left at the mountain site. The crewman collected two in his hands, turned around, took one step away from the right door and threw the slings with one hand at an angle outwards and rearward. One sling fell to the ground about 11 feet from the helicopter. The other was found approximately 115 feet in front of the helicopter at an angle to the left. A sharp noise was heard as the slings were thrown; at the same time the helicopter lifted about 4 feet and moved forward about 14 feet. It then settled vertically and started to rock both lengthways and sidewise. During the rocking motion the skids lifted about two feet above the ground.

The pilot managed to reach the fuel shut-off valve and get hold of the twist grip after about four seconds and could then shut off the engine and brake the rotor to a stop. The loading assistant, who was not employed by the helicopter company, ran away from the helicopter when he heard the sharp noise.

Analysis revealed that one of the slings had been thrown into the rotor plane and was hit by a blade. The sling was hooked by the blade near the blade tip where it was accelerated to the rotor speed and hung on the blade for about one half rotation. The force required to accelerate the sling was so large that the blade trailing edge buckled and the blade arm in the rotor head failed. The effect of this was to increase the blade angle and lift the helicopter, at the same time the blade shifted position in the rotor plane, thus moving the rotor blade's center of gravity away from the center of rotation. The large imbalance created the rocking motion.

The aircraft suffered extensive structural damage: the left main beam had been bent and the right had failed, the skin in the fuselage to tail-boom bulkhead area had buckled, the cabin floor had been displaced sidewise and downward and tubes in the engine mounting had been deformed.

The tone of the official accident report lays the blame on the pilot and the helicopter company. The crewman had not received any formal, documented training. He said that he had received oral instructions and knew of the risks connected with loading and unloading but, in his eagerness to get the slings unloaded, he forgot the risk. This does not exonerate the pilot or the company from the responsibility of correct and thorough training of all crew members.

Also mentioned in the accident report are the dangers associated with the considerable imbalance experienced:

The rocking motion created by this type of unbalance can make the helicopter roll on its side. The risk is then large that pieces of rotor blades may be thrown hundreds of yards and injure people in the vicinity. Damaged rotor blades may also cut through the pilot/passenger cabin.

There are cases on record where even very small objects hitting rotors have caused serious damage which has resulted in dangerous vibrations. It is, therefore, extremely dangerous to throw or stick even 'light' objects into turning rotors. This is not obvious. For this reason clear instructions and suitable training must be given to people engaged in loading and unloading.

Source: Swedish Board of Accident Investigation

With all investigating authorities tending to lay blame on the pilot and operator in such instances, it is vital that everyone who works around or with helicopters, even just on a single occasion, must be briefed at length. That briefing should cover: the task itself, all the peripheral dangers, and what to do if something goes wrong or something happens that is not understood. Furthermore, the person should be given a short, simple test and/or be asked to demonstrate the task and recall the important points of the briefing.

Other important procedures to follow include:

- The pilot or an experienced crew member should have some form of ability to communicate with inexperienced persons at all practicable times.
- Provide ongoing refresher training and reference material for all regular crew members.
- Position clear and prominent safety posters where they will do the most good. An example of such a poster is shown in Fig 20.1.
- If loading passengers or using inexperienced crews from the same location on a regular basis, consider making a briefing video and initiating a procedure that ensures all passengers see it prior to engine start-up and that regular personnel (including the pilot) watch it at suitable intervals.

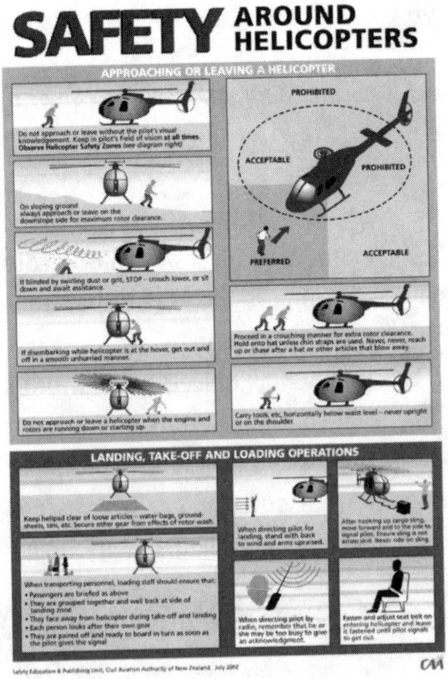

Fig 20.1 Safety posters should be prominently displayed
NEW ZEALAND CIVIL AVIATION AUTHORITY

TAIL ROTOR DANGERS

The tail rotor on a running helicopter is a constant source of danger to unwary pedestrians. And with most inadvertent contact with the fast-moving blades being from the chest up, most often the head, injuries are usually fatal.

Case study 20:2

Aircraft Accident Report 93-003

Date: February 17, 1993

Location: Glenfalloch Station, Upper Rakaia, South Island, New Zealand

Aircraft: Hughes 369 HS

Injuries: 1 fatal

The pilot in command of the helicopter was a 34-year-old with some 2900 hours of experience. His brother was his principal groundcrew or 'loader-driver' as this position is described in agricultural operations in New Zealand. The settled weather had enabled them to complete several aerial spraying jobs throughout the day in the Oamaru area.

Their father had accompanied them throughout the day's work. He did not have any tasks to perform in relation to the helicopter, equipment or chemicals, and was accustomed to keeping out of the way while the helicopter was being loaded or fueled. Essentially he was on a day out, watching the aerial spraying work in progress. He had done this on a number of previous trips and was well familiar with the operation. He had been briefed on several occasions about the hazards of helicopters on the ground and on how and when to approach the helicopter safely. His behavior around the aircraft had been in accordance with those briefings.

Mostly the father accompanied his son in the helicopter on the ferry flights between jobs, but joined his other son on the ground while the productive spraying flights were made. When they repositioned for the last job, however, he traveled in the loader vehicle by road to the new site.

The vehicle arrived first at the site, which was a field of peas to be sprayed. The vehicle was parked alongside the fence outside the field and just past the access gate. This position allowed the loader-driver to run the fuel hose for the helicopter from the vehicle through the fence into the field in preparation for refueling the aircraft when it arrived.

When the helicopter arrived at about 1950 hours, the loader-driver marshaled it to land in the field with its right side alongside the fence so that its fuel tank was accessible to the hose. In this position the helicopter was clear of and facing away from the open gateway but the tail rotor and empennage was the closest part, some 23 feet away from the gate.

As soon as the helicopter had landed, with the engine still running at flight idle, the loader-driver connected the fuel hose and started refueling it, while the pilot remained on board doing post-flight checks.

While the helicopter had been approaching to land, the father had been standing by the vehicle, outside the fence and well clear of the landing area. He had picked a handful of wild peas and was eating them while watching. During the landing, however, the rotor down-wash had blown his cap off his head and into the adjoining water race, where it was lost. This apparently disconcerted him, because shortly afterwards he walked into the field and stood about three to four yards behind the helicopter.

The loader-driver had seen him move to this position but saw him stop clear, so he continued to concentrate on fueling the helicopter.

Fig 20.2

Shortly afterwards the loader-driver glanced up to see his father walking forwards, looking down at the peas in his hand and approaching very close to the rotating tail rotor. He had no opportunity to warn him before the rotor blades struck his head. He collapsed beneath the tail of the helicopter and did not respond to immediate aid.

The helicopter was shut down and damage to both tail rotor blades was found.

The bright evening weather conditions and lighting were unlikely to have been a factor. The helicopter was on a heading of 065°M, so the approach to the tail was down-sun. While the tail rotor had normal red and white markings, it would have been almost invisible as approached, in its plane of rotation.

During the day, before the accident, 8.25 hours of flying had been done. Some fatigue may have reduced the alertness of each of the persons involved.

Fig 20.3

Source: Transport Accident Investigation Commission, New Zealand

Case study 20:3

Occurrence Brief 9502549

Date: August 10, 1995

Location: Ayers Rock
Aerodrome, Australia

Aircraft: Bell 206BIII

Injuries: 1 fatal

Four passengers had reservations for a helicopter scenic flight, and were met in the airport terminal by the ground hostess for transportation to the helipad in the company bus.

The ground hostess stated that while proceeding to the helipad she briefed the passengers about the helicopter, and the safety requirements when in its vicinity.

She parked the bus on the road adjacent to the helipad, approximately 40 feet to the right and well forward of the helicopter. The passengers were then told to remain at the bus until instructed to approach the helicopter.

Following normal practise to save engine cycles, and turn around times, the pilot left the helicopter engine running after landing, then locked the controls and got out to assist the ground hostess disembark the passengers, who were then directed to the bus. The ground hostess accompanied them as far as the edge of the main rotor disc, and then signaled the passengers (from the bus) to follow her back to the helicopter.

One passenger had expressed an interest to occupy the front left seat during the flight. This was agreed to by the others.

The ground hostess watched the passengers follow her towards the helicopter, but when she turned her head to check its proximity, the passenger who had requested the front left seat left the group to pass behind the helicopter, and walked into the tail rotor, receiving fatal injuries.

The ground hostess stated that after the occurrence she spoke to the passengers, who confirmed that they had understood her briefings, and had no idea why the other passenger had not followed her instructions.

Statements taken by the police did not address whether the passengers had received a safety familiarization briefing, but covered the last instructions given by the ground hostess concerning waiting at the bus, and approaching the helicopter. Only two of the passengers could now remember these instructions.

Reports indicated that three of the passengers were partially deaf, and the other had assisted them. Because of this it is possible they may have missed some parts of the briefing.

Source: **Bureau of Air Safety Investigation, Australia**

Ideally, pilots should shut down the engine and have nothing moving before allowing anyone to approach the helicopter. Pilots who leave machines running when there is any possibility of someone being seriously injured, or worse, must be very aware of the dangers. In case studies 20.2 and 20.3, the pilots could quite reasonably assume that everything was under control. But there are many occasions where, in hindsight, it can quite justly be said that the pilot should have known better than to leave his machine running.

Even at ground idle, painted in whatever design you care to dream up, and the subject of various warning decals plastered over the tail boom, the tail rotor is very difficult to see. This is especially true if approaching from directly in front of or directly behind the rotor. It is even harder to see if you are not looking for it.

Avoiding that area soon becomes second nature to people who regularly work around helicopters. The danger is, though, that complacency can dull intuition not only from a personal point of view, but also in failing to recognize a person who isn't aware of the danger or who had a momentary lapse of concentration during a briefing.

Steve Bone recalls:

I was flying as co-pilot on a US Navy UH-1N (Bell 212), wandering the vast Antarctic polar plateau with a group of geologists looking for meteorites. Some 9000 feet up on the white nothingness of the plateau was considered good hunting ground because the black meteorite rock contrasts with the vast white. However, I was told that you could also fly around for days and find nothing. Luck was with us and our captain spotted two small black rocks ahead of our slow-flying helicopter. He positioned the aircraft to land so that the objects were outside the disc in the 2 o'clock position. The drill was to shut down so that the geologists could investigate, measure, photograph in-situ, and so on.

Unbeknown to us, another black object had just been seen behind the aircraft by one of our passengers as we approached to land. With the aircraft still running (waiting for the two-minute engine cool-down period) the passengers were disembarked and our ecstatic rock hunters moved forward away from the helicopter. However, faster than a speeding bullet, our one overly ecstatic spotter raced straight to the rear of the aircraft. Gone.

The shut down was brought forward somewhat, and we went to investigate. Our widely grinning geologist was bent proudly over his meteorite find which was handily placed in line with our tail rotor, but about five feet adjacent. He was oblivious to the potential of his predicament.

All passengers had been fully briefed, and we had a crewman on board who took the passengers under his wing. It can still happen, and there's nothing like the helplessness you feel when you see out of the corner of your eye, someone dashing back towards your tail rotor.

Personnel who normally work around fixed-wing aircraft deserve just as much, if not more, attention than anyone else. They have a natural instinct not to go near the front of the aircraft because that's where the danger lies on most airplanes. Further, unlike the complete aviation newcomer, they are quite used to the noise associated with a running aircraft and not so overawed by their surroundings.

Many accidents that happen either to, or because of, very experienced helicopter personnel, be they groundcrew or pilot, come about due to environment and setting – the noise, the smell and the sense of purpose to not waste time. This applies particularly in a small commercial operation. Like any business, the aim of the game is to make money. But while that turbine engine is parked and guzzling costly gas at an alarming rate, nobody wants to be the reason that it's not up and producing revenue. As a result, things get overlooked, shortcuts are taken and a mishap, born of pure haste, occurs. Certainly, most of the oversights are small and insignificant, amounting to nothing at all. But just occasionally that oversight turns out to be significant and a helicopter or, worse still, a life is lost. Many of the accidents outlined in this book are examples of this.

Allowing people – be they passengers, customers, bystanders or fellow crewmen – to walk into a moving tail rotor is not always negligence as such. But somebody, other than the victim, will generally feel badly about it. Accidents like this are always tragic, emotional, and totally avoidable.

Never forget that tail rotor. Never hesitate to physically grab someone who even halfway looks like they may be headed in that direction. Never stop watching people unfamiliar with helicopters; if possible, never take your hand from their arm.

MAIN ROTOR DANGERS

The main rotor blades of a helicopter are not only very strong, but they are also quite flexible. That suppleness doesn't help much though if you happen to get hit by one that is moving. They move very fast – a tip passing the same point five times a second is doing around 700 miles per hour. Any object it contacts will suffer damage. If that object is a human being, the damage is exceedingly serious.

Case study 20:4

Date: September 6, 1993

Location: Harris Mountains, South Island, New Zealand

Aircraft Accident Report Number 93-010

Aircraft: Aerospatiale AS350B

Injuries: 1 fatal

The pilot in command of ZK-HWB was a 33-year-old commercial helicopter pilot with 4826 hours of total flight experience. At about 1615 hours the helicopter was engaged in heli-skiing operations in the Harris Mountains near Wanaka.

Four groups, each comprising four client skiers and a guide, were being transported. The operation involved flying each group to a suitable ridge or mountain-top landing site in the area. The guide would then lead his or her group to the valley bottom, generally about 2000 feet lower, where the helicopter would pick them up for a further run. As the objective was to ski over new untracked snow, routes varied on each run and this resulted in different pick-up and landing sites for the helicopter.

Coordination between the guides and the pilot was accomplished by radio, with each guide selecting the pick-up site for the helicopter to land. Pick-up sites were generally chosen on open level areas of the valley but occasionally, when a group was unable to complete a run to the valley bottom, a pick-up site would have to be chosen on a less suitable piece of terrain, such as a spur, on the side of the valley.

The procedure at pick-up was for the guide to gather all the skis and his pack together to mark the site for the pilot. The pilot would aim to place the helicopter's left skid alongside the skis so that the attached ski container was conveniently adjacent. The skiers were required to kneel together by the skis so that after the helicopter landed they had minimum distance to move before boarding. They were also required to secure all loose articles such as hats and gloves. The guide knelt one to two yards ahead of the skiers, in a position where he could monitor their behavior and maintain eye contact with the helicopter pilot during the landing. The guide would signal the helicopter to approach and land when the group was in place and ready. The skis were loaded and unloaded only by the guide, who also opened and closed the helicopter's doors and directed entry and exit from the aircraft. Disembarkation was done similarly with the skiers being required to take only two steps from the door, and then kneel together until the helicopter had departed.

The skiers in the group involved in this accident were all experienced in heli-skiing operations and had been flown directly from their Wanaka hotel to the mountain top rather than traveling by road to the forward staging area. As a result they had done more runs than other groups during the day.

During their tenth run they decided that they did not want to ski to the bottom and requested a mid-way pick-up. The guide arranged this, selecting a site on a less steeply sloping part of a spur. The skier involved in the accident was to finish skiing for the day while the others were to have one or two more runs.

The group adopted their normal positions in preparation for the helicopter's arrival and the guide signaled it to land. Just as it reached its landing site the main rotor down-wash displaced and blew away the baseball-type hat from the subsequent victim. He got up in spite of the guide's warning gesticulations to remain kneeling. The skier walked upslope, away from the left side of the helicopter. As he stood erect, the tip of one main rotor blade struck his head and fatally injured him.

The helicopter pilot did not see the accident but felt and heard the collision. He lifted the helicopter off from the site where its left skid had been placed and flew down to the valley floor where he shut down. After inspecting ZK-HWB for damage he then lifted all the skiers from the mountain and evacuated the deceased. Subsequent engineering inspection of the main rotor blade established that it had sustained damage which required replacement of the blade.

Examination of the accident site showed that it was a less steeply sloping section of a spur at about 4500 feet above mean sea level. The local slope was about 12° down to the southwest, which was beyond the helicopter's slope landing capability. It was practicable, however, for the aircraft to be hovered with the left skid on the snow surface while passengers boarded, provided they kept close to the side of the helicopter. Main rotor tip clearance above the surface on the uphill side was not measured but it was estimated to be about six feet. The normal tip clearance above a level surface was 3.5 inches lower than this but varied dynamically with the use of the cyclic control.

Normally the helicopter safety training for skiers was provided by a preliminary briefing pamphlet and then by a daily client briefing demonstration by the guides at the forward staging area. In addition each guide continuously monitored his group's behavior around the helicopter.

The accident group had been flown directly to the mountain top and had thus not participated in the briefing demonstration at the staging area. Each skier had attended briefings during the previous week as well as on innumerable occasions in previous years. The guide had found them to behave correctly on all the previous pick-ups until the accident occurred.

The victim had spent two weeks heli-skiing each year for 10 years and had probably boarded the helicopter some 700 or 800 times while doing so.

Source: Transport Accident Investigation Commission, New Zealand

Case study 20:5

Occurrence Brief Number 9500891

Date: March 27, 1995

Location: Cloncurry, Queensland, Australia

Aircraft: Hughes 269C

Injuries: 1 serious

During a stock mustering operation, the head stockman, who was a passenger in the helicopter, asked the pilot to land so that he could give instructions to stockmen on the ground. The pilot landed the helicopter and commenced to secure the controls.

The stockman left his seat, stood up on the right step attached to the skid gear, and began to signal the stockmen by waving his right arm. The stockman's arm contacted the main rotor disc causing two fingers to be severed. After first aid, the stockman was transported in the helicopter to the Cloncurry Hospital where it was discovered that he had also broken his right wrist and arm.

The helicopter was slightly damaged with a broken right Perspex panel, which was struck by the stockman's right arm as it was flung from the rotor disc.

The stockman was a very experienced helicopter passenger. He had not been specifically briefed on helicopter safety aspects for this particular flight. The pilot was concerned with securing the helicopter after landing and did not see the event. He first noticed something was wrong when he saw the stockman writhing on the ground.

Source: Bureau of Air Safety Investigation, Australia

Case studies 20.4 and 20.5 show two experienced passengers who, for a short space of time, didn't think about the main rotors for whatever reason. This is not at all uncommon.

For inexperienced passengers, the briefing is of utmost importance. The person giving the briefing must include the dangers on level ground and sloping ground, and must also include blade sailing (see pp 310–12). Illustrations from a safety chart or poster are of immense help. If you operate primarily from one base and take passengers regularly, consider making a videotape that you can run for all passengers prior to boarding.

The appointment and presence of an active ground controller, while not foolproof, is of great value. Having said that, the pilot should always be aware of persons moving in, out or under the rotor disc. Stop what you are doing and watch them safely in or out. Unless you discipline yourself to do that, you may be the person who moves the controls or the aircraft and causes the damage.

There are a million stories out there but two more will suffice to show that all an accident needs is a very brief lack of awareness:

➤ A pilot with over 2000 hours of experience left his Hughes 500 running at flight idle to attend to a task up a small adjacent rise. Seconds later his helmet was destroyed and he was seriously injured. Fortunately he made a full recovery.

➤ A Santa Claus arrived at a gathering of children and raised his hand a little too high in waving to them. The outcome was not a sight to rouse Christmas cheer!

People jumping from a height into water naturally throw their arms in the air to gain balance. Doing that from the skid of a helicopter can be very distracting and make it rather hard to swim.

Steve Bone has some words of advice:

If you are unhappy with the way a particular person or group is behaving, don't be afraid to call a halt. Shut down the machine if your location allows, and begin the briefing again. Your passengers might be annoyed, think you're pedantic and not at all politically correct. But they will take notice. And they'll be alive, which will also help your future well-being. One wise old sage I flew with on offshore operations once stopped people running around the flight line and his company's helicopter (even though it was shut down at the time) by saying, 'If I see people running around the helicopter, I go inside and have a cup of coffee'. His 'conversational' voice carried to all, and the message was not lost on the assembled, and suitably chastened, crew.

I recall being in the crew of a Hughes 500 capturing live deer. The helicopter had high skid gear and the main rotors were rarely a problem on level ground. We landed in spongy terrain to tie up and stow a live hind. I was busy doing tasks like sorting out the nets and so forth. At one stage I stood up and a fellow crewman signaled me to look up. I did and the rotor disc seemed about three inches from my nose. I instinctively ducked and he pointed again, this time to the ground. The skid had sunk a considerable distance into the swamp! It was quite disconcerting; small frights like that serve to keep you alert — which is good if you survive them all.

How do you avoid lapses of concentration, particularly in others like your passengers and crew? You can do only so much: frequent briefings, reminders and keeping an eye out whenever possible, and hoping that others are doing the same for you — if you never suffer such lapses you are a very unusual person.

BLADE SAILING

Blade sailing is the exaggerated flapping that occurs at low rotor rpm when a helicopter is starting up or shutting down in strong wind conditions (particularly when there are gusts). When a helicopter is running and the rotors are turning under power the centrifugal force to which they are subject tends to keep them horizontal to the ground. When there is no power and they slow, that centrifugal force reduces. If the helicopter is facing into the wind, the advancing blade will flap up to reach a high point in front of the helicopter. As this blade progresses on the retreating side it experiences a sudden loss of lift and will flap down rapidly, flex, and reach its lowest position to the rear of the helicopter over the tail boom.

Case study 20:6

Occurrence Brief Number 9302675

Date: August 19, 1993

Location: Western Australia, Australia

Aircraft: Bell 212

Injuries: nil

The Bell 212 was flown to the North Rankin 'A' oil platform to transport passengers back to Karratha. The pilot in command held an airline transport pilot's license (ATPL) and had 13,460 total flying hours of which 297 were on the Bell 212. The co-pilot also held an ATPL and had 6000 hours of flying, of which 350 were on Bell 212s.

When the helicopter arrived on the helideck, it was landed into a strong southwesterly wind and positioned into wind with the pilot's eyeline between the two lines painted on the helideck. The pilots elected to shut down the engines when they were told that the passengers were delayed. They did not reposition the aircraft so that it was offset from the wind, as required by company procedures, prior to shutting down the engines.

When the engines were restarted after the passengers were boarded, the blades were seen to rise during the initial couple of revolutions as they came into view of the cockpit. At about the same time a noise was heard from the rear of the helicopter and a vibration was felt. The start was discontinued and inspection found that the tail rotor drive shaft had broken at the rear end. Subsequent investigation revealed that the main rotor had struck the drive shaft, which had then failed.

Wind tunnel test results for the rig helideck platform showed that turbulent airflow over the helideck can be expected under the wind condition prevailing at the time of the accident. The pilots confirmed that the wind on the helideck was different to the free-stream airflow. They noticed that there was a strong updraft in the vicinity of the leading edge of the helideck.

Source: Bureau of Air Safety Investigation, Australia

Principles of blade sailing

The reasons for blade sailing are not hard to understand or remember. The pilot has no means of controlling a rotor blade when the engine is off and the rpm reaches into the lower bracket, as the cyclic control has little effect at such low rpm.

Knowing that takeoffs and landings are always easier facing into the wind means that helicopters are frequently parked in just that position. Of course there is no problem until the day that a strong or gusty wind arises and the blade actually strikes the tail boom.

If that is not bad enough, with the tail into the wind the lowest point that the rotor reaches is in front of the helicopter — on many models it is quite possible for the blade to strike the ground. This is incredibly dangerous for anyone under the rotor disc, no matter how low they are bending.

Fig 20.4 The rotor has been lowered just to the stops; no flexing has been introduced. Flapping could take this rotor very much lower during blade sailing STEVE BONE

To avoid blade sailing, or at least minimize the effect:

1. Make sure the collective is fully down.
2. Tilt the disc toward the wind (cyclic slightly forward and into the wind).
3. Minimize the time the blades are turning slowly (if you have a rotor brake, use it).
4. Increase rotor rpm at a faster rate than normal when starting up.

Ideally, shut down the helicopter slightly out of wind (say with the wind coming from 2 o'clock) but be very wary about any obstructions (including persons) in the 7–9 o'clock position as this will be where the blades dip the most (see Fig 20.5).

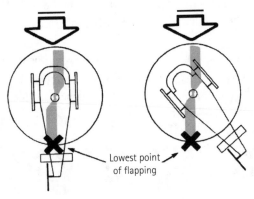

Lowest point of flapping

Fig 20.5 *VORTEX*

Resist the temptation to get behind a hangar, large wall or other building when there is a really strong wind blowing. The winds in the lee of any obstruction are generally turbulent and can cause more problems than they cure. You are usually better off out in the open.

Of course regular crews have to be very well briefed with regard to blade sailing. I had trouble with the passengers in a corporate Aerospatiale Dauphin that I used to fly. One of the Company principals in particular was always in a great rush to get away from the aircraft and get going; even while I was shutting down, before I could get the rotors stopped. Although I had told them of the dangers several times, it was not until I actually made them wait and gave them a physical demonstration of how low the main rotors can flap that I gained cooperation and had no further problems.

SOURCE: BERNIE LEWIS

DAMAGE BY FOREIGN OBJECTS

A hovering helicopter moves a lot of air, which can reach speeds of up to 35–40 knots depending on the weight of the helicopter. If a helicopter hovered just above a huge set of scales, the weight of the downward air pressure recorded on the scales would be precisely the weight of the helicopter and its cargo.

This air generally moves down from the rotor disc and then laterally across the ground (if the ground is close) for some distance before rising again. As the helicopter is driving air down through the disc, more air must be arriving above to fill the void. Some of the air that has been moved outward and upward is excess to requirements and naturally tends to return to above the aircraft and make the journey again. This, as we know from Chapter 3, is called recirculation.

A 40-knot wind affects a lot of things. In fact, if it is gusting, it can be difficult to stand. Loose articles around a hovering helicopter have a tendency to follow the airflow and end up making a very close acquaintance with the rotor disc. A foreign object that impacts with a spinning rotor blade creates massive forces. The blade may fail, strike the tail boom or cockpit, and trigger an instantaneous and disastrous chain of events. At the very least the helicopter will be substantially damaged, if not destroyed altogether. The deceleration forces convey themselves to the transmission, an intricate piece of equipment that does not cope with excessive stress.

Case study 20:7

Aircraft Accident Report Number 92-019

Date: November 20, 1992

Location: Taupo Airport, North Island, New Zealand

Aircraft: Westland Wasp NZ3094

Injuries: nil

Two Royal New Zealand Navy (RNZN) Wasp helicopters called at Taupo to refuel. They descended to hover over the runway as a pair, and then one aircraft went to the control tower to get the key to the fuel pump while the second hover-taxied to the fueling point.

The Wasp had castoring wheels, which were locked before flight, with the forward wheels parallel to the centerline and the rear wheels toed-in at 45°. Prior to touchdown it was therefore necessary for the aircraft to come to a stationary hover to ensure there was no forward or sideways motion when the aircraft touched the ground. Stabilizing the hover took, typically, about three seconds. The aircraft usually hovered at 5 to 6 feet.

Adjacent to the fueling point were two civil helicopters (a Bell 206 and a Hughes 369) which had landed there shortly beforehand and refueled, and a large powerboat (about 36 feet long) which had been put there ready to be towed away. The Wasp commander was unconcerned about the powerboat, as it was large enough to be unaffected by downwash, but he was concerned about the civil helicopters. They were much lighter than the Wasp, and it was necessary to ensure clearance between them and the Wasp's tail rotor. The second pilot, who was the pilot flying (in the right-hand seat) therefore brought the Wasp to the hover close to the powerboat, while the commander released his harness and leaned forward and down so that he could see back to the tail rotor.

The Wasp pitched forward abruptly and struck the ground. The main gearbox casing fractured and the gearbox flailed through the cockpit at the height of the tops of the seat-backs. The commander's helmet was struck but he was saved from injury by having previously released his harness and ducked down. Both pilots' flying suits were soaked in transmission oil but there was no flash fire and they were able to vacate the aircraft, which came to rest on its wheels. Neither was injured but the helicopter was substantially damaged.

Bystanders observed a light nylon cover, which had been stowed between the boat and its trailer, extracted from beneath the boat and flung into the air by the Wasp's downwash. It was then sucked down into the main rotor disc from above. Marks on the fabric and one rotor blade showed that the blade had struck the cover. Damage to the mast head bump stops, and a main rotor blade strike on the top of the tail boom, were consistent with blade flailing. This in turn would have resulted

Fig 20.6 TRANSPORT ACCIDENT INVESTIGATION COMMISSION, NEW ZEALAND

Fig 20.7 TRANSPORT ACCIDENT INVESTIGATION COMMISSION, NEW ZEALAND

in rotor disc imbalance, which was probably responsible for the main transmission casing fracture.

Parts of the Wasp were thrown a considerable distance and a bystander close to the boat was fortunate to escape injury.

A fire broke out in the vicinity of the Wasp's engine, probably due to rupture of a fuel line. The smoke was seen by a baggage handler who was also one of two volunteer firemen at Taupo Airport. He took the fire vehicle (a four-wheel-drive van with 66 US gallons of aqueous film-forming foam) and extinguished the fire just as the foam was exhausted. If the crew had been trapped, his prompt action had the potential to save lives and was commendable.

Over the preceding years, Taupo Airport had become increasingly busy and several changes had been instigated: the fuel pumps were moved, taxiing areas were redefined and facilities for fixed-wing operators were repositioned.

A helicopter operator whose hangars were adjacent to the immediate area had intended to move the boat but was called away to assist in an emergency. He knew of the imminent arrival of the two civil helicopters and considered that the boat cover may be a danger. He therefore removed the cover and stowed it under the boat. The trailer had covered sides and bottom, so this seemed to him a secure stowage.

Indeed, this proved to be the case for the two civil helicopters. As they regularly operate in an environment where foreign objects might be present, the civilian pilots were used to flying directly to the ground with minimal or no hover.

The difference in handling techniques is probably why the civil helicopters landed without disturbing the boat cover. Flying directly to the ground minimized downwash, whereas coming to a high hover, as the Wasps did, was likely to set up recirculating flow. In any event, because the Wasp was a heavier aircraft and had a smaller rotor disc, its downwash velocity would have been greater.

No aspect of the operation was regarded by the participants as being out of the ordinary and all used their best endeavors to make it safe. The question thus arose as to how the unsafe circumstances had come about. The essence of the situation was that, over the years, the area around the common fueling point had become cluttered to the extent that it was hazardous for visiting pilots.

As you can imagine, some quite in-depth questions were asked of the operator/s and aerodrome officials, but eventually the accident report merely made extensive safety recommendations with regard to aircraft parking and maneuvering areas at the aerodrome.

Source: Transport Accident Investigation Commission, New Zealand

The point is very straightforward: ensure there are no loose articles anywhere helicopters regularly land and take off. To be extra sure pilots should do a visual check of the immediate area before start-up. When approaching to land, keep a careful watch out for any loose objects.

As with many ground-related hazards, briefing of passengers and non-aviators is where it all starts. Tell them they must firmly hold caps and hats, and anything else that is likely to be picked up by strong winds, in hands; and must not discard litter. If you or your crew see someone who hasn't got this fully in mind, don't just tell them off; take the time to explain the consequences in terms of both finance and danger.

Large safety posters, which include the dangers of loose articles, will also help. Display them around your base.

EXITING AN AIRCRAFT IN-FLIGHT

Helicopters can give a false sense of security — a feeling that you are actually at home in your favorite armchair with the television remote control in one hand. It is very easy to forget you are depending on a piece of machinery and your own skill to stay aloft. The longer you have been flying, whether as a pilot or a passenger, the more blasé you can become.

But untrained crew are also at risk. A pilot once recounted an experience he had when a new crew member was hastily briefed on how to catch a deer from the skid of a Hughes 500. The procedure involved jumping on the deer when the helicopter got close enough and hanging on until someone came to help tie it up. A live animal was worth thousands of dollars at the time.

The pilot eventually found a hind and started to track it. As he looked ahead to find a likely place to get down to within a few feet of the deer's back, he was surprised to feel the helicopter lurch. The cowboy had set off on his mission at around 40 feet! Although he never caught the deer he wasn't too badly injured, having landed in thick bush. In the heat of the moment and working from such a solid and stable platform (mostly), it is quite easy to forget just where you are and what dangers exist.

Case study 20:8

Aircraft Accident Report Number 92-008

Date: March 9, 1992

Location: South Westland, South Island, New Zealand

Aircraft: Hughes 269C

Injuries: 1 serious

Early in the morning ZK-HHJ departed the township of Haast and flew into remote bush-land to recover venison. This type of operation, which usually involves flying with both doors removed, is common in New Zealand. The crewman fires a semi-automatic weapon at feral deer and the carcasses are then recovered using a sling.

The crew consisted of the pilot in command, a 46-year-old with about 8000 hours of experience, and the shooter, himself a commercial helicopter pilot experienced in the role that he was to perform that morning. Nothing was noteworthy with regard to weather.

The shooter was wearing a standard seat belt recently fitted to the aircraft. This belt had a 'flip to release' buckle. As crewmen in this role find it necessary to have some movement available, the belt is usually worn more loosely than is generally recommended.

An initial shot brought one animal down and the pilot then maneuvered ZK-HHJ in pursuit of another deer. This second deer was duly shot but, in the meantime, the first deer got up and began to run down the hill. The pilot turned, descended and approached the hillside again, coming to a hover adjacent to the steep slope, suitably positioned to enable the shooter to fire at the escaping animal.

The next moment the pilot saw the shooter 'flying out the door' and falling straight down towards the tussock- and scrub-covered slope some 30 to 40 feet below, still clutching his rifle. The shooter struck a large bush and remained motionless.

There was no landing site nearby so, after circling for several minutes and observing no movement from the shooter, the pilot decided to leave the area to seek assistance. His efforts to establish radio contact with the outside world had been unsuccessful. He landed at a nearby coastal property and awoke the property owner who agreed to return with him immediately to the accident site. His wife contacted the police to arrange for a helicopter with appropriate rescue equipment aboard to proceed to the area.

When the pilot returned to the area, he was relieved to see a flare fired from the ground. He dropped off his passenger to render assistance to the injured shooter. With no small amount of luck, the pilot then located a doctor who had been hunting in the area and brought him to the scene. Eventually the accident victim was transported to hospital. He had suffered severe injuries to his back and lower limbs. He could recall most of the morning's events apart from the actual fall from the helicopter.

The shooter was wearing a protective flight suit in which the material forming the cuff of the left sleeve was fairly hard. While the circumstances could not be established conclusively, it was likely that, as the shooter swung his body around to obtain the most effective shot, the flap of the seatbelt buckle was caught inadvertently and lifted, possibly by the sleeve of his flight suit, and the lap belt released. The momentum of the shooter's body movement was then sufficient to cause him to fall out of the aircraft.

This shooter was not the first, nor has he been the last crew member to fall from a helicopter in-flight. Several operators have tried to improve the restraint system for this particular operation by making unauthorized modifications using karabiners and the like. An official recommendation as a result of this accident suggested that a better restraint system be developed.

Source: Transport Accident Investigation Commission, New Zealand

Case study 20.8 is not an isolated event; some operations lend themselves to this type of accident. While it is rare for the pilot flying the helicopter to fall out of the machine, passengers and crew are at risk. Pilots should take every possible precaution. Manufacturing certification and general maintenance requirements should ensure compliance with the safety standards for seat restraints. It is up to pilots or other authorized personnel to ensure that the restraints are used correctly.

FAILURE TO ENGAGE CONTROL LOCKS

Leaving a helicopter idling while attending to other business can be fraught with danger, especially if the mechanism that locks the controls in place is not activated. Problems have arisen even when control locks are being used correctly. As case study 20.9 demonstrates, some control locks are more reliable than others.

Any one of a number of variables can change, opening up the possibility of damage to the aircraft and injury to nearby personnel. Remember, an idling helicopter still vibrates and generates many forces, and winds seldom blow at a constant velocity. Also, the ground on which you landed may be unstable.

Case study 20:9

Occurrence Report Number A97W0130

Date: July 9, 1997

Location: Nordegg, Alberta, Canada

Aircraft: Bell 206B JetRanger

Injuries: nil

After landing the helicopter at a private campground, the pilot rolled the throttle to idle, tightened the cyclic and collective friction controls, and briefed the fixed-wing pilot who was sitting in the left seat to hold the left-side dual controls steady. The helicopter pilot then exited the helicopter, without shutting down the engine, and ran approximately 150 feet to a recreational trailer. As he was returning to the helicopter, he stopped briefly to talk to the campground caretaker who was standing approximately 100 feet to the left of the helicopter. Immediately thereafter, a clunking sound similar to that of an out-of-balance washing machine was heard and the helicopter began to rock fore and aft on the skids. The main rotor disc was observed to be tilted to the extreme forward position.

The pilot ran back to the helicopter, ducked under the main rotor disc from the left side, ran around the front of the helicopter, and climbed into the right seat. He immediately attempted to stabilize the helicopter by applying collective and increasing rotor speed. The oscillations diminished initially and then increased dramatically when the weight was reduced on the skids. The pilot then lowered the collective and shut down the engine. When he applied the rotor brake after shut-down, a loud 'clunk' was heard before the rotor blades stopped turning. Examination determined that the swash-plate drive collar set and both main rotor pitch links were fractured. None of the four occupants on board the helicopter sustained injury.

The purpose of the flight was to transport two Transport Canada officials and a dependent around several landing strips to visit general aviation operators.

After landing, the pilot consulted with the passengers to determine if it was necessary to shut down. Since none of the passengers expressed a desire to get out of the helicopter at the campground, the pilot decided he would not shut down, as it would require only a short time to conduct his personal business and there was an experienced fixed-wing pilot in the left seat.

The helicopter pilot held a valid airline transport pilot's license (ATPL), and was certified and qualified for the flight in accordance with existing regulations. He had acquired approximately 6760 hours of flight experience, including 4700 hours in rotorcraft, and 943 hours in the Bell 206 type helicopter.

The fixed-wing pilot held a valid ATPL, and had accumulated approximately 12,000 hours of fixed-wing flying experience. He had attended Transport Canada Cockpit Resource Management training. He had never received helicopter training and, prior to this flight, had not ridden in a Bell 206 for approximately 10 years. He reported that he had pushed the cyclic forward, after the helicopter started to oscillate, in order to prevent the helicopter from becoming airborne.

The Bell 206B helicopter is fitted with an under-slung, semi-rigid, teetering, two-blade main rotor system. The teetering design allows the main rotor blades to flap to compensate for asymmetrical lift during flight. One static stop is mounted on either side of the main rotor hub to physically limit the amount of blade flapping. A condition known as mast bumping occurs if the static stops contact the mast, due to excessive blade flapping, during ground operations or in flight. During ground operations with the rotor turning, the main rotor may be affected by wind gusts and flap to its limits, resulting in a light static stop to mast contact. In such an event, mast bumping may manifest itself as a light shudder felt throughout the helicopter. The more extreme, the flapping; the more severe the shudder. Mast bumping will also occur during ground operation, if the cyclic is incorrectly positioned or is moved sufficiently to tilt the rotor disc to an extreme position. If the static stop to mast contact is severe, pronounced helicopter oscillations may develop and the helicopter can sustain substantial damage. The appropriate corrective action is to immediately reposition the cyclic, toward or near the neutral position so that the rotor disc resumes a flat position. On the ground, at idle rpm, the rotor disc is less stable and more susceptible to larger deviations due to flapping.

The Bell 206 is also fitted with cyclic and collective friction controls that allow the pilot to set the force required to move the flight controls, in order to reduce the tendency for pilot-induced oscillation or collective creep in flight. The cyclic and collective friction controls are not intended to be used as control locks. The pilot had left the hydraulics selected on prior to leaving the helicopter, and testing determined that five to six pounds of force was required at the grip to move the cyclic from a neutral to a forward position with the cyclic friction control tightened and hydraulic power applied.

The transition from fixed-wing to helicopter flying requires the acquisition of new skills and fixed-wing responses may not accomplish the desired reaction in a helicopter. The helicopter cyclic is the equivalent of the fixed-wing control wheel; however, cyclic response is extremely sensitive and rapid in all flight regimes; whereas control wheel response diminishes at low airspeeds. A fixed-wing pilot may hold a control wheel fully forward during ground operation, in strong winds, in order to keep the aircraft firmly on the ground.

Information regarding two similar Bell 206B occurrences was obtained during the investigation. In one case the helicopter was being run up by a student pilot. At about 80 percent N1 (the turbine rpm), the helicopter began to bounce and vibrate, and the rotor disc was observed to be tilted in an extreme forward position. The helicopter eventually bounced approximately 120° to the right, and a line supervisor ran to the cockpit and shut down the engine. Examination determined that the transmission isolation mount, the support plate, and the pin assembly had sustained substantial damage, and that the collar set had fractured due to overload. The second incident occurred when a trainee pilot was being checked out by a company pilot. The trainee pilot had conducted a landing at a remote staging area and the check pilot had exited the helicopter to recover some sling gear that had recently been left at the site. When the check pilot was approximately 100 feet away, the helicopter began to rock violently fore and aft. As the check pilot ran back to the helicopter, he observed that the tip path plane was in the full forward position. The check pilot entered the cockpit and immediately shut down the helicopter. Examination identified that the transmission isolation mount, the support plate and the pin assembly had sustained substantial damage. In both cases, it was determined that the cyclic had been moved to an extreme forward position.

The pilot's decision to land at the private recreational campground was based on his desire to conduct some business at his trailer. After landing, the pilot had two options with regard to exiting the helicopter. In both cases the possible outcomes were negative. One choice was to shut down the helicopter, as required by existing regulations, and lose approximately seven minutes of time. This was the less risky choice; however, the flight was already behind schedule, and the known outcome was a further loss of time. The second choice was to leave the helicopter running, have the fixed-wing pilot monitor the controls, and thereby lose less time. This was more risky in that the outcome could be far more disastrous; however, the risk probabilities were unknown. Considering that the entire trip, including stops, would have required approximately nine hours to complete, the additional time that would have been lost by shutting down was relatively insignificant.

Given that this was personal rather than official business, and that the real purpose of the flight was to visit operators, it is probable that the pilot wished to minimize any inconvenience to his passengers. The fixed-wing pilot (and the other Transport Canada official) had sufficient aviation experience to encourage a multi-person decision-making environment; however, they were relatively unfamiliar with helicopters. As well, an atmosphere of cooperation and congeniality existed and the decision to leave the helicopter running was never challenged. The pilot had the final authority in the decision-making process and, while he may have realized that option one was safer, he chose option two because the perceived outcome was more desirable. His decision to leave the helicopter in the care of the fixed-wing pilot may have been reinforced by his previous experiences of exiting a running helicopter to permit students to shut down on their own, with no adverse consequences, and by the willingness of the fixed-wing pilot to accept the responsibility of holding the flight controls. The pilot overestimated the fixed-wing pilot's helicopter skills and underestimated the likelihood of a negative outcome.

It could not be determined why the helicopter initially began to oscillate; however, it is probable that the condition occurred due to a combination of factors, including the excessive flapping of the main rotor due to a wind gust, and/or an inadvertent movement of the cyclic from the center position. The fixed-wing pilot responded by moving the cyclic forward; which induced severe mast bumping. This action was the misapplication of a normal fixed-wing corrective action. The mast bumping was sufficiently intense to damage the transmission isolation mount, the transmission mount plate and support, and the transmission pin assembly, and to permit the rotating swash-plate pitch horns to contact the surrounding cowling. When the swash-plate rotation was restricted due to pitch horn interference with the cowling, the collar set failed in overload. The main rotor pitch links subsequently failed due to overload when the flailing drive link assembly jammed and further restricted the swash-plate rotation.

Source: Transportation Safety Board of Canada

There are many cases where the pilot simply forgets to engage any control locks whatsoever. Distractions like talking on the radio, conversing with crew or passengers, answering cell phones, and so on, are normally to blame. The power creeps in, the collective control slowly rises, and there follows a few seconds of high excitement and an unbelievable bill for damage. With luck, no one is injured or killed.

During the 1980s in New Plymouth (North Island, New Zealand) engineers were working on a Hughes 500 helicopter in front of a hangar. No pilot was present but they thought it prudent to run the helicopter up to full power for some test or other. Two men were painting the roof of the hangar and weren't surprised by the sound of the helicopter taking off. Naturally, they glanced up as the helicopter passed and were shocked to see that there wasn't anyone in the helicopter – pilot or engineer! As it turned out the engineers were not attempting to install a remote control kit and, at around 50 feet, the machine inverted and destroyed itself.

How do you avoid problems? Never having a helicopter running if it is unattended by a qualified pilot is an ideal rule. But it is also an impractical rule unless you have a two-pilot operation or you work in an environment where time and money permit a shut-down every single time you leave your seat. We all have lapses of concentration sooner or later, but engaging the control locks has to be as much of a habit as pre-flight checks, fuel drains and all those other myriad of things pilots do to ensure a safe operation.

Of course, control locks should not be forgotten at the end of the landing either. Disengaging control locks before you resume your flight is just as important as engaging them when you land and leave. Flying helicopters requires skill, concentration and great attention to safety at the best of times; limit the availability of all or any one of the controls and it makes the task a great deal more difficult.

PROTECTIVE COVERS

Forgetting anything, even the most important of things, is not a hugely complicated task. That is why we have checklists. Some checklists are written down with each item ticked when complete, others are retained in the mind, possibly by using a mnemonic. Neither method is failsafe.

Case study 20:10

Report 11/76

Date: March 8, 1976

Location: North Sea

Aircraft: Wessex 60 Series 1

Injuries: nil

The helicopter was operating in the West Sole area of the North Sea ferrying gas rig personnel between various platforms. It was due to return to its shore base at Easington near Humberside when the weather deteriorated to a visibility of 1 nautical mile in heavy snow. The pilot accordingly delayed his departure from platform A until the weather improved and during the time the aircraft was parked he fitted the engine intake cover.

At about 1250 hours the weather was considered by the pilot to be suitable for flying and the passengers were called forward. The pilot ordered a further 200 pounds of fuel to be put into the aircraft and, whilst this was being done, he interrupted his external check to enter the cockpit to monitor the refueling. Whilst he was in the cockpit he also amended the passenger manifest and, as he did so, the platform superintendent indicated that all was ready for take off. The pilot then started the engines and a few minutes later took off.

The aircraft first flew to platform B and then to platform C, each flight lasting 2 and 3 minutes respectively, and embarked further passengers to a total of 13. At 1315 hours the aircraft took off from platform C for Easington and climbed to a cruising height of 300 feet in conditions of moderate snow. As the aircraft was

Fig 20.8 A Wessex 60 helicopter (not the one imentioned in this report)

CAZ CASWELL

being leveled at 300 feet, the pilot heard a slight thump and, thinking that this may have been due to ice on the rotor blades, looked outside the cockpit to check.

On resuming his instrument scan he noticed that a light on the centralized warning panel indicated that the port engine had flamed out. Assuming that this was due to snow ingestion the pilot immediately retarded the port engine speed select lever in the hope that the engine may not have fully run down. At the same time he ordered the passengers via the intercom to put on life jackets. He then turned the helicopter to port to return to platform C and, as he did so, he heard another thump followed immediately by a loss of rotor rpm.

The pilot at once initiated autorotation and within a few seconds the helicopter was close to the surface of the sea. He then flared the aircraft to reduce its speed to zero and after allowing it to drop about 15 feet, raised the collective lever fully. The aircraft dropped into the sea, which was calm, from about 15 feet. Initially the water rose to the level of the cockpit, but as soon as the flotation bags inflated, which they did immediately, the aircraft settled upright with about 12–18 inches of water in the passenger cabin.

Shortly afterwards all the occupants of the aircraft were able to disembark unhurt and board the life raft. They were picked up some 25 minutes later by a rig support vessel.

The port wheel flotation bag failed about one hour after the aircraft had entered the water, allowing it to roll over and float with the tail rotor clear of the surface. The aircraft sank later that evening in 90 feet of water whilst an attempt was being made to tow it ashore. It was successfully salvaged and brought ashore 12 days later.

The aircraft was undamaged by the ditching but was subsequently destroyed by the effects of immersion and movement on the sea bed.

The commander was a male aged 54 year with an airline transport pilot's license (helicopters) and total flying experience of 7467 hours.

The stoppage of both engines within a few seconds of each other was positively established as being due to the ingestion of the engine air intake cover, which had inadvertently been left in position by the pilot. It was remarkable that the engines apparently operated normally with the cover in position whilst the helicopter was being flown between platforms A and B and thence to C, and presumably they would have continued to do so for the remainder of the flight to Easington had the starboard securing strap not become undone. Presumably the engines were able to operate with the cover in position because sufficient air could be drawn across its unsecured upper edge.

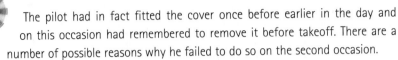

The pilot had in fact fitted the cover once before earlier in the day and on this occasion had remembered to remove it before takeoff. There are a number of possible reasons why he failed to do so on the second occasion.

Firstly, his external check was interrupted by the need to board the aircraft to monitor the refueling. It has been well established that the interruption of a check can be a common source of error.

Secondly, the cover is normally fitted and removed by groundcrew personnel during overnight stops. It is not normally fitted during transit stops unless snow is falling, as it was on this occasion. Therefore the fitting and removal of the cover by the pilot was an infrequent occurrence and the necessity to check it before each flight would not have been uppermost in his mind.

Thirdly, having entered the cockpit initially to monitor the refueling, the pilot's attention was further distracted by the requirement to amend the passenger manifest. Shortly after that, his attention was directed towards the arrival of the passengers themselves and he then received the platform superintendent's signal that all was ready for flight. By this time the pilot clearly had entirely forgotten that he had not completed the external check before boarding. The cover itself cannot be seen from the cockpit and there was nothing to indicate to the pilot that it was still in position. It is surprising, however, that none of the rig personnel noticed that it was still on. Even if it had subsequently occurred to the pilot that he had not positively checked that the cover had been removed, he would probably not have been unduly concerned since he could have reasonably assumed that the engines would not start with it in position.

The operator is now considering the provision of a suitably conspicuous hard cover for the engine air intake which would certainly prevent the engines from being started. This will doubtless prevent a similar accident happening again, though obviously in no way does it obviate the necessity for an external check.

Source: Air Accidents Investigation Branch, United Kingdom

The initial training pilots receive is hugely important. If new pilots are taught well, the important things are remembered as a matter of habit, basic checklists can be recited word perfect and self-discipline is used to successfully resist the temptation to take shortcuts.

If you are interrupted during a checklist, be it pre-flight or pre-takeoff, simply start the checklist again. Asking crew and passengers to remain quiet while you carry out your checks, or any other important process, is a good practise to adopt.

Human factors

The modern use of the term 'human factors' seems to revolve around subjects like decision making and cockpit resource management. It has given rise to an industry that deals with ideas many older pilots regarded as basic airmanship. Much of what is covered is general knowledge readily dealt with by common sense, but some warnings and precautions may be new to private pilots and newcomers to aviation.

MEDICAL MATTERS

Are you fit to fly?

I'M SAFE

I   Illness: do you have any symptoms?

M   Medication: does your family doctor know you are a pilot?

S   Stress: are you upset after a quarrel?

A   Alcohol or drugs: are you influenced by anything that will impair your judgment?

F   Fatigue: have you had sufficient rest?

E   Eating: do you have sensible eating habits?

Noise and vibration

Like sight, hearing is something we take for granted until we lose it. The ear is a complex and delicate mechanism that can easily be damaged by loud noises. 'Noise', a subjective word, can best be described as undesirable or unpleasant sound. Sound is measured in decibels (dB): the zero level is defined as the weakest sound that can be heard by a person with good hearing in a quiet location. The loudest sound that most people can bear is 140 dB, with higher levels causing pain and nausea. Public health standards define a sound level of up to 75–80 dB as acceptable and require ear protection for continuous noise levels above 85 dB.

The range of audible frequencies varies widely, but the major conversational voice frequencies are between 200 and 6000 Hertz (Hz) or cycles per second. The scale of sound measurements is logarithmic: for every 6 dB increase, sound doubles in intensity. In a library the sound level is about 20–40 dB while inside a helicopter the sound level varies from 100–120 dB. Hearing loss among pilots is of concern, but it should not be forgotten that in discos or clubs the sound level can also be damaging. In addition, many people are developing significant hearing losses from excessively loud stereos and Walkmans.

The ear

The outer ear funnels sound waves to the eardrum, a thin semi-elastic membrane, causing it to vibrate. These vibrations are conducted across the middle ear by three impedance-matching bones to the inner ear.

The inner ear, which is the size of a pea, is set in dense bone and contains the organs of balance (vestibular apparatus) and the cochlea, a snail-shaped organ that analyzes sound vibrations as they pass along a narrow 'sounding' chamber. Here they are converted to nerve impulses, which are then conducted to the brain by the auditory nerve.

Hearing loss

Obstruction of the outer ear or damage to the middle or inner ear or to the auditory nerve will cause hearing loss. The most common cause of reduced hearing is obstruction of the outer ear, usually from a collection of wax and hairs. Wax is a natural lubricant, but its quantity tends to increase with exposure to noise or vibration. The outer ear is easily cleaned, but care must be taken not to injure the eardrum.

In pilots (or divers), damage to the middle ear is often caused by pressure changes. Inflammation or infection of the middle ear (common in youth) can scar and thicken the eardrum or interfere with the mobility of the three bones that transmit sound. Damage to the inner ear and the cochlea is usually due to a sudden loud noise, which may permanently injure a segment of the cochlea, or to continuous noise at a lower dB level. Damage to the auditory nerve itself is uncommon. Hearing deteriorates with age, particularly in the high frequencies. Only in very severe cases does this affect the conversational range.

Effects of noise

A noisy environment interferes with communication and decreases intelligibility. Job-related noise, such as engine sounds, is tolerated better than extraneous noise. In its effect on work capacity, intermittent noise is more detrimental than steady noise and high frequencies are less tolerated than low frequencies. Noise increases task errors, but does not usually reduce work rate. Noise can be stressful and distracting, and prolonged noise is a significant cause of fatigue. Aside from hearing damage, noise can cause irritation and frustration.

Noise protection

Aircraft noise varies widely by type. In rotary-wing aircraft, for example, there are two peaks. The gear box and drive trains emit frequencies between 400 and 600 Hz and the rotors a lower-pitched frequency of 12–125 Hz. The older piston aircraft, which often had noise levels between 100 and 130 dB, caused damage at 4096 Hz, which was a harmonic of their engines' sounds. Modern jets are somewhat quieter, except on takeoff and landing when levels of up to 120 dB may be measured in the cockpit.

Noise protection generally involves preventing sound waves from entering the outer ear or dampening their vibrations. This is accomplished by ear plugs, headsets or helmets. All of these methods of sound protection require well-fitted equipment. Placing cotton wool or ill-fitting pads in or on the ears is ineffective and can be dangerous. Active noise-canceling earphones are now available, which include a microphone that picks up extraneous sound and separates it from the signal being received from the radio or intercom. The extraneous noise is subtracted by feedback. These are extremely effective but, at present, expensive.

However, the importance of hearing protection to the pilot, both on and off the job, can scarcely be overestimated. Nobody employs deaf pilots!

Vibration

Sound is a vibration whose frequency lies within our range of hearing. Any form of motion that repeatedly alternates in direction is a vibration. The range is enormous, varying from an earthquake to the infrasonic (inaudible to the human ear) vibrations of our vocal cords or chest. Vibrations particularly affect mobile body parts such as the head, shoulder girdle and limbs, and the large organs such as the liver. Intense vibration interferes with visual acuity and neuro-muscular control, particularly of fine movements. It is a significant cause of fatigue.

Helicopter pilots experience vibration created by the rotors at 4 Hz and its harmonics and from the engines at 60 Hz: these problems increase with air speed and are worst at transition to hover. Noise, vibration and turbulence combine to cause serious fatigue problems in helicopter pilots, and vibration is also a factor in the development of chronic back pain, common amongst helicopter pilots. Vibration can rarely be completely controlled, but it can be dampened. Dampening consists of putting a barrier between the pilot and the source to reduce the effects: shock absorbers on a car are an example. Dampening can be accomplished by reducing the source of vibration or by modifying the transmission pathway. Much work is being done on dampening helicopter seats; unfortunately, modifications often interfere with the crashworthiness of the seats. Pilots who think they are doing themselves a service by using a big foam pad as a seat cushion are in fact exposing themselves to the risk of serious back injury in a high-vertical-speed impact.

Colds

We have all suffered from the common cold. When you have the sniffles and are just getting a cold, should you go flying? The symptoms of an oncoming cold can develop into the symptoms and associated problems of a serious cold in a matter of hours. It's called a 'common' cold for good reason — they are common. So common that nobody attaches much seriousness to them. It's quite easy to ignore them and often pilots get away with this . . . but the potential is there for things to go seriously wrong in more ways than one.

Case study 21:1

Occurrence Brief Number 199400960

Date: April 15, 1994

Location: south of Brisbane, Queensland, Australia

Aircraft: Robinson R22 Beta

Injuries: nil

The pilot substantially damaged the aircraft when he was seen by witnesses to land in the corner of a fenced paddock. A short time later the helicopter took off again but after climbing to about 10 feet above the ground, it began rotating and gyrating erratically, contacting the ground a number of times. It came to rest in an upright position but with the tail boom severed and damage to the main rotor assembly and gearbox.

The pilot reported that he had recently recovered from a viral complaint that was characterized by severe coughing bouts, but had been free of these symptoms for a few weeks. As he flew over the area of the accident, however, he had experienced the incipient stages of a coughing fit so he landed the helicopter, shut down the engine, and walked around for a short time until the symptoms disappeared. He then reboarded the helicopter to continue the flight but shortly after lift-off was overcome by a severe coughing fit. This caused him to partially lose control of the helicopter and it contacted either the fence or the ground.

Source: Bureau of Air Safety Investigation, Australia

Other hazardous symptoms

A cold reduces tolerance to fatigue, to cold stress, to hypoxia; and increases susceptibility to decompression sickness. A cold can lead to pressure vertigo, symptoms from self-medication or extremely painful ear or sinus block.

Sinuses are holes in the skull. They are there to reduce the weight of bone and to provide resonance for your voice. They are located between and above the eyes and nose (frontal), in the area adjacent to the nose and below the eyes (maxillary), two very small ones between and just below each eye (ethmoid), and down the line of the nose (either side of the septum).

In normal circumstances the interior of the sinuses is lined with a mucous membrane and vents to the nasal cavity. Each sinus contains air, which is affected by changes in outside air pressure. When venting normally, we don't even notice it, but if you have a cold or flu, the mucous membrane can swell and block or impede normal venting. Now the air is trapped in the sinus. When climbing or descending in an aircraft with a pressurized cabin, as the trapped air pressure changes, it cannot escape or increase and this causes pain. Sometimes (particularly during descent) the pain can be so intense and distracting that it can dangerously affect a pilot's ability to concentrate on flying. Helicopter pilots tend to think that they don't often fly high enough for this to be a problem, but it can be, even on descent from as low as 3000 feet. A bit of discomfort at that altitude can amplify to excruciating pain at sea level (for instance, when you're on short final). Blood vessels in the cavity can burst and fill the sinus with blood, vision is blurry or double and someone is driving a red hot spike into your head and your head feels like it's going to explode. With all that happening are you really flying safely? Will you notice that you are slowing and losing translational lift? Can you remember whether that light wind is a tail wind? Who put that hangar there?

Other air traps

There are other air traps too — not related to a cold perhaps, but it is worth mentioning that you should avoid flying for 24 hours after having a tooth filled — there may well be air trapped under the new filling!

A big danger is the eardrum — perforation is possible. The pain and/or discomfort even when perforation doesn't occur is easy enough to imagine. If you cannot clear your eardrums using the Valsalva method — block your nose, close your mouth and 'blow' — then don't go flying. If you are actually flying, descending, feeling discomfort in the ears and Valsalva doesn't work, you may be in trouble.

The Eustachian tubes between your ears and your throat become blocked by secretions and inflammation. If they are totally blocked there is no way to equalize the pressure; the pressure on the outside of the eardrum is much higher than that on the inside. One way for natural equalization to take place is for the eardrum, which is only a few cells thick, to perforate and let air in — ouch! What normally happens is that the inner ear fills with fluid (particularly blood) until the inside air is at the same pressure as the outside air. There is a large loss of hearing, and the fluid takes at least two weeks to dissipate, during which time you cannot fly. Permanent damage may result in the form of scarring to the tiny bones in the inner ear.

So, there you are — descending, unable to clear your eardrums and uncomfortable with 4000 feet to go — what do you do? If you are short on fuel, accept what is going to happen, grit your teeth and fly the aircraft to the ground, descending as slowly as possible. But don't let your discomfort distract you from maintaining control. An aircraft crash can be fatal; a blocked ear is usually not.

If you have fuel to burn, climb until you are comfortable, force a Valsalva, descend slowly and keep forcing the equalizations (almost constantly) as you descend. Bear in mind what kind of torture an autorotation would mean in this situation and the unlikelihood of its success — don't run out of fuel!

Be very careful when taking medicines to prevent 'hay fever' or relieve cold symptoms. Many adversely affect flying; some of them mean you should not be flying at all. If you are thinking about using medication, play it safe and consult a designated aviation medical examiner for advice.

There are thousands of aviators out there who are reluctant to report in unfit for duty when they have a cold coming on or are actually suffering from an active cold. There are a few hundred who have had very bad experiences and now don't hesitate to report in sick. There are a few who will die because they don't report in sick, and dozens who will infect fellow pilots because they do go to work, forcing those pilots to then make the call for themselves. What do you think your call should be?

Intoxication

There can be no doubt that consumption of alcohol or cannabis causes thought and decision-making processes to deteriorate. The rate of deterioration is rapid and directly related to the rate and amount of consumption. Great amounts at high rates can cause unconsciousness. But armed with a weapon like a helicopter, even a small amount can have catastrophic results. Decision making and overall performance are not the only things adversely affected; intoxicated pilots are much more susceptible to disorientation (see pp 357–77).

Case study 21:2

Occurrence Brief Number 199903335

Date: November 3, 1999

Location: Ross River Homestead, Northern Territories, Australia

Aircraft: Robinson R22 Beta

Injuries: 1 fatal

The helicopter had been engaged in cattle mustering operations on the day of the accident. Late in the afternoon the pilot invited one of the stockmen to accompany him on a flight to a nearby tourist resort to purchase bread for the stock camp. They arrived at Ross River Homestead resort at about 1700 CST and some time later decided to remain overnight at the resort. Witnesses reported that the pilot had consumed a quantity of alcohol during the course of the evening.

At about 2345–2400 hours witnesses heard the helicopter engine start and run for a period of time before the helicopter was seen to take off and depart in a northeasterly direction. It climbed steeply to about 600 ft above ground level, after which the engine noise appeared to change and the aircraft descended quickly until impacting with the terrain. Searchers found the wreckage soon after first light the next morning on a flat area of land between hills, approximately 850 yards from the resort.

Witnesses reported that there was no wind at the time of the accident. There was high level cloud; overcast and very dark conditions. Examination of the astronomical ephemeris (a table of the moon's position), confirmed that the moon did not rise until approximately three hours after the accident. The helicopter was not equipped for flight under the instrument flight rules.

The evidence showed that the helicopter impacted the terrain banked to the right, in a nose-low attitude, and at high forward and vertical speeds. Impact forces destroyed the forward right and central cockpit area of the aircraft. The investigation could find no evidence of pre-existing damage to any of the helicopter's flight control systems. The type of damage to the main and tail rotor blades indicated low power and low rotor rpm at impact. Examination of the engine indicated that it was either at idle or a very low power setting at impact. The investigation determined that sufficient clean fuel of the correct grade was on board the helicopter at the time to power the engine. No defect was identified that may have influenced the circumstances of the accident.

Due to the severity of the impact, the accident was not survivable. The pilot had 1550 hours of flight logged.

Source: Bureau of Air Safety Investigation, Australia

In case study 21.2, the decision to fly a non-instrument flight rules equipped helicopter and the pilot's increased susceptibility to disorientation proved fatal. The answer is simple: avoid the problem by not flying after consuming alcohol.

Blatant disregard for the dangers of alcohol, prior to or while flying, has been the cause of many accidents. Just drinking, resting, and then flying is very risky. Individuals who believe their skills to be either enhanced or unaffected by drinking should be stopped, using physical force if necessary, from flying. Their passengers may not be aware of the danger to which they are about to expose themselves.

If you have drunk alcohol and are planning to fly, looking for an instant recovery is not wise. Various 'cures' for a hangover have been mooted and some may indeed lessen one symptom or another, but there are none that will immediately restore sound thinking and good judgment.

The general guideline is:

➤ If you consume one or two alcoholic drinks, don't fly for another eight hours.

➤ If you consume more than two standard drinks, don't fly for another 24 hours.

Even though our livers and other bodily functions have expunged the alcohol from our bloodstream, the effects on the brain continue for much longer as there are still traces there and in the nervous system.

Intoxication from illicit drugs is no less dangerous than alcohol. Even some legal non-prescription and prescription medicines can affect pilots' ability to fly to their usual capabilities.

Mixing alcohol and drugs, prescription or otherwise, produces effects that are far worse than taking one on its own. If you're taking strong medication for a cold, smoking cannabis regularly and have a night on the alcohol, go to bed. Walking onto an apron to start a helicopter is like handing someone a gun and asking them to shoot you in the head — same result, less mess.

Flying and intoxicating drugs of any description are not a good mix, and the inappropriate actions of some individuals are to the detriment of the aviation industry as a whole. Be proactive about encouraging others not to fly when they are anything less than 100 percent capable.

Fatigue

There are two types of fatigue: acute (short term) and chronic (long term).

Acute fatigue can be the result of a long hard day — an early start, a hectic schedule, flying around adverse weather, a groundcrew who weren't as efficient as usual, and difficult passengers. Or perhaps you were giving lesson after lesson in hot weather to students who weren't concentrating at the ground briefing and were having trouble performing.

After such stress acute fatigue is perfectly natural. You can't concentrate on monotonous tasks and can't think clearly enough for the complex jobs. The only cure is a good night's sleep.

Chronic fatigue is a lifestyle issue. Work is a part of your lifestyle, as are relaxation, diet and exercise. Consider how much stress you are under; chronic fatigue can be a direct result of chronic stress. If you have good nights of sleep but consistently wake up still feeling tired, you may need to change your lifestyle.

Sometimes chronic fatigue is a symptom of a more serious illness (psychological or medical). A designated aviation medical examiner could be helpful. Alternatively, there are now several sleep clinics with specialists to diagnose sleep disorders. A surprisingly common problem is sleep apnea, a life-threatening condition whether or not you are a pilot.

As a pilot, you need to always be very aware of fatigue; and you must accept that responsibility for avoiding it is in your hands. The same can be said for any condition that robs you of your ability to concentrate and perform.

Stress

Stress is such a common word these days that, like the cold, we accept it as a concept but are far too busy to worry about whether or not we are suffering from it. Well, we are all stressed. Yes, even you. Put simply, stress is the physical state of arousal of our bodies. Your body is even stressed while you are reading this book. If you have no stress at all, don't call a doctor; call an undertaker. You have no brain activity or blood flow.

What is of vital concern, though, is the level of stress. Like fatigue, excess stress can be either acute (a sudden crisis) or chronic (a slow build-up over weeks, months, even years).

There is a chain reaction in the body with acute stress. The brain tells the hypothalamus — the bit that dictates your body's subconscious reactions — what is going on. The hypothalamus stimulates the pituitary gland to release a hormone that tells the adrenal gland to send out adrenalin and other hormones. This prepares the body for maximum performance, all in the space of a few seconds.

It is important to remember that it is the subconscious running a lot of the shop here. You may be sitting in a theatre watching a horror film and terrified by the image on the screen. Your brain knows that, barring earthquake or fire, you are in no danger at all, but your subconscious assesses what you are seeing as a real and present danger. The result? Adrenalin flows and your body gets a dose of heavy stress — and you go there for fun?

So what is happening during this adrenalin rush? Most of you will know of the 'fight or flight' response — either requires maximum performance; to win the fight or get the heck out as fast as possible. So the heart beats faster, blood pressure rises and blood is redirected away from the skin and digestive system and into the muscles. Breathing increases and the air passages dilate to allow maximum flow of oxygen to the lungs. Maximum energy is helped further by a release of more glucose into the blood.

How much benefit is all this when you are flying a helicopter? You require refined and precise control movements (as always) in the face of any emergency, but here your body is poised with so much power and energy that you feel you should change clothes in a telephone booth! There is also an element of rage involved with the fight-flight reaction, and rage is hardly a suitable emotion for a pilot battling an emergency situation.

How do you control this acute stress response? Fact of the matter is that, in essence, you cannot totally control it. But you can moderate its effects. Be aware of your breathing rate and try to slow it down. Try to relax a little and release the tension in your muscles. Above all, be aware of your 'self-talk', those internal thoughts that are racing through your mind at what seems like 10 gigabytes a second. Assess the situation, be honest about any risks involved, and direct positive and creative thought toward a solution and recovery.

I once made a horrendous decision to fly into a lowering cloud base and rising terrain in a Piper Warrior on Visual Flight Rules. The worst did come to the worst and I found myself flying in formation with a set of high tension wires. I could just see from one pylon to the next. As a frequent reader of aircraft accident reports, my mind formulated the wording of one that was going to feature me: 'Injuries, crew — 1 fatal'. My rage was directed at myself; I was going to be thought of as an idiot by my peers for making such a stupid decision. It was more pride than survival instinct that made me determined to get through the situation one way or the other. I controlled everything pretty well; tried things like sunglasses on or off for better visibility. I was alone, the load was light and, with two notches of flap, I was confident that 75 knots was the best trade-off not to stall but to see things in time to react. By the time I got through the high ground and back into some clear air (around 10 minutes), I was pretty exhausted, elated and disgusted all at the same time. A valuable lesson learnt but the fee was almost exorbitant.

It is all about ensuring that the pilot in command is also a pilot in control — and not just of the aircraft. Training for emergencies is routine, repetitive and often ho hum, but it does hone your instincts and reflexes. If you get it 100 percent right 90 percent of the time in practise, you should get it 80 percent right in the stressful environment of reality and, with luck, that will make the difference between walking away and not walking away.

Acute stress is natural and quite healthy in small doses; some daredevils even chase it. Chronic stress is quite another matter. It is negative and potentially destructive. You can get stress from work, from home, from within (attitudes) and from social settings. It can be cumulative and it needs to be controlled. Unfortunately, the human being is a complex beast and it is very difficult to assess just how much stress an individual (including yourself) is under.

There are four groups of indicators that may reveal stress levels:

➤ Mental — suffering from frequent mental blocks, reluctance to decide (procrastination), difficulty in concentrating on one task, having a reduced attention span, and forgetting things like names and places that were once familiar.

➤ Physical — high pulse rate, dry mouth, heart disease, headaches (sometimes severe), ulcers, rashes, asthma, hot flushes, sweating, insomnia and nightmares.

➤ Emotional — mood swings, short temper, weariness, apathy/depression and anxiety.

➤ Behavioral — daydreaming at inappropriate times, restlessness, loss of job satisfaction and enthusiasm for work, doing things to excess (drinking, smoking, eating or drug-taking on a regular basis).

There are three identifiable stages of coping with chronic stress:

1. Alarm — acute stress.
2. Resistance — acceptance of the situation and a resolve to battle through. Behavior outwardly shows acceptable performance but all the while stress is actually accumulating.
3. Collapse — the point where the body, brain or both simply say, 'Enough is enough', and shut down to some extent, often in a spectacular fashion. All reserves have been burnt out. At this stage, the road to recovery can be long.

Many pilots resign, take early retirement or lose their medical clearance when they reach or are near a state of collapse. This is a huge shame as there is a lot that can be done to alleviate stress and control it to an acceptable level that offers quality of life. We hear of people being 'in denial', refusing to accept that they are under any excessive stress, internalizing their feelings, often subconsciously. It is ironic that there is the perception of a stigma in admitting stress levels are 'getting to you'. And the threat of the stigma is another source of stress in itself. You need to quantify things; get them in perspective. Write down all your 'stressors' (things causing your stress) on a piece of paper and think about them individually. Just how great is the threat? What are the real consequences? Is your subconscious exaggerating them?

Discuss your stressors with someone you trust; this alone can often cut short the resistance stage and give you a positive outlook. Change some things in your life; perhaps your attitude, your fitness level, your method of relaxation, or a stressor that you have identified as unacceptable. There are as many different ways to relieve stress as there are individual people. If you are having trouble sorting out why you are stressed or what it is best to change, consider getting some expert advice — most such professionals don't claim to be able to fly a helicopter; most pilots aren't fully conversant with managing stress.

Hypoxia

Lack of oxygen is the greatest single danger to humans at high altitudes. Shortage of oxygen in the human body results in a condition called hypoxia, which simply means oxygen starvation. When a pilot inhales air at high altitudes, there is not enough oxygen pressure to force adequate amounts of this vital gas into the bloodstream, so it can be carried to the tissues of the body. The function of various organs, including the brain, is then impaired.

Unfortunately, the nature of hypoxia makes you, the pilot, the poorest judge of when you are its victim. The first symptoms of oxygen deficiency are misleadingly pleasant, resembling mild intoxication from alcohol. Because oxygen starvation strikes first at the brain, your higher faculties are dulled. Your normal self-critical ability is out of order. Your mind no longer functions properly; your hands and feet become clumsy without you being aware of it; you may feel drowsy, languid, and nonchalant; you have a false sense of security; and the last thing in the world you think you need is oxygen.

As the hypoxia gets worse, you may become dizzy or feel a tingling of the skin. You might develop a dull headache, but you are only half aware of it. Oxygen starvation gets worse the longer you remain at a given altitude, or if you climb higher. Your heart races, your lips and the skin under your fingernails begin to turn blue, your field of vision narrows, and the instruments may begin to look fuzzy. Hypoxia is a grim deceiver. It makes you feel confident that you are doing a better job of flying than you have ever done before. But regardless of acclimatization, stamina, or other attributes, every pilot will suffer the consequences of hypoxia when exposed to inadequate oxygen pressure.

What do you do about it? Leaving aside the regulatory requirements for the moment, there is one general rule. Do not let hypoxia get a foot in the door. Carry oxygen and use it before you start to become hypoxic. Do not gauge your 'oxygen hunger' by how you feel. Gauge it by using your altimeter and your watch. The light aircraft pilot flying non-pressurized types need remember only the following:

1. Oxygen must be used by all crew members for any period of flight in excess of 30 minutes above 10,000 feet and up to 13,000 feet pressure altitude.

2. Oxygen should be administered to passengers whenever it is evident that there are indications of hypoxia.

No one is exempt from the effects of hypoxia. Everyone needs an adequate supply of oxygen. A few pilots may be able to tolerate a few thousand feet more altitude than others, but no one is really very far from average.

Remember this: serious trouble lies in wait for pilots who try to test themselves to prove how much higher they can fly or how much longer they can function without supplemental oxygen. Pilots who are older, fatter, out of condition, or smoke heavily should limit themselves to a ceiling of 8000 to 10,000 feet unless oxygen is available.

Hyperventilation

Some people believe that breathing faster and deeper at high altitudes can compensate for oxygen lack. This is only partially true. Such abnormal breathing, known as 'hyperventilation', also causes you to flush from your lungs much of the carbon dioxide your system needs to maintain the proper degree of blood acidity. The chemical imbalance in the body then produces dizziness, tingling of the fingers and toes, sensation of body heat, rapid heart rate, blurring of vision, muscle spasm and, finally, unconsciousness. The symptoms resemble the effects of hypoxia and the brain becomes equally impaired.

You are most likely to hyperventilate while flying under stress or at high altitude. For example, the stressful feeling of unexpectedly entering instrument conditions, noting both fuel gauges bouncing on empty, or developing a rough-running engine over water or mountainous terrain may make you unconsciously breathe more rapidly or more deeply than necessary.

A pilot who suffers an unexpected attack of hyperventilation, and has no knowledge of what it is or what causes it, may become terrified, thinking that he or she is experiencing a heart attack, carbon monoxide poisoning or something equally ominous. In the resulting panic and confusion, they may lose control of the aircraft; exceed its structural limits and crash.

A little knowledge is all you need to avoid hyperventilation problems. Since the word itself means excessive ventilation of the lungs, the solution lies in restoring respiration to normal. First, however, be sure that hyperventilation, and not hypoxia, is at the root of your symptoms. If oxygen is in use, check the equipment and flow rate. Then, if everything appears normal, make a strong conscious effort to slow down the rate and decrease the depth of your breathing. Talking, singing or counting aloud often helps. Normally paced conversation tends to slow down a rapid respiratory rate. If you have no one with you, talk to yourself. Nobody will ever know.

Normal breathing is the cure for hyperventilation. The body must be allowed to restore the proper carbon dioxide level, after which recovery is rapid. Better yet, take preventative measures. Know and believe that over-breathing can cause you to become disabled by hyperventilation.

Hypoglycemia

Fuel for the pilot is just as important as fuel for the aircraft — and neither can operate without an adequate supply of the right kind.

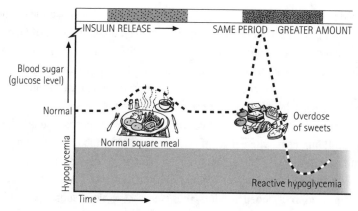

INSULIN RELEASE ⟶ SAME PERIOD – GREATER AMOUNT

Blood sugar
(glucose level)

Normal

Overdose
of sweets

Normal square meal

Hypoglycemia

Reactive hypoglycemia

Time ⟶

Fig 21. 1 Blood sugar levels *NEW ZEALAND FLIGHT SAFETY*

Hypoglycemia is literally low blood sugar: 'Hypo' means low, 'glyc' sugar, 'emia' in the blood. In aviation, where human performance is crucial, we must be concerned about situations in which the blood sugar is in the 'normal' range, but either it is lower than desirable for the activity engaged upon, or it is at a 'normal' level but being kept there by intensive body system effort.

For want of a better term, we will discuss this aspect as 'hunger stress'. Either decreased intake or increased consumption of sugar, or a combination of both, can cause the blood sugar to fall. In healthy individuals, except in the type of case described later, blood sugar rarely falls below the normal range. Additional stresses, however, such as those experienced in aviation, require additional metabolic effort to keep enough sugar circulating to meet the needs of the brain and body. This extra effort reduces performance and increases the frequency of errors, both minor and major, sufficiently to increase the risk of accidents.

It may be claimed that humans weren't designed to eat three regular square meals a day — our ancestors, whose knuckles really did scrape along the ground, ate as and when they could get a meal. This often meant they gorged themselves fit to bust when they caught something, and later tightened their belts until they

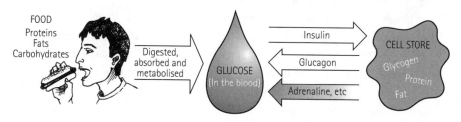

FOOD
Proteins
Fats
Carbohydrates

Digested,
absorbed and
metabolised

GLUCOSE
(In the blood)

Insulin

Glucagon

Adrenaline, etc

CELL STORE

Glycogen
Protein
Fat

Fig 21.2 Food assimilation *NEW ZEALAND FLIGHT SAFETY*

managed to nab their next meal. So how did they manage? Just like the dromedary with its hump, we have stores of energy that are laid down in times of plenty, and broken down to make sugar in times of famine. This process is going on all the time in between meals to keep the blood sugar within quite tight minimum and maximum limits. The process is controlled by hormones circulating in the blood, which are released in response to a rise or a fall in the blood sugar.

Insulin is one such hormone. It controls the entry of glucose (sugar, the basic energy unit) into cells, where it is either burned in the metabolic cycle or converted into stores of energy — either as glycogen (strings of glucose molecules joined together) or as fat. Increasing the amount of circulating insulin causes more glucose to be converted to glycogen and fats, thus causing the blood sugar level to fall.

Glycogen forms a readily available store of glucose energy, which is released rapidly when the glycogen is broken down. The agent which controls this breakdown, and thus causes blood sugar level to rise, is another hormone — glucagons. There are others, however, such as the 'injury hormones' and the 'stress hormones' (like adrenalin) which also break down the glycogen in order to ensure that plenty of sugar is available at the time of greatest need.

As long as there is a reasonable store of energy; the hormonal control of blood sugar levels is functioning normally; and glucose energy is required in moderate quantities and at a slow rate of delivery, then this system will keep the blood sugar from dropping too far – but at a price.

The first penalty is hunger, a sensation of stomach emptiness and a desire to eat. This is pleasant when mild, but as the pangs increase hunger rapidly becomes a strain, a distraction, and eventually a frank discomfort. The second penalty is the effect of the increased levels of stress hormones that have been released to carry out the conversion of stores to energy. These increased hormonal levels cause feelings of anxiety, fear, excitability, tremulousness and difficulty in concentration. All of these will impair aircrew performance and safety, even if the blood sugar levels are reasonably normal. Tolerance to fatigue, hypoxia, and heat stress (either hot or cold) is reduced. Tolerance of high-G maneuvers is also reduced but this is not of great concern to most helicopter pilots.

Under certain circumstances, 'real hypoglycemia' (less than minimum blood sugar levels) can occur and the commonest of these is self induced. It often occurs at the time of least resistance — disrupted programs, short and long term fatigue, uncertainty about weather for the flight, missing meals that you usually take (for example, a hearty breakfast), and operating in the early morning when the blood sugar is lowest anyway because of the body's biological rhythms. Drinking alcohol in abundance the night before also lowers your blood sugar, as well as your fluids.

Now, if you are due to take a meal at this time but have to skip it to catch the weather, or for some other reason, then your blood sugar will be low. But, you can really ensure that it will drop into your boots by doing what seems to be the sensible thing — swallow two cups of really strong coffee, laden with sugar, in quick succession and then chew a pocketful of chocolate bars while filing your flight plan and walking round the pre-flight checks. Initially your gut gets a huge dose of sugar, which gets absorbed rapidly and the blood sugar rises above normal limits — so far so good. You've kicked the tires, lit the fires and you're dialing up the pressure on the altimeter feeling good.

But then your body scuppers everything. Your brain detects the rapid massive rise in blood sugar and switches on the release of insulin to prevent this becoming excessive, and both the blood sugar and you level off. Good. But large amounts of sugar normally enter the body in meat-and-two-vegetable sized packages and are broken down slowly and absorbed over longer periods. Your system is used to this and continues to release insulin to meet the anticipated extra input. The problem now is that you have, in a very short space of time, taken on board all the sugar that the candy and sweet will give you, and the high level of insulin will now cause the blood sugar to plummet to levels when real hypoglycemia will occur. This is known as 'reactive hypoglycemia'.

What symptoms do you typically get from real hypoglycemia?

➤ Extreme hunger and fatigue, sensation of worry, feeling of intoxication or confusion, increased self criticism, headache.

➤ Trembling and difficulty in fine control movements.

➤ Sweating and flushing.

➤ Eventually loss of consciousness, convulsions, and potentially death.

You may occasionally get similar hypoglycemia following violent exercise, having skipped a meal, when there is a rapid demand for glucose energy before the glucose used up can be replenished.

You can avoid hypoglycemia:

➤ Be aware of the factors that can cause or aggravate the effects of reduced blood sugar:
   — missed meals
   — too much sweet food consumed too quickly
   — period of intense workload

- heat or cold stress

- hypoxia

- illness (like a cold or stomach upset)

- fatigue

- alcohol or drugs, both prescribed and recreational.

➤ Eat sensibly and do not miss meals that you normally take. If you want to fly, never ever miss more than one mealtime, even if this is one that you routinely miss (for example, people who skip breakfast). If you are on a weight-reducing diet, drop the calories after flying, not before.

➤ If you really must miss a meal, do not 'load up' with sugar, but have at least some moderately sweet coffee and a bun before you fly and, if the flight is intended to be longer than a local sortie, carry something to eat along the route to top up. In these circumstances, also always ensure that your fluid intake is adequate.

➤ If you intend high-G maneuvers; or you're likely to get thermal stress in flight; or you're flying at night or early morning; make sure that you get a square meal and adequate fluid before the flight.

➤ Eat to live so that you can live to eat!

Dehydration

This section has been adapted from the article 'I need a drink', *Vortex*, 3/2002.

The classic image of a dehydrated pilot is one stumbling through the desert in a torn flight suit, a tattered parachute in tow, kicking up dust with every scuff of a pair of worn-out boots. Sweating bullets, the pilot is trying to suck the last drop out of a dirty canteen, and gasping, 'Water! Water!' The sun is a huge fireball in the sky, and the heat is shimmering off the ground like it's alive. While this version is great for movies and beer commercials, the dehydrated pilot can take on a much less dramatic profile; one you probably see more often than you think.

Helicopter pilots can be exposed to climatic extremes on a regular basis. Many pilots spend hot summer days inside a poorly vented, greenhouse-like structure, with brief periods outside to load cargo, throw around a few 100-pound nets, or roll some fuel drums across a soft bog, or in snow; and in winter there is the added pleasure of putting the aircraft covers on in the evening. This is best accomplished in a nice 35-knot breeze.

Much of the work pilots do, and the environments in which they work, exposes them to a greater than average rate of dehydration. The following material contains a discussion on how and why the body uses moisture, the mechanics of dehydration, what effect it has on performance and health, and how it can be prevented in our working environment.

The human body is primarily composed of water, about 70 percent, in fact. To put this in perspective, a 182-pound man is likely to contain over 11 imperial gallons, or 50 liters of water in and around all the cells of his body, and in his bloodstream. This water is used for virtually every function the human body performs: regulating its temperature, eliminating waste, digestion, transporting nutrients and disease-fighting agents, and even has a role in neurological functions and the thought process. It is truly what keeps us alive.

Water enters our bodies when we eat and drink, and exits in three primary ways: urination, perspiration and transpiration (breathing). The body can also lose fluids through vomiting, diarrhea, excessive mucus production, and so on. Under normal circumstances, a body will lose between 2 and 2.5 liters of water in a 24-hour period, or about 2 to 3 percent of total body weight.

As mentioned earlier, water is used in virtually all of our metabolic processes. Urination is one method the body uses to dispose of its waste material, and urine is a good indicator of our state of hydration. Normally, it should be relatively clear with a yellowish tint, whereas darker yellow and often pungent-smelling urine is a signal that you need more water. This will often be your first sign of dehydration, even before you get thirsty. Use of diuretic substances (those which increase the production and excretion of urine) like alcohol, caffeine and many carbonated soft drinks, will have a major impact on dehydrating the body. In these cases, the urine may not appear dark in color, but fluid levels are being significantly depleted. Less frequent, but still very significant, are the excessive fluid losses through diarrhea and vomiting. The average body loses approximately 0.2 liters of fluids per day through normal bowel movements. This can literally become liters with diarrhea.

Perspiration, one of the body's ways of temperature regulation, is another large consumer of fluid reserves. Our brain senses an increase in body temperature and takes action in two ways to bring the core temperature down. The body secretes sweat, which cools the surface of the skin through evaporation. At the same time, there is increased blood flow to the skin as the blood vessels have expanded (vasodilatation). The temperature of the blood is lowered as it flows through the cooled skin, then the newly cooled blood returns to the body's core. There, it picks up excess heat, and the process continues until a satisfactory temperature is reached. Temperature alone can cause the body to heat up and perspire, but performing work will exacerbate the creation of heat, hence we sweat more when working hard.

Some facts about perspiration:

➤ It works only when the sweat can evaporate, so tight clothing, or clothing which doesn't 'breathe', will hinder cooling.

➤ When perspiration drips off or is wiped away, it has afforded no cooling benefit and is wasted.

➤ In high humidity, less evaporation occurs; therefore less cooling. This is why we feel hotter on a muggy day.

We also lose water when we breathe, as evidenced by breathing on your sunglasses before cleaning them, and by the ability to see your breath when it is cold. In fact, it is during cold dry weather that we lose the most moisture through transpiration. When cold dry air is inhaled, it is warmed by body heat. While the air is in our lungs, it absorbs moisture, which in turn leaves the body when exhaled. In a harsh winter, this can be a significant cause of dehydration, made even more insidious by the fact that we don't usually feel thirsty when it is cold.

So what happens when the lost water is not replaced? The onset of dehydration appears in stages. After the loss of approximately 1.5 liters of water (about 2 percent of the total body weight), we get thirsty. At the 3-liter mark, we are getting sluggish and tired, maybe nauseated and irritable. This is a very dangerous level for pilots, as this is where your faculties start to become affected, but you may not be aware of the deteriorating performance.

United States army experiments on helicopter pilots clearly indicate that self-reporting is notoriously inaccurate, even at relatively early stages of dehydration. Aircrew who felt no adverse effects had clear objective difficulty with cognitive tests.

By the time the body has lost 4 liters, it has entered a stage of clumsiness. Headache is likely, along with an increase in core temperature, heart rate and breathing rate. At this level or beyond, you have ceased to function effectively, as your body copes with the fluid loss.

One serious episode, or several repeated moderate episodes of dehydration can result in kidney stones, which are stone-like masses of mineral salts. They can cause intense, incapacitating pain when passing through the urinary tract, and other symptoms like fever, chills, blood in the urine, nausea and vomiting. There are several manners in which kidney stones may be treated. Shock wave lithotripsy

Water lost (Liters)	Symptoms
1.5	Thirst
3	Sluggishness, fatigue, nausea, emotional instability
4	Clumsiness, headache, elevated body temperature, elevated pulse, elevated respiratory rate
5	Dizziness, slurred speech, weakness, confusion
6	Delirium, swollen tongue, circulatory problems, decreased blood volume, kidney failure
9	Inability to swallow, painful urination, cracked skin
12	Imminent death

(breaking stones) literally pulverizes them into smaller pieces that can be passed more easily. Other methods are a little more invasive. Diagnosis of a kidney stone may have an affect on pilot license privileges. Anyone who has ever had kidney stones will tell you, you don't want to be dealing with one of these things while flying an aircraft; the pain can be overwhelming and debilitating. Thankfully, simply keeping the body well hydrated can prevent stones in most people. A friend who suffered a bout of kidney stones a number of years back is rarely seen without his water bottle these days. I took it as a lesson learned.

How then do we prevent dehydration? Well, the simple answer is drink plenty of water; but we can do better than that.

➤ Recognize environments where the risk of dehydration is increased.

➤ When changing environments, like going from a cold climate to a warm climate or vice versa, the body can take up to two weeks to acclimatize. During this time, it may use more fluid reserves than it normally would.

➤ Do not rely on thirst to be the signal that you need water. By that time, you are already on your way to dehydration. In addition, drinking a small quantity of water, insufficient to rehydrate, may fool the thirst mechanism.

➤ Carry a container or bottle that allows you to monitor how much fluid you drink.

➤ Avoid excessive use of diuretics like caffeine, alcohol, and so on.

➤ Monitor activities such as exercise or heavy work, and rehydrate accordingly.

➤ Monitor your health state. Vomiting, diarrhea, fever, and many illnesses like influenza or the common cold, will cause the body to lose fluids at a much greater rate than normal.

➤ As a guide, drink enough water throughout the day to keep the urine relatively clear.

➤ Plan to carry sufficient water, and ensure it is readily available. This can mean up to eight liters per day in some environments, or even more in extreme cases.

➤ If water doesn't do it for you, try a sports drink. They are sold everywhere these days, and are great for replenishing fluids and restoring electrolyte levels, which may become depleted in episodes of heavy perspiration. If you find these drinks a tad expensive or unavailable in your bush camp, make your own. Here's a recipe that works well.

Vortex-ade
1 liter water
$\frac{1}{3}$ cup sugar
$\frac{1}{2}$ tsp salt
$\frac{1}{2}$ cup unsweetened orange juice, or add lime juice,
 lemon juice or lime cordial to taste.

As we can see, preventing dehydration is important all year round. Cold dry winter weather can sneak up and deplete the body of vital fluid without our even knowing it. In summer, keeping the body's fluid reserves topped up goes a long way to help prevent another killer — heat exhaustion.

Transporting patients

Modern trends have seen more and more patients and accident victims being transported by helicopter. Some pilots are gaining a huge amount of expertise and knowledge about differing conditions of patients and how they can be affected by air transport.

The primary change is, of course, air pressure. As an example, someone with a pneumothorax (trapped air in the lung that cannot be ventilated) is obviously going to be adversely affected by a change in air pressure.

Nobody expects a helicopter pilot to be a doctor or trained paramedic but it is important to carry a crew that is appropriately trained and have an aircraft that is properly equipped. There is then oxygen at hand and someone who knows how to put a drain into the lung prior to the flight.

But what of pilots who are not in a specialist medical role and find themselves faced with a situation where the obvious thing to do is to put an injured person in the helicopter and get them to hospital? It's quite simple really – do your best; it's an emergency. Knowing that the change in air pressure is the single most potentially problematic difference, you might fly at the lowest safe height.

Shock is a serious concern and, if the patient has little experience in helicopters (some people actually believe they are at risk flying in one), soothing words and comfort may be needed. It sounds simple but is often difficult to achieve. Remember, trauma and shock can kill quickly and effectively so keep the 'drama factor' as low- key as possible. If there is anyone nearby with advanced medical training or qualification, 'abduct' them for the flight.

Shade patients' eyes from direct sunlight and, at night, protect them from strobe or rotating beacon reflection; they may be susceptible to epilepsy.

Diving and flying

Picture a nice cold bottle of pop; just the thing on a hot day. Look at those bubbles fizz. Ever stopped to wonder where they come from? The manufacturer pumps gas in under high pressure during bottling. The moment you depressurize (take the cap off) it all comes out with bubbles forming from the gas that had been dissolved in the drink under pressure.

The 'manufacturer' of the human body pumps gas (nitrogen) into our bodies; this time at just one atmospheric pressure. It gets into the blood and other body fluids, into the brain, into joints, into the spine. And what a fizzer that can be if we have the pressure taken off!

Under normal atmospheric pressure the human body contains about $1\frac{1}{2}$ pints of gaseous nitrogen dissolved in body fluids; even more than that at altitude. So what happens is that, if you take the pressure off, some of that gas comes out of solution and bubbles form in the bloodstream or in the tissues. Luckily the amount of depressurization needed is great before large numbers of bubbles form. A scientist by the name of Haldane reckoned that the ambient pressure had to fall to half that normally experienced (that is, pressure under which nitrogen was forced into the body) before the bubbles formed. So for someone dwelling at sea level, bubbles would form at 18,000 feet, at half atmospheric pressure.

Decompression sickness doesn't occur immediately on exceeding 18,000 feet — it usually takes 20 minutes or so for enough bubbles to form to cause symptoms, and it may take some while for these to go away during the descent.

While 18,000 feet seems a good rule of thumb, various studies show that bubbles are detectable above 12,000 feet cabin pressure, but are too few in number to cause symptoms unless there are predisposing factors.

So at sea level there is only 1½ pints of nitrogen; if you go diving for any length of time the amount of nitrogen taken on board (inhaled) during the time at depth will increase significantly. The ambient pressure increases one atmosphere for every 35 feet of water so that for a 100-foot dive you have four times as much nitrogen as you normally have. If you depressurize that lot, you've got a problem. Luckily nitrogen is not very soluble in body fluids, even blood, and so you are unlikely to take on the full nitrogen dose until you have been down there for eight hours or more, but you will absorb enough to aggravate the problem. So decompression sickness need not be isolated to above 18,000 feet. It may occur at 9000 feet or even sea level!

It is a function of how high you are and how long you have been at that altitude; for people who have previously been scuba diving, 3000 feet may be okay for 15 minutes, but not for 30 minutes. Going to altitude, descending and then regaining altitudes does not by itself increase your risk of getting decompression sickness; but, if it increases your total time at altitude, it will.

The names for the symptoms for decompression sickness sound rather quaint now, as they arise from old-fashioned navy terms:

The bends

This is a pain in a joint which usually starts out as a sore nagging ache, but if you move that joint around or rub it, it will soon get much, much worse. Joints commonly affected are big joints (elbows, knees, shoulders, hips, hands and feet). However, the bends may focus on joints affected by a previous injury where the joint has been dislocated or involved in a break.

The creeps

Ever see the film where the hero is staked out, with ants crawling over him and biting him? Well you've got the feeling! Usually known as 'formication' (no, the spelling is correct), the prickling and itch is associated with a skin rash from the nitrogen bubbles tracking under the skin. Usually affects the trunk and thighs in areas covered by clothing.

The chokes

This is a feeling that you can't get your breath, with wheezing, tightness and pain in your chest and stomach. Taking a deep breath becomes impossible. Painful coughing is often set off by attempts at taking deeper breaths; the coughing becomes more difficult to stop and may terminate only as you pass out.

The staggers

A general term for any damage to the brain or nervous system caused by nitrogen bubbles. This tends to fall into three groups:

1. In the brain — bubbles can cause severe headache, confusion, double vision, loss of vision in whole or part of the visual field, flashing lights, symptoms similar to a stroke and eventually loss of consciousness with convulsions (primary collapse).

2. In the inner ear — bubbles may form in the fluid of the balance organs, causing intense vertigo and disorientation that is impossible to overcome in flight.

3. In the spinal cord — damage to arteries by bubbles causes serious, often irreversible damage to the peripheral nervous system; symptoms are numbness, weakness and loss of control (bodily or aircraft). Often the legs feel heavy and numb, like a leg 'going to sleep'.

Certain things make you more prone to decompression sickness:

➤ Scuba diving — this can be very significant at altitude.

➤ Age (over 30) — the older you are, the more likely you are to get decompression sickness.

➤ Obesity — fat contains lots of nitrogen that readily comes out of solution.

➤ Cold — extremes of cold increase the formation of bubbles.

➤ Hypoxia — if you are not breathing oxygen, it is probably nitrogen.

➤ Activity — general activity, moving or rubbing limbs increases the bubbles.

➤ Previous illness or injury or hangover.

All of the symptoms of decompression sickness may be serious in themselves (as they may not get better after recompression) or they may cause a pilot to lose functional ability. The real danger is that of a secondary collapse; for example, a pilot who has had a bend pain in the knee at altitude and has then not descended as advised, collapses as more bubbles form and collect in key body areas. Death may follow as the result of decompression sickness per se or from consequent loss of aircraft control. Prompt action is necessary!

Decompression sickness — vital actions

> ➤ Be suspicious. If you are above 18,000 feet, or at any altitude if you have been diving, decompression sickness is possible; and if you cannot blame how you are feeling satisfactorily on any other cause, assume decompression sickness until proven otherwise, and act accordingly.

> ➤ Descend immediately because of the hazard of secondary collapse; this is vital. Fly to the nearest place where medical assistance is available.

> ➤ Declare an emergency. Let air traffic control know that you need immediate clearance to sea level. If you do collapse your location will be known, and if you do perish, at least others will know why.

> ➤ If possible, transfer control of the aircraft.

> ➤ Keep warm. Turn the cabin heating up to maximum as cold increases the likelihood of further bubbles forming.

> ➤ Keep any affected joints still. You probably won't want to move the joints anyway as they will be painful.

> ➤ Apply pressure to affected joints. Wrapping a bandage around a painful joint will locally pressurize the body fluids in that area, causing the size of bubbles to reduce and symptoms to abate.

> ➤ Use oxygen. Pure oxygen contains no nitrogen. Breathing pure oxygen after the event is a late gesture, but any action to reduce further nitrogen absorption in a case of the bends is worth it.

> ➤ Arrange a doctor to meet you on landing. There is a phenomenon known as post-descent collapse where, four hours after landing, a pilot who successfully recovered from the bends at altitude during the descent suddenly gets unwell. Only medical help can prevent or treat this condition. You may need therapy at a decompression chamber ('pot'). If equidistant from two main centers, choose the larger center or, better still, that nearest to a nautical or naval unit. But do

not fly to a more distant airfield simply because it has a 'pot' — getting back to sea level may be enough.

Should a passenger appear to have decompression sickness, the actions described above are still necessary (and possibly lifesaving) even though loss of control of the aircraft is not a concern.

Little has been said about how the bubbles cause their damage — primarily because the mechanism is not clearly understood. It may be direct irritation by gas bubbles; or the gas bubbles may cause the blood to clot in crucial body centers, or may act through a hormonal intermediary. What we do know is that nitrogen has extremely serious and unpleasant consequences at altitude.

If you have experienced any symptoms while diving that could have been decompression sickness and then plan to fly, you must consult a doctor who knows about diving medicine. This applies even if the gap between diving and flying is weeks. The chances of reactivation of symptoms are significant if the symptoms occurred at sea level.

Diving and flying

Nature of dive	Don't fly until rested at sea level for
Decompression stops not required	4 hours
Decompression stops required — duration of dive less than 4 hours	12 hours
Decompression stops required — duration of dive more than 4 hours	48 hours

Head injuries

Case study 21:3

Accident Brief

Date: mid 1990s

Location: unknown, Canada

Aircraft: Bell 206

Injuries: 1 fatal

It was to be a simple and very enjoyable trip. The pilot was to ferry a machine from the company maintenance facility to Grizzly Falls and bring another one back for maintenance. The trip was going to take two days outbound and two coming back, with two stops (with company approval) at his parents' farm. The weather was forecast to be excellent for the next week with scattered clouds and light winds. Pre-flight planning was completed with the pilot's usual care and concern for detail. He hadn't accumulated over 10,000 hours of accident and violation free flying by cutting corners and being reckless. He was the company safety manager and led by example.

The first leg went smoothly, and he arrived at his parents' farm just before supper. He secured the helicopter for the night and went inside for what was to be a very enjoyable visit, except for that nagging sinus headache that had started about a week before. He took a couple of pain killers, which seemed to help a bit, and spent an enjoyable few hours with his parents and his younger brother and sister.

He awoke the next morning at 6 a.m., fairly well rested, but he still had a sinus problem. Along with the headache he noticed a tingling in his left hand — he must have slept on it, he thought and it quickly disappeared. After a very thorough pre-flight, he telephoned in his flight plan, said goodbye to his family, and took off on the final leg of his trip. About an hour out he was abeam Rutland Airport, and he called the Rutland flight service station (FSS) to give a position report — his sinus headache was a bit worse, but he felt that was because he had not taken, for safety reasons, any pain killers before takeoff. He was only 30 miles from Rutland but it took him five calls to get his message through — the FSS kept asking him to repeat, remarking that his transmissions were strong but very garbled and unreadable. He thought that his calls sounded a bit strange but attributed the problem to a fault in the radio. Thirty minutes later his headache had gotten worse and he thought about taking a pain killer, but decided not, again for safety reasons. At the same time he noticed a tingling sensation in the fingers on his left hand and he wondered if perhaps he had been gripping the collective a bit too hard.

At 11.43 Eastern daylight time, in clear weather conditions, a gentle 5-knot wind out of the northwest, at 1000 feet above ground, cruising at 110 knots, his head slumped forward — he was held upright by the shoulder harness. He lapsed into unconsciousness, a condition he would never recover from. He was alive, but unconscious and unable to control his helicopter. For some reason he continued to hold the cyclic, actually bringing it back a small amount — enough to start a gradual deceleration. The little bit of collective tension he flew with prevented the collective from lowering rapidly. The collective started down slowly, as did the helicopter. Speed on impact with the small trees and brush was about 65 knots. The helicopter was 5° nose up with 55° of right bank when it struck the ground. The first hit caused the machine to partially right itself, almost returning to a level attitude, and it removed the rotor, the transmission, and part of the tail boom. The fuselage missed direct contact with any trees, and, when it came to a stop, it was basically intact and upright. Damage to the cockpit area was confined to a broken left front windshield, left chin window and left door. There was no post-crash fire. The emergency locator transmitter (ELT) was activated on impact and the COSPAS-SARSAT system had the area located in minutes. A short while later the C-130 Hercules aircraft that had been dispatched by the rescue coordination center homed in on the ELT and located the wreckage on the first pass. The search and rescue (SAR) technician team on board the C-130 parachuted to the crash site and, shortly after landing, reported back to the SAR aircraft that the lone occupant, the pilot, was dead.

What happened to the pilot? Did he have a heart attack or a stroke? Did he faint or have some type of seizure? Did he choke on something? None of these were the reason for him becoming unconscious, and the real answer required some tenacious investigating on the part of an investigator. The Transportation Safety Board sent a team to the crash site and quickly determined that the crash was very survivable, based on the complete integrity of the cockpit, the fact that the pilot was wearing both a helmet and the installed four-point harness, and the relatively low forward and vertical speed on impact. The question then was: what had killed the pilot? He was unmarked, had no broken bones and did not appear to be bleeding externally or internally. In addition, there was the baffling question: why had a perfectly serviceable helicopter in the hands of a very skilled and safety conscious pilot crash on a superb day for flying?

The investigator wanted answers, so he requested a complete autopsy, which included, by law, a full cranial examination, and with it came the shocking answer. The cranial examination revealed that the loss of consciousness was caused by pressure on the brain exerted by a subdural haematoma, which had most likely been caused by a blow to the head that occurred days, or even months, before the accident. With this discovery came more questions: What had caused the blow to the head? Was there a safety message for the aviation community?

The 'what' was answered quickly when the investigator spoke to the pilot's wife. He asked her if her husband had hit his head in the past weeks or months and she said that she didn't think so but to be sure she would ask her children. The final piece of the puzzle fell into place when she spoke to their 12-year-old son. The boy recalled that he had been playing catch with his dad in the backyard about three weeks before the accident and that his father had tripped trying to catch a ball and had fallen, hitting his head. He hadn't thought anything more about it because his father was up immediately, saying he was fine. They both forgot about the incident and continued playing ball. The minor blow to the head ruptured veins in the skull and the blood pooled, forming a haematoma. The haematoma grew in size as the bleeding continued, causing pressure on the brain. The pressure on the brain caused the headache and, as the pressure increased, the tingling sensation and, ultimately, the loss of consciousness. As the pilot sat in the wreckage the pressure increased even further, forcing the brain to shift downward. Three hours after he lapsed into unconsciousness, the area of the pilot's brain responsible for involuntary response shut down and the pilot's heart stopped beating.

Source: *Vortex*, 2/99

This accident was tragic: a wife lost a husband; two children lost a father; parents lost a son; and the accident was totally preventable. The blow to the head was, in itself, not cause to rush off to the hospital but the 'signs' certainly were.

A subdural haematoma is a brain condition involving a pooling of blood in the space between the inner and outer membranes that cover the brain. Symptoms usually develop within a short time after a head injury. Subdural means 'below the dura' – the dura is the outer membrane covering the brain. A subdural haematoma develops when veins that are located under the dura, between the dura and the inner membranes covering the brain, leak blood as the result of an injury to the head. This could be as obvious as a traumatic accident or as trivial as a bump to the head from diving for a football, and does not require initial loss of consciousness. Blood collects, forming a mass (haematoma) that presses on the brain. Pressure can damage the brain and causes loss of function that may get progressively worse as the haematoma grows and pressure within the head increases. The injury and the resulting collection of blood cause inflammation of the brain, which leads to swelling (cerebral edema). Cerebral edema further increases pressure on the brain, resulting in a loss of consciousness and, if massive enough, causing herniation of the brain stem and death.

Some, but by no means all, of the possible symptoms to watch for are: persistent headache; impaired vision; decreased sensation or numbness (sensory deficit); confusion, delirium; changes in personality, irritability, apathy, decreased memory, slowed thought processes; slurred speech; withdrawal from social interaction; and absent sweating on one side of the forehead. This is not to suggest that you rush off to the hospital every time you bump your head; but it is a reminder of how fragile we actually are. If you hit your head, tell your spouse or partner what you did. Then when you pass out just before you get into your helicopter and they rush you to hospital, the doctors will have some information to work with.

SPATIAL DISORIENTATION

Non-flying folk often find it difficult to understand how a trained pilot cannot fly an aircraft without reference to the horizon or other visual cues. They do not appreciate the training required to fly using only instruments, and that without such training pilots can lose control of their aircraft. Even when a pilot has been trained to fly on instruments, it can be a skill easily dulled by lack of use. Instrument ratings must be kept current. Disaster can strike quickly once orientation is lost.

To most people, orientation means being aware of their position in space (and time), but sometimes just knowing which way is up is a real problem. We rely on three systems for most of our orientation. These are in the kinesthetic sensors (muscle-bone-joint sense), vision and the vestibular (labyrinth) organs in the inner ear. Vision is the dominant sense and is integrated with the vestibules whose prime role is to coordinate eye movement with body movement. This is achieved by multiple connections in the brain between the nerves that control the eyes and the organs of balance.

On the earth's surface, our sense of orientation is remarkable, for even with the eyes closed we are (subconsciously) aware of the position of all our body parts. Standing, we can sense the nature of the surface underfoot. Sitting or lying, we sense the texture of surfaces and are able to make rapid movements to maintain balance, should circumstances change.

In flight, orientation is more difficult because we are routinely exposed to forces other than gravity. In flight we produce our own gravity, called 'G' force, and quite often it is not in the same direction as the earth's gravitational pull. A lot of training and experience is needed to develop the stored mental images required for flight. Often, the senses we have learned to trust on the ground give us unreliable information in the air. In Instrument Meteorological Conditions we must rely on instruments instead of our instincts or we may become the victim of illusions (false impressions) and suffer disorientation. Understanding the causes of disorientation will not prevent its occurrence but does demystify it and allow us to cope with the consequences.

Kinesthetic illusions

Pilots use the phrase 'flying by the seat of the pants' to describe the (subconscious) position sense used in flight on most days. When peripheral vision is limited, however, this sense becomes dangerously unreliable. Experience has taught us that gravity acts toward the center of the earth and that the gravitational pull is 'down'.

In an aircraft making a coordinated turn, the force we feel is actually centrifugal, acting out from the radius of the turn. In a looping maneuver, the situation is even more bizarre because at the top of a loop the blindfolded pilot would sense the earth's pull as being up, not down. We need to develop a mental encyclopedia of unusual positions before we can begin to analyze correctly our kinesthetic sensations; even then they are often wrong. The wise pilot knows that when kinesthetic sensations and the instruments disagree, the instruments are, most of the time, correct!

Visual illusions

Vision may be separated into central and peripheral, although the two are always intimately connected. Central or focused vision is used for object recognition, but peripheral vision is our main source of spatial orientation. Central visual illusions are usually misunderstandings of what we see; peripheral illusions are false impressions of movement or rotation. Central visual illusions are often affected by expectancy. A pilot's judgment may be biased by previous experience and preconception. A helicopter pilot who is accustomed to working in very tall trees, for example, may have difficulty in areas where the trees are small or non-existent.

Another central visual illusion is what a pilot sees on a normal 3° approach to two identical runways, one of which has a 2° uphill and the other a 2° downhill slope. The uphill slope represents a larger (taller) image at the retina, which is interpreted as being high: the tendency is for a flat approach to be flown, and the aircraft may make contact before the roundout has been completed. The downhill slope gives the impression of being too low: a steep approach is likely, and roundout may be made too high. For the helicopter pilot, change 'runway' to read 'open field', and the impression is the same.

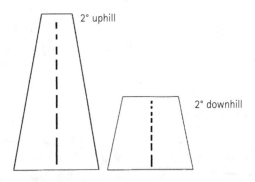

Fig 21.3 Normal 3° glideslope *VORTEX*

Whiteout and black holes

Whiteout and black holes, both due to lack of contrast, cause many accidents.

In whiteout, a layer of fresh snow on the ground merges with a white sky and indistinct horizon to make depth perception practically impossible. A similar effect may be caused by blowing or recirculating snow. This is an all-too-common occurrence if a helicopter is brought to the hover over an abundance of loose powdery snow. Under whiteout conditions, experienced pilots have flown aircraft into the ground while maneuvering at low levels. Even with fair visibility the lack of contrast can disguise contour changes.

Float planes have similar problems making landings on glassy water. It is common practise for them to set up a constant low-rate descent under these conditions rather than try to estimate the height above the water.

Night operations into (or out of) dark, featureless areas, such as unlighted water or woods, are difficult. Such areas are commonly referred to as black holes. On approach to a lighted runway where the surroundings are not visible, the runway may appear to move or tilt as the aircraft attitude changes. If there is a lighted town some distance beyond the runway at higher elevations, another problem arises; pilots tend to keep the lighted area in the distance in the same relative position as they descend. This may take them below the correct glide path, and a premature landing can occur.

False horizons

False perceptions of the horizontal can be confusing. Lining-up with sloping cloud tops, particularly between layers, is not uncommon. At night, in flight over sparsely populated areas, ground lights and stars may be confused, giving a feeling of tilt or nose-high attitude. A dimly lit, straight road in the distance can be mistaken for the horizon. During takeoff into a black hole, the receding shoreline may be mistaken for the horizon, with disastrous results. In these cases, however, awareness of the illusion and good instrument checks will prevent problems.

Vectional illusions

Illusions of false movement are common and difficult to ignore. Most car drivers have experienced a common vectional illusion: at a traffic light the neighboring car creeps forward and you appear to be slipping backward. Or how many of you have been in your car in the automatic car wash and been sure that you were moving forward when in fact it was the wash apparatus moving past you to the rear? Many of us have jumped on the brakes with this sensation. Helicopter pilots, trying to hold a hover over water, feel they are moving because of the motion of the waves.

Over fields, the movement of the grass in the rotor wash or blowing snow creates a similar illusion. Illusions of vertical movement can be caused by raindrops running down the windscreen of an aircraft in clouds or in marginal visual meteorological conditions.

Angular vection

Angular vection occurs when there is rotation in the field of peripheral vision. Full-vision simulators make use of this phenomenon by moving the scene outside the windscreen: the viewer's sensation is that the simulator is rotating, not the scenery. The astonishing power of these illusions can be felt at an IMAX cinema as the viewer drops from dizzying heights to the earth while firmly seated.

Autokinesis

A small, fixed light, viewed steadily at night, appears to move. The movement is actually caused by the eyes losing fixation, drifting away, and then jumping back to the target. Pilots, however, have altered course to avoid a collision with stationary lights, believing them to be moving aircraft. The feeling can be overcome by deliberately looking away from the light and then back again. If lights are multiple, bright or large this illusion is uncommon.

Vestibular illusions

Vestibular illusions are the most complex and dangerous. The labyrinth contains two related organs; the otoliths, sensitive to linear acceleration; and the semicircular canals, sensitive to angular acceleration. Although both organs are similar in function, they will be described individually for simplicity.

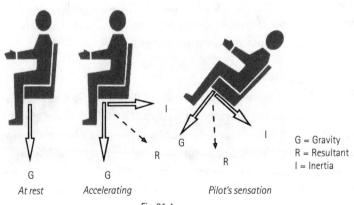

At rest   Accelerating   Pilot's sensation

G = Gravity
R = Resultant
I = Inertia

Fig 21.4 VORTEX

Linear accelerations

There are two otoliths in each inner ear, and they are set at right angles to each other. One records accelerations in the horizontal plane, the other in the vertical plane. They are located in the common bulbous portion at the base of the semi-circular canals and consist of hair-like fibers tipped by tiny calcium stones that project into the fluid (endolymph) filling the vestibular system. These hair fibers sway like weeds in a river current, swept by movements of the endolymph caused by acceleration forces. The movements generate nerve impulses which the brain interprets as changes of head or body position in the linear plane.

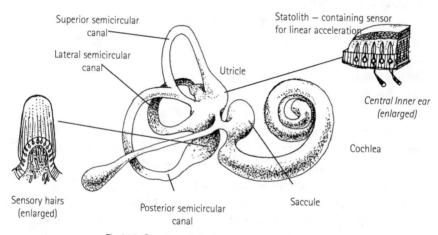

Fig 21.5 Structure of the inner ear NEW ZEALAND FLIGHT SAFETY

On the ground this system is accurate, but under flight conditions the otoliths can give rise to incorrect information. The pitch-up or somatogravic illusion is a good example. When an aircraft is stationary on the tarmac or in a stable hover, the otoliths sense gravity acting downward. When the aircraft accelerates for takeoff, a new force is sensed as the hair cells are swept backward by the inertia of the fluid. The brain resolves the two forces (gravity and acceleration) as a single resultant force acting downward and backward. We have learned to interpret such a force as the head being tilted backwards, so the pilot feels that the nose of the aircraft is pitching up. In normal conditions the sensation is corrected by vision, but when a takeoff is being made at night from a well-lit field into a black hole, it is difficult to ignore. It can easily occur during a missed approach as the aircraft is accelerated from approach speed to a climb configuration.

A recent accident in a Fairchild Metro III is a good example. The aircraft commenced what seems to have been a normal missed approach when, for no apparent reason, it nosed over and impacted the ground. The normal reaction to pitch-up is to push forward on the stick. Accident reports from such cases often state, 'The aircraft struck the ground at a steep angle on the runway heading'.

With deceleration a similar but opposite (pitch down) illusion occurs. Sudden decelerations, such as those caused by deploying the speed brakes of lowering the flaps and landing gear, sway the otoliths forward, and the pilot feels that the nose of the aircraft is dropping. This illusion is most likely to occur on final approach at slow speed, and the reaction of pulling back on the stick may cause a stall.

Angular accelerations

The semicircular canals are responsive to angular accelerations. Each canal is filled with a viscous fluid (endolymph) into which sensitive hair cells, similar to those in the otoliths, project. These hair cells are sensitive to fluid movement. There are three canals in each (inner) ear, lying in planes roughly corresponding to pitch, roll and yaw. Movement of the fluid in the canals is recognized as rotation.

When a glass of water is rotated, inertia causes the water to move after the glass. In the same way, as we enter a turn, the fluid in the canal system lags behind the bony canal so that the hair cells are displaced, telling us we have entered the turn. As the turn is continued, the fluid begins to move and, after 10–30 seconds, the movement is synchronized when the walls of the canal and the deflected hair cells return to a neutral (upright) position. The feeling of turning will then disappear but, when the turn is completed and the aircraft levels out, inertia will cause the endolymph to continue to flow, although the canals are now still. The hair cells will be momentarily swept in the opposite direction and a sensation of an opposite turn will be felt, which may last for 10–20 seconds. This is the opposite turning illusion.

Graveyard spiral

The opposite turning effect can cause a very dangerous illusion. Say an inexperienced pilot enters a spin under instrument conditions. After two or three turns, the initial sensation of spinning may disappear. When the appropriate corrective action is taken and the spin stops, the pilot will experience a sensation of turning in the opposite direction. If this is reacted to, the spin will be re-entered, probably with disastrous results. Similarly, in a spiral dive, pulling back on the stick without correcting for bank feels much like a full recovery, and the inexperienced pilot, whose sensation of turning is absent, may believe that the aircraft is level and that a recovery has been made. This illusion has been called the graveyard spiral.

Coriolis illusion

The Coriolis (excess G) illusion, caused by inappropriate head movements, is the most confusing and possibly the most dangerous of the vestibular illusions. Because the semicircular canals are inter-connected, movement of fluid in two canals at the same time can cause fluid movements in the third canal. The following is an actual case.

The pilot takes off and enters cloud. While climbing and accelerating, a left turn is initiated and at the same time the pilot turns his or her head quickly downward to the right, say to locate a switch. The canal, which sensed the acceleration, was repositioned by this head movement, and a second canal in a new plane is stimulated. The combined effects cause a movement of fluid in the third canal, and the pilot experiences a violent sensation of tumbling. Because the vestibular organs also stabilize the eyes by reciprocal connections in the brain, focused vision is affected, and the whole scene, including the instruments, appears to rotate.

Under such conditions it is extremely difficult to maintain control of the aircraft. Turning the head sharply while in Instrument Meteorological Conditions, particularly if the movement is against the direction of turn, is extremely hazardous.

Picture yourself as the sole occupant in a light helicopter. The weather is not very good, but you have only 50 miles to go, so you press on (helicopter Visual Flight Rules). In the back of your mind you have a plan if you should stumble into Instrument Meteorological Conditions. You steal a split second of your straight-ahead scan to check your instruments, and when you look ahead again, you see nothing but gray and white. Your head snaps to the right in an attempt to regain visual reference, and at the same time you drop a bit of collective, ease back on the cyclic and also start a right turn. You don't find any visual references out the right window, and by now you have induced some mind-numbing illusions. The rest is in the accident report.

The leans

The otoliths are very sensitive, and in fact changes in acceleration as small as 0.01 g per second can be detected and misinformation sent to the brain. The semicircular canals are less sensitive and, if the pilot is distracted, roll rates of up to 3° per second may go unnoticed. For example, a pilot, flying straight and level, is studying a chart or talking while the aircraft slowly drops one wing 15°. Becoming aware of the incorrect attitude, the pilot makes a quick recovery. Since the brain did not sense the original bank but now senses the correction, the pilot will feel that the aircraft is in a 15° bank in the direction of the recovery, even though the instruments clearly indicate level flight. The feeling is so compelling that the pilot leans toward the opposite side of the aircraft to maintain a feeling of balance. The feeling is disturbing rather than dangerous, but extremely common. Usually it is short-lasting, but one well-documented case in cloud lasted for over an hour. If this happens in a helicopter, close to the ground, the results can be disastrous.

General factors

It is extremely difficult to mimic these conditions in the air, even under the hood, but most can be demonstrated in simulators. The head-turning (excess G) phenomenon can be demonstrated by spinning a blindfolded subject on a piano stool while head movements are made. This must be demonstrated with great care, as the subject can easily be thrown from the stool by the body's righting reaction.

Disorientation is not a disease or an illness; and even experienced pilots suffer from it. Fatigue, inattention, alcohol and hangover all make disorientation more likely to occur. Flying with a cold can also cause problems if one ear clears before the other in a descent or ascent. The sudden difference in pressure in the two inner ears may produce a short-lasting but acute sense of vertigo (spinning), which is known as 'alternobaric vertigo'.

Motion sickness

Airsickness is a form of motion sickness, and all are susceptible, from the new trainee to the long-time pilot. Experienced aerobatic pilots, for example, can become seasick in small boats. Space sickness has been a problem with over 75 percent of the astronauts.

Motion sickness appears to be caused by a conflict between what the eye sees and the motion receptors report. For example, experienced pilots in a simulator may become ill because the movement seen on the instruments is not accompanied by the expected G forces of an airplane. Inexperienced pilots do not have this problem.

Usually pilot trainees adapt quickly and motion sickness is temporary, but it can be prolonged by anxiety or fear. Reassurance by an instructor is extremely important. Passengers are more likely to experience this condition than drivers, and giving control to a student pilot will often be enough to relieve the symptoms. Those prone to motion sickness should not eat or drink immediately before flight, should avoid greasy or other hard-to-digest food, and should not smoke. Several effective medications are available, but at this time there are no motion sickness drugs that are safe or authorized for use by a pilot in command.

The following case studies show how the various aspects of spatial disorientation conspire to trap helicopter pilots.

Case study 21:4

Accident Report ANC99FA073

Date: June 9, 1999

Location: Juneau, Alaska, United States

Aircraft: Eurocopter AS 350B

Injuries: 7 fatal

The air-tour helicopter, with the pilot and six passengers, departed Juneau for a 50-minute flight over mountainous glaciers. About 10 minutes after making a normal radio transmission, the helicopter was located by the pilot of a second company helicopter who was also conducting a tour. The accident helicopter had impacted a nearly level, snow-covered glacier, and all occupants received fatal injuries. The leading edge of the impact crater was angled between 30° and 45° below the horizon, and the face of the airspeed indicator gauge had a needle slap mark at 130 knots. The pilots of the only two helicopters near the accident site at the time of the accident said 'flat light' made the featureless, snow-covered terrain difficult to discern from the indefinite, overcast ceiling. Photographs taken within one hour of the accident showed that the mountain pass the pilot was attempting to fly through was difficult to distinguish from the clouds, the glacier surface, and the surrounding snow-covered terrain.

A post-accident inspection discovered no evidence of any pre-accident mechanical anomalies. At the time of the accident, the pilot had accrued 7.9 hours in AS 350 helicopters, and a total of 37.5 hours in turbine helicopters, all with the accident company. This was the second day the pilot operated the AS 350 by himself. The pilot did not hold any instrument certificates, nor was he required to. The pilot received no emergency instrument training from the company, nor did the company require him to demonstrate the ability to control the helicopter solely by reference to the installed flight instruments. The helicopter was required to, and did have a gyroscopic pitch and bank indicator installed. The pilot's previous helicopter piloting experience (as a student and as a flight instructor) was in Arizona and California. The pilot stated on his company resume that he had accrued 891 hours of helicopter flight experience at the time of his employment. The National Transportation Safety Board investigator in charge, and the Federal Aviation Administration, estimated the pilot actually had 612 helicopter flight hours when hired. The company had not received background checks for the pilot before the accident occurred, and was allowed by the Pilot Records Improvement Act to use a pilot for 90 days prior to receipt of background information. The pilot had expressed to a previous employer, and a previous instructor, that he was uncomfortable with company pressure to fly tours in bad weather.

The National Transportation Safety Board determined the probable cause/s of this accident as follows:

The pilot's continued Visual Flight Rules flight into adverse weather, spatial disorientation, and failure to maintain aircraft control. Factors associated with the accident were pressure by the company to continue flights in marginal weather, and the 'flat' lighting leading to whiteout conditions. Additional factors were the pilot's lack of instrument experience, lack of total experience, inadequate certification and approval of the operator by the Federal Aviation Administration, and the Federal Aviation Administration's inadequate surveillance of the emergency instrument procedures in use by the company.

Source: National Transportation Safety Board, United States

Case study 21:5

Accident Report FTW98FA256

Date: June 5, 1998

Location: La Gloria, Texas, United States

Aircraft: Eurocopter AS 350BA

Injuries: 3 fatal

The 2905-hour instrument rated commercial helicopter pilot encountered an area of limited visibility during a dark night flight over an unlit, very sparsely populated, rural area while en route to evacuate a truck driver injured in a highway accident. The helicopter crashed 19 miles west of the truck accident site, indicating that the pilot failed to recognize his intended destination and flew past it. The helicopter impacted trees and terrain in a left turn in an 85° to 95° nose down attitude. The pilot had accumulated a total of 4 hours of actual and 45 hours of simulated instrument flight time, none of it within the 90 days preceding the accident. The pilot was reported to have been concerned about night flights in the area due to the lack of lights on the ground to maintain visual reference.

Another helicopter pilot stated that 'at night the area west of the highway is a big black hole'. No discrepancies were found that could have prevented normal flight operations. The pilot who flew the helicopter prior to the accident flight stated that 'the aircraft flew well and responded normally'. The visibility had been severely restricted by thick smoke from fires in Mexico. No distress calls were received from the helicopter. There were no reported eyewitnesses to the accident.

The National Transportation Safety Board determined the probable cause/s of this accident as follows:

The pilot's continued flight into adverse weather conditions resulting in a loss of control due to spatial disorientation. Contributing factors were the dark night illumination, the lack of visual cues, the pilot's lack of total instrument time, and the pressure induced by the medical emergency to complete the medical evacuation.

Source: National Transportation Safety Board, United States

Case study 21:6

Accident Report ANC02FA057

Date: June 21, 2002

Location: Central, Alaska, United States

Aircraft: Hughes OH-6A

Injuries: 1 fatal, 3 serious

About 2319 Alaska daylight time, a helicopter was destroyed during an in-flight collision with terrain while maneuvering about 15 miles west-southwest of Central. The helicopter was being operated as a Visual Flight Rules personal flight when the accident occurred. The private pilot/operator sustained serious injuries, two of the three passengers aboard were seriously injured, and the third passenger was fatally injured. Instrument Meteorological Conditions prevailed in the area of the accident, and no flight plan was filed. The flight originated at Eagle Summit, elevation about 4400 feet, and was bound for Circle Hot Springs, Alaska, about 8 miles southeast of Central.

On-site interviews were conducted by the National Transportation Safety Board investigator in charge on June 22. Witnesses said an engaged couple was transported to Eagle Summit in the helicopter for a mountain-top wedding ceremony. Witnesses and wedding guests said when they arrived on Eagle Summit in their personal vehicles, it was windy but visibility was good, and the clouds were high above the summit. They said they could see worsening weather approaching from the north, and shortly after the helicopter arrived, the weather closed in and it began to snow. They said visibility was reduced to 100 to 300 feet in wet heavy blowing snow. The wedding ceremony was concluded quickly and the pilot voiced concerns about having to leave the helicopter on the mountain-top over night. The pilot and his wife boarded the helicopter and occupied the front two seats and the bride and groom occupied the two rear seats. Witnesses said the helicopter lifted off the ground, made a right turn and disappeared into the blowing snow en route to Circle Hot Springs. They said they had to remove an accumulation of wet heavy snow from their cars prior to descending along the summit access road to reach the Steese Highway en route to Circle Hot Springs. About two-tenths of a mile east of the summit access road, on the Steese Highway, the wedding guests spotted the crashed helicopter. It was lying on its belly, facing the highway, in an open, down-sloped area, about 60 yards north of the highway. The tail boom, tail rotors and main rotor system were separated from the helicopter, and the engine was still running. The occupants of the helicopter were evacuated from the helicopter by the wedding guests and moved to the highway. Due to the remoteness of the area, the occupants were transported to the hospital by private automobile. The witnesses said the helicopter engine was still running at a high rpm when they left the area.

During an interview with the investigator on July 12, 2002, the pilot said when they left Circle Hot Springs for Eagle Summit en route for the wedding ceremony the visibility was greater than 10 miles, and the ceiling was about 8000 feet. When they landed on Eagle Summit the winds had died down somewhat from his earlier trip to the summit (to check the location). While they waited for all the guests to arrive, a line of dark clouds descended on them from the north; as the guests arrived it started to snow and the ceiling and visibility decreased. He said the ceremony was concluded rather quickly because of the deteriorating weather conditions.

As the helicopter lifted off for departure the ceiling was about 600 feet and the visibility was two miles. It had stopped snowing. The pilot recalled taking off, making a right turn, and heading downhill toward the road that would lead them back to Circle Hot Springs. He said he could not recall any of the circumstances just prior to, or related to, the accident. He said there were no mechanical problems with the helicopter.

The pilot held a private pilot certificate with Federal Aviation Administration ratings for airplane single-engine and helicopter. He did not hold an instrument airplane or instrument helicopter rating, and was issued a Federal Aviation Administration third class medical certificate on July 3, 2001. According to information received from the pilot, he had accumulated 502 total flying hours, 126 of which were in helicopters. All of the 126 helicopter hours were in the make and model of the accident helicopter. The pilot had accumulated 2.2 hours of simulated instrument flying time, and 0.2 hours of actual instrument flying time. The pilot did not provide information with respect to the date of his last Federal Aviation Administration flight review.

The accident helicopter was an ex-military OH-6A, which received Federal

Aviation Administration certification on March 21, 2001. The helicopter was not equipped for instrument flight.

The closest official weather reporting station is at Fairbanks, Alaska, about 130 miles southwest of the accident location. The Fairbanks Next Generation Weather Radar loop with a start time of 020621/23:01 and an ending time of 020622/01:02 shows a line of heavy precipitation that transits the area of the accident from west-northwest to east-southeast at the time of the accident.

A witness, who was not a party to the wedding, was on the summit at the time the helicopter took off. During an interview with the investigator he said that he was a pilot and when the helicopter landed visibility and sky conditions were good, but the winds were picking up and gusting. He estimated gusts of over 20 knots. He watched as the wedding progressed and the weather deteriorated; after the wedding ceremony the wedding party got into the helicopter. He assumed they got in the helicopter to get out of the weather, which was strong winds, blowing snow, sky obscured, and visibility of less then 100 feet. He was then surprised when the helicopter's engine started, but thought they were just going to turn the heater on. He was more surprised when the helicopter picked up into a hover. From inside his camper, about 50 yards from the helicopter, the witness could see only the red beacon light on the tail, and could make out the faint outline of the helicopter. He said the helicopter made a hovering right turn and then started forward and disappeared northbound into the blowing snow. The witness provided the investigator with a video tape of the wedding, the deteriorating weather conditions, and the helicopter's departure.

The terrain sloping down from Eagle Summit to the Steese Highway varies in pitch from gently sloping about 10° to a moderate slope of about 30° in places. The slope is covered with tundra, sparse low growing shrubs about three feet tall, and punctuated with low rock outcroppings. The view from the summit to the highway is unobstructed. The Steese Highway cuts east to west along the north side of the summit hill about 300 feet below Eagle Summit, and extends eastbound along the top of a ridge. The highway is a two lane gravel roadway with shoulders, which is about 30 feet wide. The north edge of the roadway drops off abruptly about 40° for about six vertical feet, and then decreases in pitch angle to about 20°, and then to about 10° where the helicopter came to rest.

Source: **National Transportation Safety Board, United States**

Case study 21:7

Accident Report SEA02FA008

Date: October 18, 2001

Location: Anchorage, Alaska, United States

Aircraft: Bell 206L

Injuries: 3 fatal, 2 serious

N400EH impacted the waters of Cook Inlet about six-tenths of a mile west of the shoreline off the approach end of runway 06 at Ted Stevens Anchorage International Airport. The pilot, who held a commercial pilot's certificate, expired as a result of the accident sequence. Two passengers were fatally injured and two other passengers received serious injuries. The aircraft was on a Visual Flight Rules on-demand charter flight. The flight was on a company Visual Flight Rules flight plan.

On the morning of the accident, around 0915, the subject pilot flew five individuals from the south ramp of Anchorage International Airport to Fire Island, and then returned without passengers to the company facility. Around 1300, he again flew to Fire Island, and landed in a cleared area adjacent to the Federal Aviation Administration navigational radio facility near Big Lake. After shutting down the aircraft, he entered the facility building, and waited for the Federal Aviation Administration technicians to complete their work. By about 1515, the technicians were finished with their tasks at the facility, and were ready to depart the island. Federal Aviation Administration certified audiotapes that recorded radio communications audible at the Anchorage Air Traffic Control Tower (Local Control) show that at 1531, the pilot requested a special VRF clearance back to Anchorage International Airport with a landing at the south airpark. The tower advised the pilot to hold on the ground for a couple of minutes because there was inbound traffic to runway six left. The pilot acknowledged this transmission, and then waited on the ground until 1540 when Anchorage tower issued him a special Visual Flight Rules clearance to enter the Anchorage class C airspace. That clearance included instructions to maintain 1100 feet or lower, and to proceed to the south airpark. The pilot then read back the clearance and indicated that, 'we're on our way; thank you very much'. About two minutes and thirty seconds after the pilot indicated he was en route, the tower asked him what altitude he was at. He responded with '400 is at 50 feet'. About 15 seconds after this transmission, Anchorage tower directed Dynasty 212 Heavy to maintain 2000 feet and to continue to fly its present heading. At the end of that transmission, after what sounds like the click associated with the tower controller releasing his transmit button, there is what appears to be a separate sound that lasts for about five-tenths of a second. That sound, which appears to be the end of a transmitted spoken word, was followed immediately with a one second-long squealing static sound that is consistent with the termination of a radio transmission overlapping the tower's transmission to Dynasty 212. After waiting for about five seconds for a response from Dynasty 212 Heavy, the tower repeated the instruction for Dynasty 212 to maintain 2000 feet, but changed the heading to a right turn to 200°. The flight crew of Dynasty 212 responded with a confirmation of the clearance, and then asked, 'Did you copy aircraft going into the water?' Tower asked Dynasty 212 to 'say again', and at 1543:17 Dynasty 212 responded with, 'Did you copy? It sounded like [company name] said he was going into the water'. There were no further transmissions or responses from the subject helicopter, and approximately two minutes later both the Ted Stevens Anchorage International Airport Police and the National Guard's 11th Rescue Coordination Center were notified that the helicopter had probably entered the water. The two survivors were rescued by an Air National Guard HH-60 helicopter, and were transported to Alaska Regional Hospital in Anchorage by approximately 1645. Due to weather conditions, the fire department rescue boats and the national guard search helicopters were recalled at approximately 1730.

During the investigation, the investigator in charge conducted multiple interviews with the two surviving passengers. According to the passenger in the right rear seat (facing forward), the helicopter had been sitting on the landing area adjacent to the facility for at least two hours before he and the other passengers boarded for the flight back to the south ramp of Anchorage International Airport. After the aircraft's engine was started, it remained on the ground with the engine running for what seemed to him to be about five to 10 minutes. He said that by the time the helicopter lifted off, there was a considerable amount of condensation on the inside of the side windows in the area where he was sitting. After it lifted off, the helicopter entered a hover over the landing area, and remained in that position for about one minute. At the end of that time, the pilot set it back down on the landing area, but kept the engine running.

Soon after the helicopter set back down on the landing area, this passenger saw the pilot look at the left front seat passenger and point toward the left side of the front windscreen. Immediately thereafter, the passenger in the front left seat started wiping the inside of the windscreen with his glove. The rear seat passenger believed that the pilot's pointing action had been an indication to the front seat passenger to assist in removing the condensation from the inside of the front windscreen. The rear seat passenger then heard the pilot say 'Can you hear me now' over the intercom. The passenger responded that he could, but he did not hear the front seat passenger, who was the only other passenger with a headset on, respond to the pilot's question. Soon thereafter, he heard the front seat passenger ask the pilot if they were on 'weather hold', to which the pilot responded, 'Yes'. About five seconds later, after being on the ground for what seemed to the passenger to be about two minutes, the helicopter took off a second time. After takeoff, the helicopter flew 'low over the trees', on a route that was more to the north than what the passenger had expected ('toward Palmer'). Upon reaching the shoreline of Fire Island, the helicopter descended until it was 'very low over the water'. When the investigator in charge asked the passenger for an estimate of the aircraft's height above the water, he said that it was really hard to tell, but that he thought it was about 10 to 15 feet. It appeared to the passenger that, as the pilot continued on, the helicopter descended lower and lower, and was eventually flying at what seemed to be about five feet above the water. At that time, he could see white chop and 'a lot of spray' being created where the wind from the aircraft's main rotor was impacting the surface of the water. He said that he was 'real uncomfortable' with the low altitude, and wanted to say something to the pilot, but did not because he was concerned that he would break the pilot's concentration. According to this passenger, who was looking through the right side window of the helicopter, the pilot continued on at this low altitude for a period of time, and then appeared to enter what seemed like a hover. It appeared to the passenger that the aircraft was in a hover for about a minute, and then it seemed to start moving forward again.

As it appeared to start moving forward, the helicopter seemed to descend a few more feet, and then about three to five seconds after descending, the skids dragged through the water. Immediately thereafter, the aircraft pulled rapidly up from the water and climbed quickly for a few seconds to an undetermined altitude. As it leveled off, the helicopter began to rock sideways, and seemed to move erratically in many different directions. It then descended for approximately five seconds and impacted the water 'very hard'. Soon after the bottom of the helicopter impacted the water, it began to roll to the right, and the main rotor blades started to hit the surface of the water on the right side of the aircraft. The helicopter then slowly began to sink.

After exiting the cabin and coming to the surface of the water, the passenger tried to make a call for help on two cell-phones he had in his possession, but he was not successful. He then attempted to sit on a portion of the helicopter that was on the surface of the water, in order to get his personal flotation device repositioned and connected correctly before inflating it. But, since he found this exacerbated the pain from the back injury he sustained during the impact, he slid back into the water. Once he was back in the water, he made the adjustment of his flotation device and pulled the inflation lanyards. The vest then successfully inflated. Soon after inflating his vest, he realized that he could see a shoreline, which at the time he thought was Fire Island, and he started swimming in that direction. After swimming in that direction for a while, he discovered that he was making little progress against the receding tide, so he stopped swimming and floated until being picked up by a national guard helicopter.

According to this passenger, soon after the accident helicopter departed the shoreline of Fire Island, he was no longer able to see any land or inlet shoreline. The water was calm and flat, and it was very hard to tell exactly how high above the surface they were. In addition, he said that it was snowing at the time, and that everything, including the reflection on the water, was kind of grayish white. He could not see any clearly defined horizon, and he felt there were no clear visual clues as to where they were going. Although he could not ascertain the condition of the front windscreen of the helicopter, because most of it was not in his line of sight, he reported that he had to repeatedly wipe off the side window in the passenger cabin with his glove in order to see out.

The helicopter did not make any sudden or unexpected rolling or pitching movements prior to the skids contacting the water. Although he was uncomfortable with the aircraft flying at what looked to him to be just above the surface of the water, there was otherwise nothing unusual about the movements or the flight path of the helicopter until the skids touched the water.

This passenger did not see any lights come on or flash inside the helicopter, nor did he hear any horns or beepers prior to the initiation of the dragging of the skids. He did not hear anything that sounded like a change in the engine or rotor rpm, and that there were no other unusual noises.

The account of events from the other surviving passenger was similar in all respects. It appeared to him that 'the pilot just flew it into the water'.

The pilot held a commercial pilot certificate, and was rated to operate helicopters and single-engine airplanes. His commercial license was first issued in October of 1969 and, according to the company pilot experience record that he completed in early July 2001, he had accumulated over 10,000 pilot hours in helicopters, 8000 of which were accrued in the Bell 206. Of his total pilot time, 50 hours were logged as offshore operations. The pilot did not hold an instrument rating, nor had he been issued an instructor rating in either airplanes or helicopters. His most recent second class Federal Aviation Administration medical was completed on April 19, 2001. A search of the Federal Aviation Administration's accident/incident data system and the enforcement information system revealed no accident/incident or enforcement records.

Source: **National Transportation Safety Board, United States**

These accidents show that spatial disorientation strikes quickly, often at the time Instrument Meteorological Conditions are encountered. A popular theory, no doubt derived from accident sequences and testing, suggests that the life expectancy of a Visual Flight Rules pilot after entering cloud is a mere 178 seconds!

Avoidance

Avoidance of spatial disorientation is obvious and simple — always fly in Visual Meteorological Conditions with clear visual cues. There is no guarantee that Instrument Meteorological Conditions can be avoided completely, but the chances of encountering them can be minimized by thorough weather planning. To achieve this, the pilot must be fully conversant with charts, reports, forecasts and local weather trends.

When their mind is set on an objective, many pilots push ahead no matter what the risks. This 'get-there-itis' is generally caused by pressure (self, company or passenger induced) to reach a destination or complete a task. Proving they are equal to any situation is a strong motivator for many pilots. Often this is hard to fathom in the harsh light of post-accident common sense, and has caused much head shaking at funerals. Unbelievably, often within a relatively short time, it is one of the head shakers who meets the same fate!

It is a psychological pitfall that defies logical explanation. All experienced pilots know that spatial disorientation regularly kills pilots, sometimes close colleagues and friends. Yet accident reports clearly show that a wait of less than one hour often turns weather conditions from marginal or atrocious to clear or nearly clear. It is like a motorist deciding to drive the wrong way on a motorway and then, to really spice things up, do it blindfolded. Actually, since the on-coming drivers are not blindfolded, the driver's odds would be better than that of the Visual Flight Rules pilot entering Instrument Meteorological Conditions.

Instead of perceiving and accepting a challenge when that feeling of unease emerges as the ceiling lowers and visibility reduces, pilots should sense danger. Consider a precautionary landing rather than risk inadvertent Instrument Metrological Conditions or 'scud-running'.

Scud-running is a term that describes the situation where pilots push the capabilities of themselves and their aircraft to the limit by trying to maintain visual contact with a terrain without making physical contact with it. Don't play Russian roulette with five bullets and one empty chamber. The healthy decision is to land the aircraft while you can still see what you are doing.

Perhaps you are a proficient instrument-rated aviator who has filed a Visual

Flight Rules flight plan and have, unexpectedly, encountered Instrument Meteorological Conditions. Not to worry, you think, climb through it and change the flight plan to Instrument Flight Rules. The only shortfalls in this plan are obstructions such as towers, pylons, hills and mountains (because it is unlikely you were at the minimum safe altitude for the area). What do you do if the minimum safe altitude is well above the freezing level and you can't get there without icing up? Then there is always the possibility of a mid-air collision with another aircraft flown by a pilot who couldn't see Instrument Metrological Conditions in the pre-flight planning either.

Federal Aviation Administration Advisory Circular 60-4A, *Pilot's spatial disorientation*, reported that, 'during a recent five-year period, there were almost 500 spatial disorientation accidents in the United States. Tragically, such accidents resulted in fatalities [more than] 90 percent of the time. Tests conducted with qualified instrument pilots indicate that it can take as much as 35 seconds to establish full control by instruments after the loss of visual reference with the surface.' It must be noted that these tests were conducted in fixed-wing aircraft (inherently stable). Helicopters are inherently unstable. Gaining control could take considerably longer.

If you are going to fly Visual Flight Rules, fly in Visual Meteorological Conditions. If you are going to push the limits, sooner or later you are going to end up in cloud or fog with no visibility. Do you have a plan? Do you have the skill to carry out the plan? When did you last think about your plan or do any practise to give you the ability to effect it? Much of the answer is in training and proficiency. If you are trained, proficient and have a plan of action when you inadvertently enter Instrument Metrological Conditions you are far more likely to survive.

The four Cs technique
Many schools teach the four Cs technique: control, climb, course, communicate.

Control
The foremost priority is controlling the aircraft. Remember the 35 seconds mentioned above. Initially devote your full attention to the instruments and begin your scan as you have been trained, vital attitude instruments first. Believe your instruments; they are rarely wrong and the seat of your pants will be telling lies to your brain. Don't worry about anything else until you are firmly confident that you have the aircraft under control.

Climb

When you feel you have your machine under control, climb. You were probably at low altitude and you don't want to descend into the ground or fly into an obstruction. The climb you initiate should slow you to a speed (say 60 knots) that will delay the arrival of any obstruction ahead of you. Don't let the speed get too low; the air speed indicator won't tell you you are flying backwards at 30 knots. Also, you don't want to get into any of the aerodynamic hazards that can be encountered at low airspeeds when you're unexpectedly flying on instruments; you have enough problems as it is. If you know you were surrounded by hills 4000 feet high, climb to 4500 feet. Your miraculous ability to control the aircraft solely using the instruments as a reference will be wasted unless you are clear of the terrain.

Course

Next consider your course. What should you be on? If you were surrounded by 4000-foot high hills but to the east of your last known position was a 10,000-foot high mountain, don't use an easterly heading. Start by heading in the direction that your intuition tells you is where the clearer weather lies, so long as it's not toward a large obstruction.

Communicate

Lastly, communicate. Tell someone, preferably air traffic control, your situation (including current height, heading and last known position) and your intentions. If you think they can do something for you, don't hesitate to ask. They are well trained and know that your 178-second clock is ticking! If you have a transponder, tell them and make sure it is turned on. They may or may not have you on radar. If they tell you to turn one way or the other or to climb (or descend), do it. They will know the safe minimum height for the area and, depending on the information available to them, will tell you the best heading to avoid obstructions. They will try not to ask you to climb or descend in a turn or to make any abrupt turns.

Stay as calm as possible, stay in control and, if you make it, never forget how you felt. Remember it really well next time you are flying near marginal weather and whatever you do, don't think you beat it last time — you just got lucky. Next time could be very different.

Brownout and whiteout

Blowing dust (brownout) or blowing snow (whiteout) are very similar; it is only the color that is strikingly different. The hazard is very easy to understand. A helicopter is in clear air, perhaps in total visual conditions, and approaching to land. As soon as ground effect is encountered (roughly $1\frac{1}{2}$ times the rotor disc diameter) there is suddenly a wind on the surface. Any dust or light snow on the ground is whipped up, increasing in both speed and quantity as the helicopter gets closer to the ground. Often this can cause total loss of horizon and reference while the helicopter is still airborne, leaving the pilot completely reliant on instruments with little airspeed and very close to the ground. Because of recirculation the problem is not likely to resolve itself.

Both options available to pilots have their own set of hazards: to carry on down means risking a dynamic rollover or skid collapse; to try a 'go-around' means moving laterally, blind to any obstructions and reliant on the instruments to maintain control of the helicopter, without the 35 seconds available to gain complete control. Of course, the same hazard arises when taking off.

This trap has snared many. It has been, and still is, a huge problem for military pilots in the Middle East, where often they are trying to land multiple helicopters in the same dusty location. Imagine being the pilot of the fourth or fifth machine to approach!

To avoid whiteout, don't take off when it is snowing heavily. Even if you can see sufficiently when you are sitting there with the collective hard down, that may all change quickly when you apply some pitch to the rotors and the downdraughts begin.

There are two techniques you need to be especially proficient at if you are to successfully handle brownout and whiteout:

> Maximum rate takeoffs that are steeper than normal.

> Straight to the ground approaches and landings.

Any time spent mucking around hovering and dithering in the ground effect area is time where the murk is forming very, very rapidly.

Many commercial pilots, particularly owner-operators, learn early that they don't tend to make a lot of profit by hovering before landing. So they become adept at the straight in, no-hover landing. It is a skill that is not too hard to master with practise and it is invaluable when facing a possible whiteout or brownout. Foresee the problem and react accordingly.

FLICKER VERTIGO

The pilot turned left about 30° on to a westerly heading; he was on the last leg of a two-hour trip to a mining site in northern Alberta, Canada. It was mid-afternoon, and the sun was right on the nose about 30° above the horizon. This was a bit bothersome as the light blinked through the rotor, but the pilot had a good pair of sunglasses, and there was only 15 minutes left before landing.

He blinked and shook his head as the sun suddenly appeared to increase in brightness. The light got wider, and then almost instantly, grew to a tremendous size and brightness. The pilot let out a gasp as his arms began to jerk about. His mouth twitched as he fought to speak, and still the light grew.

The urge to escape the light, now huge and dazzling, brought an inner command to dump the collective, cyclic hard over, kick pedal — anything. Fully conscious, wanting desperately to do anything to escape the light. For a brief moment the pilot felt paralyzed.

Within an instant, his back arched in one long agonizing spasm. His muscles jerked, twitched and convulsed, driven by unknown and puzzling commands. His head snapped wildly from side to side without purpose or intent. Yet, through all this, his faculties remained.

The light — the light — it had to be the light! He squeezed his eyes shut as tightly as possible. That single act of self-defense signed his own death sentence. Closing his eyes changed the light to the orange-red of his own blood. His eyelids restricted all incoming light to that one color, that diffused red.

That was all that he would ever know. His muscles were now rigid, commanded by an unknown force, his body frozen and unmoving, his mind a total blank.

The helicopter continued its rolling, uncontrolled spiral from 2000 feet. As it accelerated through 190 knots, still rolling violently, with the pilot's hands frozen on the controls, the main rotor separated. A microsecond before smashing into a granite ridge line the tail rotor wrenched free. The fuselage crashed vertically at about 210 knots, and as vaporized fuel from the ruptured cells ignited into a giant fireball, torched by the hot engine, the pilot's life was terminated by the transmission as it hurtled forward through the cockpit.

The killer was unpredictable, unknown to many and, in almost every respect, beyond control once the attack was started. Flicker vertigo had claimed another victim.

Flicker vertigo is also known as 'photic stimulation' and 'flicker unconsciousness'. It comes in a wide variety of forms, strikes with varying reactions, attacks pilots of a wide variety of aircraft from light singles, helicopters and even, under certain circumstances, the crews of large transport aircraft. It has also claimed drivers of cars and trucks. In other words, if the circumstances are right it can strike anywhere.

Flicker vertigo is one of the least known hazards to flight, but the fact is that flicker unconsciousness has been known for thousands of years. Some 2000 years ago, Roman slave masters attempted to induce epileptic seizures, with some success, by forcing their slaves to look downwards into a fire above which a spoked wheel was spun at a steady rate. It was a common belief that one in every 10 adults was, under certain conditions, prone to convulsions, coma and violent reactions. This percentage is consistent with modern-day test results. In the most violent cases, a single burst of light is enough to send a person into convulsions or unconsciousness.

If there is any trace of epilepsy in the individual's history, even though it has not previously appeared, it is easily triggered. An individual can experience the physical convulsions and be fully aware of what is happening but be unable to control his or her actions.

In the 1950s, the American Federal Aviation Administration became very interested in flicker vertigo. What they had was a series of crashes that escaped reasonable explanation, and all but one of them were fatal. As the reports were compared, a pattern began to emerge. All had been in single-engine aircraft landing to the west into the setting sun. The sole surviving pilot remembered descending on final with the power back — when the sun suddenly became very bright and was expanding rapidly. That was all that he remembered till he regained consciousness in hospital. The pilot was later tested in the same light conditions and very quickly passed out. When he regained consciousness he could remember only the bright expanding light.

Fig 21.6 Flicker vertigo *NEW ZEALAND FLIGHT SAFETY*

It now appears a number of accidents which were previously recorded as 'unknown', 'equipment failure' or 'pilot error' could in fact been caused by flicker vertigo. Pilots who crash and die, of course, cannot tell what happened. On the other hand, there have been numerous accidents which were recorded as 'cause unknown' where the pilots could not remember what had happened. Unfortunately, in many cases, no one thought to ask where the sun was at the time of the accident.

What happens when an aircraft is flying with the sun shining through the rotor of a helicopter or the propeller of a fixed wing is a direct function of the rpm and the number of blades. Pilots subjected to between 10 and 30 flickers of light per second are in a potentially hazardous area. If the figures are between 16 and 20 per second then the potential for vertigo is extreme.

A four-bladed rotor system rotating at 300 rpm would produce 20 flickers per second. What is most insidious at 16 to 20 flickers per second is that the pilot sees the flashing for only a second or less and then the light appears to be from a steady source. A steady light is seen in much the same way that the images from a motion picture film running at a steady rate through the projector is perceived as motion rather than individual frames.

If there is a hint of susceptibility to this flash pattern (not everyone is susceptible) the brain may react violently in the form of instant unconsciousness, body lock-up that completely paralyzes, or violent convulsions. All the time this is going on the person may be fully conscious.

Can pilots recognize and escape the effects of the flicker vertigo? There's no hard and fast answer to this question, since the onset varies from individual to individual. If they react in what would be considered a normal fashion, by immediately closing their eyes tightly, the onset may be much more rapid and devastating. Medical tests have shown that the physical and mental condition of the pilot is a factor in how severe the reaction will be. Persons who are fully alert and in good physical and mental condition may escape the worst symptoms but might be thrust into physical exhaustion, a sense of their eyes being glazed over and malaise. They may remain fully conscious, however, and in control of their faculties. Fatigue, emotional distress and excess use of caffeine can turn a survivable encounter to complete helplessness.

If you have the presence of mind, don't close your eyes. Closing your eyes will almost guarantee the onset of vertigo. Turn away from the light source and, if you can, shield your eyes. If you are down low, attempt to gain some altitude and try, try to stay composed. Perhaps a tall order, but there isn't really anything else you can do. It's always associated with a light source and pattern.

Who's susceptible? This question has not been fully satisfied and research continues. There is some indication that epilepsy-prone people may be more susceptible, and so people with a family history of epilepsy should take notice. In most cases there must be a flickering of the light source for a few seconds, but there have been cases were a single short burst of light triggered a reaction.

While not everyone is prone, if you do encounter flicker vertigo, the chances of you being affected again increase dramatically. It can also happen without the sun.

If you are in cloud and your beacon, strobes or navigation lights are on; they can set up a pattern on the clouds that could induce vertigo.

It would seem that the biggest defense against light-induced vertigo is to be aware of the hazard and to be healthy and well rested.

A few years back the pilot of a medium twin helicopter was on final after a long hard day, when his co-pilot noticed that the normally smooth, accurate flying was starting to deteriorate. The co-pilot thought the pilot was just tired and did or said nothing.

A second or two later the pilot pushed the cyclic hard over to the left and started shaking and twitching violently. The co-pilot grabbed the cyclic and for the next few seconds (to the co-pilot, hours) fought his pilot for control of the helicopter. Fortunately for all concerned, there were five people in the back and they came to the co-pilot's rescue, subduing the pilot and dragging him into the back. He soon lapsed into unconsciousness and the co-pilot made an uneventful landing.

The pilot was rushed to hospital, and when he regained consciousness he had no recollection of the incident. He was tested with a flickering light pattern and his reaction was identical. Guess what? The approach was on a westerly heading, the sun was about 30° above the horizon and on the nose, and the pilot was tired.

A study by the US navy showed that, out of 102 aviators, 22 were susceptible in varying degrees, but in only one incident was an impending accident a likely outcome. The conclusion was that such 'photic stimulation' tests were not sufficiently reliable to become standard requirements for flight crew.

Many pilots will become quite uncomfortable in the 'risk flicker range' and simply make a change to alleviate the problem, sometimes without even realizing precisely what has prompted them to do so. They will change heading, change aircraft attitude; in a fixed wing they might change the rpm.

So the main things to remember are:

➤ Know when your flight characteristics could set up the required conditions.

➤ Try to be in good condition and well rested, as pilots should do in any event.

➤ If you are suddenly feeling off-color, change something as soon as you can; if you are in a two-pilot operation, tell the other pilot.

CREW RESOURCE MANAGEMENT

The following material comes from an article published in *Helicopter Safety*, Vol. 22, No. 2 (1995) by the Flight Safety Foundation. It was written by Joel S Harris, then of FlightSafety International. The Flight Safety Foundation is dedicated to the promotion of flight safety through information and education, and is an excellent resource for pilots. You can find out more through the webpage www.flightsafety.org.

The single helicopter pilot can use crew resource management (CRM) techniques to improve communication, decision making, workload management, stress management and other skills for improving the safety margin.

Training to refine the human-factors aspects of crew operations can be traced back to the early 1970s, when KLM Royal Dutch Airlines introduced a program for its flight crews. In 1979, the US National Aeronautics and Space Administration (NASA) suggested that business managerial concepts could be applied in the cockpit to reduce the high number of 'human factors' accidents occurring to airlines. Within 10 years, cockpit resource management – later expanded conceptually to crew resource management – was included in training worldwide at most major airlines, and the US air force had begun full-scale CRM training of all crews of multi-person aircraft. Today, there is sufficient evidence that CRM training and practise have improved safety.

CRM skills are no substitute for technical proficiency. Nevertheless, high technical proficiency cannot guarantee safe operations without effective resource management.

Not all helicopter pilots and operators agree, however, on the value of CRM training in the single-pilot cockpit. To some, CRM is about multi-pilot crews learning to work together, and has little relevance in the single-pilot cockpit. Although it is true that traditional CRM training emphasizes interpersonal behavior to improve group dynamics among crew members (for example, junior crew members are encouraged to be more assertive toward veteran superiors, and veterans are encouraged to give more attention to input from less senior crew), this is only one aspect of CRM. The US Federal Aviation Administration's advisory circular on CRM includes a training curriculum that recommends training in such areas as teamwork, communication skills, decision making, workload management, situational awareness, preparation and planning, cockpit distractions and stress management.

The concepts must be applied somewhat differently to the single-pilot operator, and the advisory circular instructs the reader to 'customize the training to reflect the nature and the needs of the organization' to increase CRM's relevance.

By definition, CRM is the effective management of all resources available to the pilot. The captain of a multi-crew aircraft can and should delegate many tasks to the first officer – for example, checklists, radio calls, approach briefings, and so on. The single pilot simply has a different set of resources and therefore must manage them differently. Nevertheless, cockpit management for the single pilot may be more demanding than for pilots in a multi-person crew.

According to Lonney McCann, director of training at Indianapolis Helicopter Corporation, Indianapolis, Indiana, 'CRM is more critical in single-pilot cockpits than in those with multiple crew members. In a multiple-crew cockpit, even a poor manager will benefit from the self-preservation instincts of the other crew members. The single pilot does not have the luxury of checks and balances provided by other crew members. He or she must invest an even greater effort in organizing the cockpit and thought processes to accomplish all of the same tasks that are required of a multiple-crew operation.'

The Federal Aviation Administration makes the following observations:

➤ CRM is a comprehensive system of applying human-factors concepts to improve performance.

➤ Human factors is a field devoted to optimizing human performance and reducing human error.

➤ CRM embraces all operational personnel.

➤ CRM can be blended into all forms of training.

➤ CRM concentrates on attitudes and behaviors, and their impact on safety.

➤ Success of CRM training programs depends on check airmen, instructors and supervisors who are highly qualified and specially trained in CRM.

➤ CRM training requires commitment from all managers, starting with senior managers, to be received positively by operations personnel.

➤ CRM training should be customized to reflect the nature and needs of the operation.

➤ CRM training should include initial indoctrination, recurrent practise, feedback and continuing reinforcement.

➤ CRM should be an inseparable part of the organization's culture.

The topics outlined below are included in the FAA advisory circular, but have been interpreted to more readily fit the single-pilot operator.

Teamwork

CRM includes the effective use of human resources. Human resources includes groups that routinely work with the pilot and are involved in operating a flight safely. These groups can include, among others, dispatchers, medical attendants on emergency medical service flights, maintenance personnel, air traffic controllers, other pilots and management.

One advantage of teamwork is that it facilitates the transfer of information. Many accidents and incidents were precipitated by a lack of knowledge. The ideal flight operation might function in the same way as a sports team: each member has a particular assignment and is responsible for fulfilling it, while at the same time never losing sight of the ultimate goal. In sports, that goal is winning. In aviation it is the safe and efficient operation of an aircraft.

The team concept must begin with management and be constantly reinforced. The FAA recommends that instructors and check airmen use team concepts as grading criteria during training and checking.

Communication skills

Miscommunication with air traffic control is often cited as a causal factor in aircraft accidents and incidents. Good communication skills are essential in developing teamwork. Pilots increase the probability of a safe flight by learning to effectively seek and evaluate information, overcoming barriers to communication and being assertive at appropriate times.

Decision making

The single pilot makes decisions with fewer resources. Nevertheless, fewer does not mean none. One frequent error observed in simulator training is the failure of a captain to seek all relevant information before making an important decision; an error made by single pilots too. The single pilot's resources may include;

➤ Other persons aboard the aircraft, such as passengers and medical crew members.

- Aircraft gauges and sensations – vibrations, sights, smells, and so on.

- Air traffic control.

- Other pilots who can be contacted by radio.

- Flight manuals, checklists and other documentation.

- Ground-based support personnel.

To effectively use a resource, a pilot must know that the resource exists, understand its use and limitations and – most importantly – the pilot must ask for assistance. Some pilots feel a self-imposed pressure to perform without outside assistance. For example, a single pilot was flying a turbine helicopter that experienced a tail rotor malfunction. Because the pilot was new to the aircraft, he was not sure of the appropriate emergency procedure. Rather than instruct a passenger to read him the procedure from the flight manual (his hands were busy controlling the aircraft), he elected to execute an unapproved procedure. Fortunately, he was successful, but this pilot's unwillingness to use every resource available to the pilot at a critical moment could have resulted in an accident.

Workload management

Captains of multi-pilot crews are encouraged to manage the cockpit workload by sharing duties with other crew members. Workload management is even more important to the single pilot, because he or she has fewer resources and therefore must carefully prioritize tasks.

Learning to prioritize, while avoiding distraction from the primary duty of flying the aircraft, requires training and practise. Yet learning to prioritize tasks often receives too little emphasis during training.

The single pilot can also benefit by sharing tasks: requesting air traffic control assistance during heavy workloads; asking for help from the company dispatcher; prudently using automation such as an autopilot; and, in the case of emergency medical service operators, involving trained medical personnel in certain flight duties.

Situational awareness (SA)

Situational awareness is defined as the accurate perception of the factors that affect the aircraft and crew during a specific time period. Or, put more simply, it is knowing what is going on around you. Helicopters normally operate in airspace that is considered hazardous by many jet crews — below 10,000 feet mean sea level. SA becomes imperative in this environment. Maintaining a high level of awareness is more challenging for a single pilot, and automation and other resources should be used effectively.

One method to help maintain a high level of SA is simply remaining aware of SA's importance. For example, in a recent study of fatal turbine-helicopter accidents, six (7.1 percent) were mid-air collisions. All six accidents occurred during daylight and in the immediate vicinity of an airport. Five occurred at uncontrolled fields and each of these involved one aircraft climbing after takeoff. The lesson learned in this study is that, when operating in an area where takeoffs and landings are occurring, special vigilance must be maintained. Other required tasks — such as entering navigation coordinates and communicating with base operations or with passengers — may need to be delayed, to focus full attention on the hazards of other air traffic.

Preparation

For the single-pilot operator, pre-flight preparation is of special importance. No first officer is available to confirm radio frequencies and make radio calls, fix positions and call out checklists. The single-pilot operator may commit to memory information, such as frequencies and emergency procedures, that otherwise might be difficult to confirm in an emergency or unexpected high-workload situation.

Careful cockpit organization also simplifies many tasks and allows the single pilot to do more with less.

Cockpit distractions

The NASA aviation safety reporting system studied pilot distraction, based on reports submitted by crew members and controllers. The study's conclusion was that human susceptibility to distraction is one of the most frequent causes of hazardous events in air-transport operations.

If this is true in the multiple-crew air-transport cockpit, how much more adept at dealing with routine distractions must a single crew member be? The study showed that cockpit distractions are often the result of routine cockpit tasks interrupting or preventing the pilot's performance of other routine tasks. A single pilot must recognize and prevent distractions, while performing the primary task of the flying of the aircraft.

22

Training mishaps

There will always be times when an instructor has to reach deep into his or her bag of skill to extricate the aircraft and its occupants from an unintended adverse situation. Some do it better than others.

Case study 22:1

Accident Report NYC01LA204

Date: August 5, 2001

Location: Islip, New York, United States

Aircraft: Robinson R22

Injuries: nil

A Robinson R22 was substantially damaged when it impacted terrain during a practice autorotation at Long Island MacArthur Airport, Islip, New York. The certificated flight instructor and the private helicopter pilot under instruction were not injured. Visual Meteorological Conditions prevailed at the time.

According to the flight Instructor, the private pilot under instruction had completed 10 straight-in autorotations, and then flew five 180° autorotations. As the helicopter approached the ground during the last practice autorotation, the flight instructor noted that its nose was too low. He took the helicopter's controls and tried to level it, and was able to bleed off some of the vertical and forward airspeed. However, the left skid touched down and dug into the turf, and the helicopter tumbled forward and came to rest on its side.

The pilot under instruction had 1540 helicopter flight hours, with 7 hours in the R22. The flight Instructor stated that, although he was on or near the controls at all times, he had been confident in the pilot under instruction's ability to 'keep it safe'. He also noted that prior to touchdown, the helicopter's nose attitude 'changed rapidly for the worse', and that he had little time to take over and complete a safe landing.

Source: National Transportation Safety Board, United States

Case study 22:2

Accident Report LAX01LA190

Date: May 23, 2001

Location: Murietta, California, United States

Aircraft: Robinson R22

Injuries: nil

The instructor had just completed 30 minutes of hover practise with the student. The student landed the helicopter and the instructor deplaned to allow the student to execute his first solo flight. As the instructor watched, the student began to lift off and immediately rolled to the right and crashed. According to the instructor, the helicopter never left the ground. The student had logged approximately 29 hours of dual instruction in the R22 prior to the accident. The instructor stated that he had done many pick-ups and set-downs with the student just prior to the accident. He emphasized the differences in flight characteristics that the student would experience when the instructor was not in the helicopter. He explained that he had covered dynamic rollover situations in ground school sessions, including a Robinson factory training videotape on the subject.

The instructor normally occupied the left seat of the helicopter during dual lessons. At the time of the accident, the student was occupying the right seat, and the left seat was empty. The solo flight was conducted on an asphalt helicopter landing pad. The helipad was inspected and found to be clean, smooth and free of obstructions. The instructor weighed 185 pounds, and the helicopter had consumed 30 pounds of fuel prior to the attempted solo flight. This made the helicopter 215 pounds lighter than the student was accustomed to. If the student did not exercise extreme caution to lift off very slowly and deliberately, with the cyclic centered, a dynamic roll over would be very likely to occur. With an excessive input of collective, the helicopter at a light takeoff weight and the cyclic even slightly off center, the roll would extremely rapid and nearly impossible to arrest.

Source: National Transportation Safety Board, United States

Case study 22:3

Accident Report CH197LA277

Date: August 31, 1997

Location: Council Bluffs
Airport, Iowa, United States

Aircraft: Hughes 269A

Injuries: 2 serious

A Hughes 269A with a designated pilot examiner acting as pilot in command during a commercial rotorcraft-helicopter practical test was destroyed when it collided with terrain while on short final to runway 13 at Council Bluffs Airport. The examiner and commercial applicant sustained serious injuries. Visual Meteorological Conditions prevailed at the time of the accident.

After oral examination with the examiner, the student prepared for the flight portion of the check ride. The examiner said that, during the oral portion of the test, he discussed who would be responsible for the flight and who would be acting as pilot in command. He stated that he would be 'acting as a passenger'. He also stated that he reviewed the procedure for transfer of controls.

After the oral portion of the test, they departed Council Bluffs Municipal Airport at approximately 1100 CDT. The examiner stated that 'various tasks were completed by reference to [applicant's name] pre-arranged plan of action satisfactorily and without incident'. Transfer of controls occurred two times prior to the accident.

Approximately one hour into the flight, when they were hovering in a confined space, the examiner gave the student the instruction to 'climb out, turn to the right, fly over the airport and head back towards the college'. After departing the confined space and at an altitude of 1700 feet above mean sea level (approximately 450 feet above ground level) at 50 knots indicated airspeed and in a westerly heading over the runway, the examiner instructed the student to perform a 'simulated engine failure', and reduced the throttle to the idle position.

The helicopter began to immediately turn left to parallel the runway. The student did not adjust the collective to the appropriate down position and the aircraft was not appropriately adjusted to the recommended 50 knot attitude. The examiner recognized a rpm decrease in the main rotor. He then took control of the aircraft and 'decreased the collective to the full down position, increased the throttle to regain rotor rpm, and applied slight forward cyclic to lower the nose to increase airspeed'. The aircraft began a right turn and the examiner 'manipulated the throttle as required to maintain optimum power and acceptable yaw'.

At approximately 10 feet he 'attempted to level the aircraft into a landing attitude and rapidly increased the collective'. The aircraft impacted in a slightly nose-low attitude, bounced, and came to rest on its left side with extensive damage.

A witness, who was taxiing a Cessna 310 to the ramp at the time, stated that he saw the helicopter throughout the approach. He thought the 'helicopter made a very stable, rectangular approach' but just after it crossed the 13 threshold, 'their forward motion suddenly stopped and the helicopter rapidly entered a nose down, very steep, very tight spiral descent into the ground'. The witness estimated the altitude of the helicopter to be 50 to 75 feet off the ground when this occurred. After departing his airplane and calling for help, the witness ran to the accident site. The witness heard the student ask the examiner what had happened and if something failed. The examiner's response was, 'No. It was me; I failed'.

Source: National Transportation Safety Board, United States

Instructors are generally on their guard when a student has the controls, and none more so than an R22 instructor. The instructor, by virtue of experience and self-preservation, will normally see situations or trends developing and will decide when to intervene either verbally, or physically. After all, jump in too early and the student will never learn anything; too late, and it may be too late to avert disaster. Where even the most experienced instructor can be caught out is when the student is either experienced or an above average 'safe' student. Here is when the instructor's vigilance can be low and the unexpected happens quickly.

One of the major revelations when analyzing training accidents is that it is rarely the student's fault. The common theme in most mishaps is lack of judgment or lack of adequate supervision on the part of the instructor. Also, many accidents happen during the practise of an emergency procedure where there is a fine line between effective learning and disaster.

However, spare a thought for instructors as they attempt to ensure a realistic and effective learning environment, with the fine judgment that that often requires:

➤ When to send a student on the first solo.

➤ What weather limits to impose on a student for a particular solo exercise.

➤ When to intervene verbally.

➤ When to take control of the aircraft.

But the fact remains: the student is entrusted to the care of the instructor. It is unfortunate that many training accidents can be summed up as:

The instructor allowed the student to get the aircraft into a situation from which the instructor could not recover.

Further reading

Bradbury, Tom, *Meteorology and Flight: Pilot's Guide to Weather.* A & C Black, 2000 (3rd edn), ISBN 0713642262.

Buck, Robert N., *Weather Flying.* McGraw-Hill Professional, 1997 (4th edn), ISBN 007008761X.

Holleran, Renee S., *Air and Surface Patient Transport.* Mosby, 2002 (3rd edn), ISBN 0323017010.

Johnson, Wayne, *Helicopter Theory.* Dover Publications, 1994, ISBN 0486682307.

Leishman, J. Gordon, *Principles of Helicopter Aerodynamics.* Cambridge University Press, 2003, ISBN 0521523966.

Machado, Rod, *Rod Machado's Instrument Pilot's Survival Manual.* Aviation Speakers Bureau, 1997, ISBN 0963122908.

Reinhart, Richard O., *Basic Flight Physiology.* McGraw-Hill Professional, 1996 (2nd edn), ISBN 0070522235.

Stepniewski, W.Z., *Rotary-Wing Aerodynamics.* Dover Publications, 1984, ISBN 0486646475.

Wagtendonk, Walter, *Principles of Helicopter Flight.* Aviation Supplies & Academics, 1996, ISBN1560272171.

Index